T0216264

Octave and MATLAB for Engineering Applications

Andreas Stahel

Octave and MATLAB
for Engineering Applications

 Springer

Andreas Stahel
Engineering and Information Technology
Bern University of Applied Sciences
Biel, Bern, Switzerland

ISBN 978-3-658-37210-1   ISBN 978-3-658-37211-8   (eBook)
https://doi.org/10.1007/978-3-658-37211-8

© The Editor(s) (if applicable) and The Author(s), under exclusive license to Springer Fachmedien Wiesbaden GmbH, part of Springer Nature 2022
This work is subject to copyright. All rights are solely and exclusively licensed by the Publisher, whether the whole or part of the material is concerned, specifically the rights of translation, reprinting, reuse of illustrations, recitation, broadcasting, reproduction on microfilms or in any other physical way, and transmission or information storage and retrieval, electronic adaptation, computer software, or by similar or dissimilar methodology now known or hereafter developed.
The use of general descriptive names, registered names, trademarks, service marks, etc. in this publication does not imply, even in the absence of a specific statement, that such names are exempt from the relevant protective laws and regulations and therefore free for general use.
The publisher, the authors and the editors are safe to assume that the advice and information in this book are believed to be true and accurate at the date of publication. Neither the publisher nor the authors or the editors give a warranty, expressed or implied, with respect to the material contained herein or for any errors or omissions that may have been made. The publisher remains neutral with regard to jurisdictional claims in published maps and institutional affiliations.

This Springer imprint is published by the registered company Springer Fachmedien Wiesbaden GmbH part of Springer Nature.
The registered company address is: Abraham-Lincoln-Str. 46, 65189 Wiesbaden, Germany

Preface

About This Book

These lecture notes grew out of lectures at the Bern University of Applied Sciences (BFH). The main goal is to familiarize students with *Octave* or MATLAB, such that typical engineering problems can be solved. It is not an introduction to programming, using MATLAB/*Octave*. The reader is assumed to be familiar with basic programming techniques. The key part of this book is chapter 3 with many engineering applications of MATLAB and *Octave*.

The book consists of three chapters.

- The first chapter is an introduction to the basic *Octave*/MATLAB commands and data structures. The goal is to provide simple examples for often used commands and point out some important aspects of programming in *Octave* or MATLAB. The students are expected to work through all these sections. Then they should be prepared to use *Octave* and MATLAB for their own engineering projects.

- The second chapter presents a few commands for elementary statistics, illustrated by short sample codes. This chapter was never presented in class, but handed to the students as an aid to perform elementary statistical tasks.

- The third chapter consists of applications of MATLAB/*Octave*. Most topics were part of a Bachelor or Master thesis project at BFH-TI (Bern University of Applied Sciences, School of Engineering and Computer Science). In each section the question or problem is formulated and then solved with the help of *Octave*/MATLAB. This small set of sample applications with solutions should help the reader to solve **his/her** engineering problems. In class I usually selected a few of those topics and presented them to the students. As an essential part of the class the students had to select, formulate and solve a problem of their own.

First versions of these notes were based on *Octave* only, but by now (almost) all codes works with MATLAB too. Wherever possible I attempted to provide code working with *Octave* and MATLAB. Most of the codes for this book are available on Springer Link or on my web page on Github at https://AndreasStahel.github.io/Octave.html, together with some more information on my contributions to *Octave*. A list of errata for this book will be available too.

There is no such thing as *"the perfect lecture notes"* and improvements are always possible. I welcome feedback and constructive criticism. Please send your observations and remarks to Andreas.Stahel@gmx.com.

Acknowledgments

This book would not have been possible without the help of many students, either as participants of the class or as authors of their thesis. The required software to work with this book is either `MATLAB` (developed and supported by MathWorks Inc.) or the free software *Octave*. The main author of *Octave* is John Eaton, and there are many contributors, see the section `Acknowledgments` in the manual of *Octave*. This book would not be possible without *Octave*/`MATLAB`.

Finally I wish to thank Eric Blascke from Springer Nature for the excellent support while setting up this book.

To my beloved wife and life companion
Ludivica Baselgia Stahel

Port, Switzerland
February 21, 2022   Andreas Stahel

Contents

Chapter 1

Introduction to *Octave*/MATLAB

The first chapter, consisting of six sections, gives a very brief introduction into programming with *Octave*. This part is by no measure complete and the standard documentation and other references will have to be used. Here are some keywords presented in the sections of this chapter:

- Remarks on MATLAB and pointers to documentation.

- Starting up an *Octave* or MATLAB work environment.

- How to get help.

- Vectors, matrices and vectorized code.

- Script files and function files.

- Data types, functions, control statements, conditions.

- Data files, reading and writing information.

- Solving equations of different types.

- Create basic graphics and manipulate images.

- Solve ordinary differential equations. Include C++ code in *Octave*/MATLAB.

1.1 Starting up *Octave* or MATLAB and First Steps

The goal of this section is to get the students started using *Octave*, i.e. launching *Octave*, available documentation and information. Some of the information is adapted to the local setup at this school and will have to be modified if used in a different context. *Octave*[1] is very similar to MATLAB. If you master *Octave* then MATLAB is easy too. *Octave* is developed and maintained on Unix systems, but can be used on Mac and Win* systems too.

---

[1] http://www.gnu.org/software/octave/

Supplementary Information The online version contains supplementary material available at [https://doi.org/10.1007/978-3-658-37211-8_1].

© The Author(s), under exclusive license to Springer Fachmedien Wiesbaden GmbH, part of Springer Nature 2022
A. Stahel, *Octave and MATLAB for Engineering Applications*, https://doi.org/10.1007/978-3-658-37211-8_1

There is a number of excellent additional packages for *Octave* available on the internet at Octave Forge[2].

For most tasks MATLAB and *Octave* are equivalent

References

- For quick consulting there is a reference card for *Octave*. It should come with your distribution of *Octave*.

- David Griffiths from the University of Dundee prepared an excellent set of short notes on MATLAB [Grif18].

- The *Octave* manual for the most recent version is available on the *Octave* web site at https://www.gnu.org/software/octave/support in the form of HTML files or as PDF file. This provides basic documentation of all *Octave*–commands. Almost all of the commands apply to MATLAB too.

  - Access the manual of *Octave* at https://octave.org/doc/latest/ as HTML files.
  - Access the manual of *Octave* at https://www.gnu.org/software/octave/octave.pdf as a single PDF file.
  - The *Octave* packages are documented on the web site for Octave Forge, available at http://octave.sourceforge.io/.

- The book [Hans11] by Jesper Hansen is an elementary and short introduction to *Octave*.

- A good reference for engineers is the book by Biran and Breiner [BiraBrei99].

- Another useful reference is the book by Hanselman and Littlefield ([HansLitt98]). Newer versions of this book are available. As an introduction to MATLAB and some of its extensions you might consider [HuntLipsRose14].

- On the *Octave* web page there is a Frequently Asked Questions (FAQ) page: http://www.gnu.org/software/octave/FAQ.html

- Find a wiki for *Octave* at http://wiki.octave.org/ with useful information.

- There is an active mailing list for *Octave*, avalable on Nabble to be found at octave.1599824.n4.nabble.com/.

- The book [Quat10] is considerably more advanced and shows how to use *Octave* and MATLAB for scientific computing projects.

---

[2]http://octave.sourceforge.io/

Since *Octave* and MATLAB are very similar you can also use MATLAB documentation and books.

- The on-line help system of MATLAB allows to find precise description of commands and also to search for commands by name, category or keywords. Learning how to use this help system is an essential step towards getting the most out of MATLAB.

- As part of the help system in MATLAB two files might be handy for beginners:

    - GettingStarted.pdf as a short (138 pages) introduction to MATLAB.
    - UsingMatlab.pdf is a considerably larger, thorough and complete documentation of commands in MATLAB.

    The above documents are also available on the web site of MathWorks at
    http://www.mathworks.com/access/helpdesk/help/techdoc/matlab.shtml

> One of the most important points when using advanced software is how to take advantage of the available documentation.

Notations in these notes

In these notes we show most *Octave* or MATLAB code in a block, separated by horizontal lines. If input (commands) and results are shown in the same block, they are separated by a line containing the arrow string --> , short for "leading to".

```
                                    ── Octave ──
code
-->
results
```

Individual commands may be shown within regular text, e.g as plot(x,sin(x)).

1.1.1 Starting up *Octave*

There are many command line options when startin *Octave*. Type octave --help in a terminal or examine Section 2.1.1 Command Line Options in the *Octave* manual.

Working with the *Octave* GUI

Starting with version 4.0.0 of *Octave* has a GUI (Graphical User Interface) as interface, see Figure 1.1. To start *Octave* with the GUI use

- your mouse to click the menue entry on your desktop environment, e.g. Xfce, Gnome, Mac OS*, Win*

- type octave& in a terminal with versions 4.0 and 4.2

- type `octave --gui&` in a terminal with version 4.4 or newer.

Within one window frame you can

- execute commands and observe their results in the **Command Window**.

- edit code segments and run the directly from the **Editor** window with one key stroke (F5).

- gather information on all variables in the current workspace in the **Workspace** window.

- work with the built-in **File Browser**.

- read the standard *Octave* documentation in the **Documentation** window.

- read and change the current directory in the top line of the GUI.

- Starting with version 4.4.0 *Octave* has a **Variable Editor**. By clicking on a variable in the workspace the name and the value(s) will be displayed in the variable editor, where you can display and change the value(s). To show and modify the variable a use the command `openvar('a')`.

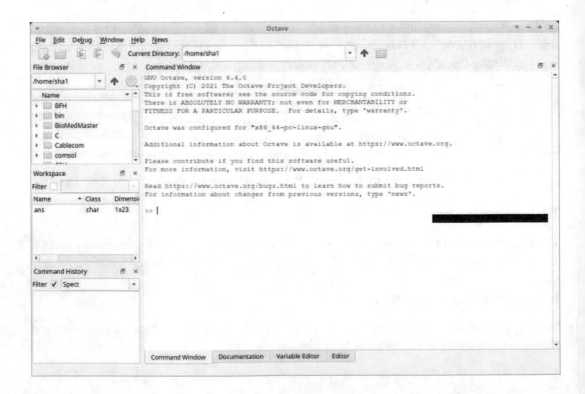

Figure 1.1: The *Octave* GUI

Figure 1.1 shows a typical screenshot of the *Octave* GUI. There are several advantages using the GUI:

- The built in editor has good highlighting and syntax checking of *Octave* code.

- In the editor you can set break points and step through your code line by line.

- You can detach some windows from the main frame. I most often use the editor in a separate window.

- You can move in the directory tree using the top line of the GUI.

- Graphics generated by *Octave* will always show up in separate windows.

Working with the CLI (Command Line Interface) of *Octave*

To start *Octave* without the GUI type

- `octave --no-gui` in a terminal with versions 4.0 and 4.2

- `octave` in a terminal with version 4.4, 5.1 and beyond.

A working CLI environment for *Octave* consists of

1. A command line shell with *Octave* to launch the commands.

2. An editor to write the code. Your are free to choose your favorite editor, but editors providing an *Octave* or `MATLAB` mode simplify coding.

   - This author has a clear preference for the editor `Emacs`, available for many operations system. On Linux systems you might want to try `gedit`.
   - For WIN* systems the editor `notepad++` might be a good choice.

3. Possibly a browser to access the documentation.

4. Possilbly one or more graphics windows.

Thus a working screen might look like Figure 1.2. Your window manager (e.g. Xfce, KDE or GNOME) will allow you to work with multiple, virtual screens. This is very handy to avoid window cluttering on one screen.

The startup file `.octaverc`

On startup *Octave* will read a file `.octaverc` in the current users home directory[3]. In this file the user can give commands to *Octave* to be applied at each startup. You can add a directory to the current search path by adding to the variable `path`. Then *Octave* will search

---

[3]On a Windows10 system in 2017 the file `octaverc` was located in the directory `C:\Octave\Octave-4.0.3\share\octave\site\m\startup`. On your system it might be in a similar directory.

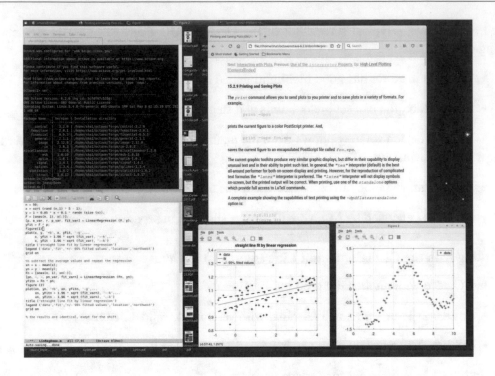

Figure 1.2: Screenshot of a working CLI *Octave* setup

in this directory and all its subdirectories for commands. Thus the user can place his/her script and function files in this directory and *Octave* will find these commands, independent of the current directory. My current version of the startup file is

```
                                    .octaverc
pkg prefix ~/octave/forge ~/octave/forge;
addpath(genpath('~/octave/site'))
set (0,'DefaultTextFontSize',20);% capitalization of letters is ignored
set (0,'DefaultAxesFontSize',20)
set (0,'DefaultAxesXGrid','on')
set (0,'DefaultAxesYGrid','on')
set (0,'DefaultAxesZGrid','on')
set (0,'DefaultLinelineWidth' ,2)
set (0,'DefaultFigureColormap',jet(64))
more off
```

With this initialization file I configure *Octave* to my desire:

- The packages are installed and searched for in the directory `~/octave/forge` .

- *Octave* will always search in the directory `~/octave/site` and its sub-directories for commands, in addition to the standard search path.

- I choose larger default fonts for the text and axis in graphics.

- For any graphics I want to show the grid lines, by default.

- I choose a larger line width by default for lines in graphics.

- Use the colormap `jet` as default.

- The pager `more` is turned off by default.

If you want to ignore your startup file for some special test you can use a command line option, i.e. launch *Octave* by `octave --no-init-file`.

1.1.2 Packages for *Octave*

As an essential addition to *Octave* a large set of additional packages[4] is freely available on the internet at https://octave.sourceforge.io/. An extensive documentation is given by the option `Packages` on that web page. If you find a command in those packages and want to make it available to your installation you have to install the package once and then load it into your *Octave* environment.

How to install and use packages provided by the distribution

On Win* systems

- On a recent (2021) win* system with *Octave* 6.3.0 many packages came with the distribution. You have to make the prebuilt packages available. To be done once: launch the command `pkg rebuild` to make the packages available.

- The Win* package for *Octave* has many packages prebuilt and installed.

- Use `pkg list` to generate a list of all packages. Those marked by * are already loaded and ready to be used.

- Use `pkg load image` to load the image package and similar for other packages.

On Linux/Unix systems   Most Linux distributions provide *Octave* and the most of Octave Forge with their distribution. We illustrate the use of the package manager by Debian, also used by Ubuntu and it derivatives.

- To install *Octave* use a shell and type `sudo apt-get octave`.

- To install most of the documentation use the command `sudo apt-get install octave-doc octave-info octave-htmldoc`

- To install a few packages (image, optimization and statistics) use `sudo apt-get install octave-image octave-optim octave-statistics`.

---

[4]MATLAB does not have packages, but toolboxes. They are installed when you install your version of MATLAB.

- If you plan to compile the packages on your system (see below) you also need the header files and some libraries. To install those use a shell and `sudo apt-get install liboctave-dev`.

- To list the installed packages use the *Octave* prompt and type `pkg list`. Those marked by ∗ are already loaded and ready to be used.

- Use `pkg load image` to load the image package and similar for other packages.

How to install a package from Octave Forge

On the web page https://octave.sourceforge.io/ current versions of the packages are available and can be installed manually.

- To be done once:
  - Decide where you want to store your packages. As an example consider the sub directory `octave/forge` in your home directory. Create this directory, launch *Octave* and tell it to store packages there with the command

  ```
  ┌─────────────────── Octave ───────────────────┐
    pkg prefix ~/octave/forge ~/octave/forge
  └───────────────────────────────────────────────┘
  ```

  Since the tilde character ~ is replaced by the current users (in this case `sha1`) home directory the packages will be setup in `/home/sha1/octave/forge` and its sub-directories. This command can and should be integrated in your **.octaverc** file.

- To be done for each package:
  As example we consider the image package, aiming for the command `edge()`. There are different options to install the image package, containing this command:

  - It might already be installed, try `help edge`. If *Octave* knows about the command, the package is installed and probably loaded. If it is not loaded yet, type `pkg load image`.
  - On a Linux installation use your favorite package manager to install the image package of *Octave*, e.g. Synaptic or `apt-get ...`
  - Let *Octave* download, compile and install the package.

  ```
  ┌─────────────────── Octave ───────────────────┐
    pkg install -forge image
  └───────────────────────────────────────────────┘
  ```

  - You can download, compile and install step by step, as outlined below.

  - Go to the web page https://octave.sourceforge.io/ and select the menue entry `Packages`. Then search for the package `image` and download it to your local disk and store it in the above directory.

– Launch *Octave* and change in the directory with the package. Then install the package with the command

```
──────────────────────────── Octave ────────────────────────────
  pkg install image-2.12.0.tar.gz
```

From now on the commands provided by the package are available. To use them you still have to load the package, e.g. `pkg load image`.

– You may also locate the source for the command `edge.m` by

```
──────────────────────────── Octave ────────────────────────────
  which edge
  -->
  'edge' is a function from the file
                 /home/sha1/octave/forge/image/edge.m
```

Having the source code may allow you to adapt the code to you personal needs.

• Loading a package: if a package is installed it will show up with the command `pkg list`, but it is not loaded yet. To load a package use `pkg load image`, then the package `image` will be loaded and listed with a star.

```
──────────────────────────── Octave ────────────────────────────
  >> pkg load image
  >> pkg list
  Package Name    | Version | Installation directory
  ----------------+---------+-----------------------
          control |   3.3.1 | ~/octave/forge/control-3.3.1
         femoctave |   2.0.1 | ~/octave/forge/femoctave-2.0.1
        financial |   0.5.3 | ~/octave/forge/financial-0.5.3
          general |   2.1.1 | ~/octave/forge/general-2.1.1
          image *|  2.12.0 | ~/octave/forge/image-2.12.0
               io |   2.6.3 | ~/octave/forge/io-2.6.3
    miscellaneous |   1.3.0 | ~/octave/forge/miscellaneous-1.3.0
              msh |  1.0.10 | ~/octave/forge/msh-1.0.10
            optim |   1.6.1 | ~/octave/forge/optim-1.6.1
           signal |   1.4.1 | ~/octave/forge/signal-1.4.1
          splines |   1.3.4 | ~/octave/forge/splines-1.3.4
       statistics |   1.4.2 | ~/octave/forge/statistics-1.4.2
           struct |  1.0.17 | ~/octave/forge/struct-1.0.17
```

• Commands to maintain the packages:

– To show a list of all packages use

```
──────────────────────────── Octave ────────────────────────────
  pkg list
```

– To make the additional commands unavailable you may unload a package, e.g.

```
——————————————— Octave ———————————————
  pkg unload image
```

– You can load an already installed package, e.g.

```
——————————————— Octave ———————————————
  pkg load image
```

– To update all installed packages using the Octave Forge site.

```
——————————————— Octave ———————————————
  pkg update
```

1.1.3 Information about the operating system and the version of *Octave*

When working with *Octave* you can obtain information about the current system with a few commands. Obviously your results may differ from the results below.

- The command `computer()` shows the operating system and the maximal number of elements an array may contain.

```
——————————————— Octave ———————————————
  [C, MAXSIZE, ENDIAN] = computer()
  -->
  C = i686-pc-linux-gnu
  MAXSIZE =  2.1475e+09
  ENDIAN = L
```

The resulting maximal array size of $2.1475 \cdot 10^9 \approx 2^{31}$ is a consequence of 32-bit integers being used to index arrays.

- With the function `uname()` some more information about the computer and the operating system is displayed.

```
——————————————— Octave ———————————————
  uname()
  -->
  ans =  scalar structure containing the fields:
      sysname = Linux
      nodename = hilbert
      release = 5.4.0-89-generic
      version = #100-Ubuntu SMP Fri Sep 24 14:50:10 UTC 2021
      machine = x86_64
      sysname = Linux
```

- With `version()` the currently used version of *Octave* is displayed, e.g. in November 2021 the current version was 6.4.0.

```
──────────────────────────── Octave ────────────────────────────
version()
-->
ans = 6.4.0
└──────────────────────────────────────────────────────────────┘
```

With the command `ver()` the versions of *Octave* and all installed packages are displayed.

```
──────────────────────────── Octave ────────────────────────────
ver()
-->
----------------------------------------------------------------
GNU Octave Version: 6.4.0 (hg id: 8d7671609955)
GNU Octave License: GNU General Public License
Operating System: Linux 5.4.0-89-generic #100-Ubuntu SMP
                  Fri Sep 24 14:50:10 UTC 2021 x86_64
----------------------------------------------------------------
Package Name  | Version | Installation directory
--------------+---------+-------------------------
       control |   3.3.1 | ~/octave/forge/control-3.3.1
      femoctave |   2.0.1 | ~/octave/forge/femoctave-2.0.1
      financial |   0.5.3 | ~/octave/forge/financial-0.5.3
        general |   2.1.1 | ~/octave/forge/general-2.1.1
          image *|  2.12.0 | ~/octave/forge/image-2.12.0
             io |   2.6.3 | ~/octave/forge/io-2.6.3
  miscellaneous |   1.3.0 | ~/octave/forge/miscellaneous-1.3.0
            msh |  1.0.10 | ~/octave/forge/msh-1.0.10
          optim |   1.6.1 | ~/octave/forge/optim-1.6.1
         signal |   1.4.1 | ~/octave/forge/signal-1.4.1
        splines |   1.3.4 | ~/octave/forge/splines-1.3.4
     statistics |   1.4.2 | ~/octave/forge/statistics-1.4.2
         struct |  1.0.17 | ~/octave/forge/struct-1.0.17
└──────────────────────────────────────────────────────────────┘
```

The star on the line `image` indicated that the package `image` is not only installed, but also loaded.

- With the commands `ispc()`, `isunix()` and `ismac()` you can find out what operating system is currently used. These commands are useful for code depending on the OS, e.g. the exact form of file and directory names.

```
──────────────────────────── Octave ────────────────────────────
[ispc(), isunix(), ismac()]
-->    0   1   0
└──────────────────────────────────────────────────────────────┘
```

The results shows that currently a Unix system is running, in this case Linux.

- To verify that your system conforms to the IEEE standard for floating point calculations use the command `isiee()`.

```
────────────────────────── Octave ──────────────────────────
octave:16> isieee()
--> 1
```

- With the command `getrusage()` you can extract information about the current
  Octave process, e.g. memory usage and CPU usage.

- With the command `nproc()` the number of parallel threads to be used by *Octave* is
  returned. The results depends on the hardware and the operating system. By setting
  an environment variable in Linux the number of parallel threads can be limited, e.g.
  `export OMP_NUM_THREADS=1`. On many systems *Octave* uses the OpenBLAS
  libraries for basic matrix and vector operations. By using the environment variable
  `export OPENBLAS_NUM_THREADS=1` the number of threads to be used can be
  limited. Usually this should be left at the default value.

1.1.4   Starting up `MATLAB`

A working environment for `MATLAB` consists of

1. a command line shell with `MATLAB` to launch the commands

2. an editor to write the code

3. possibly a browser to access the documentation

4. possibly one or more graphics windows.

To start up `MATLAB` on a Unix system you may proceed as follows:

- Open a shell or terminal.

- Change to the directory in which you want to work, use `cd`.

- If you type `matlab &` then `MATLAB` will launch with the flashy Java interface.
  By using the ampersand `&` it is started in the background and you can launch other
  commands in the same terminal.

- Type `matlab -nojvm &` to launch `MATLAB`. The option `-nojvm` launches it
  without the splashy Java interface and thus uses a lot less memory, which might be
  a precious resource your system. On newer versions of `MATLAB` some of the features are not available without the Java interface, this concerns mainly the handling
  of graphics. Recently (2020) one may use `matlab -nodisplay` to achieve the
  above.

- You may also use the GUI of your operating system, locate the menu entry for MATLAB, click on it and the program will start.

  - After a short wait a flashy interface should appear on the screen.
  - On the right you find the command line for MATLAB. Elementary commands may be entered on this line.
  - On the left you find a history of the previously applied commands
  - On the top you can choose the working directory and a few menus
  - When you launch MATLAB with the interface you will use considerably more memory for the interface (Java) and Greek characters will not show on the screen. You will have to set the working directory with a cd command.

- When typing the command edit an elementary editor will show up. It might be useful to type longer codes and store them in a regular text file with the extension .m , the standard for any MATLAB file. You are free to use your favorite editor to work on your codes.

- On startup MATLAB will read a file startup.m in a subdirectory matlab of the users home directory. Often it is in ~/Documents/MATLAB/. In this file the user can give commands to MATLAB to be applied at each startup.

```
────────────────────────── startup.m ──────────────────────────
set (0,'DefaultAxesXGrid','on')
set (0,'DefaultAxesYGrid','on')
set (0,'DefaultAxesZGrid','on')
addpath(genpath('/home/sha1/Documents/MATLAB'))
```

1.1.5   Calling the operating system and using basic Unix commands

	Unix	*Octave*
show current directory	pwd	pwd
change into directory MyDir	cd MyDir	cd MyDir
change one directory level up	cd ..	cd ..
list all files in current directory	ls	ls
list more information about files	ls -al	ls -al
remove the file go.m	rm go.m	delete go.m
create a directory NewDir	mkdir NewDir	mkdir NewDir
remove the directory NewDir	rmdir NewDir	rmdir NewDir

Table 1.1: Basic system commands

Within *Octave* or MATLAB it is possible to launch programs of the operating system. This can be very useful, e.g. to call external image processing tools, see page 96. The details for the functions obviously depend on the underlying operating system.

- system() : execute a shell command and return the status and result of the command. As a simple example call the command whoami which returns the current users name.

```
unix('whoami');
-->
sha1
```

- unix() : execute a system command if running under a Unix-like operating system, otherwise do nothing. dos() : execute a system command if running under a Dos or Windows-like operating system, otherwise do nothing.

With MATLAB the three commands system(), unix() and dos() are interchangeable. With *Octave* their result depends on the operating system. It is a good idea to use the command system() only.

When working on a computer system some basic commands might be handy. Table 1.1 shows a few useful Unix commands. Some of the commands also work on the *Octave*/MATLAB command line. The behavior on a Win* system might be different.

1.1.6   How to find out whether you are working with MATLAB or *Octave*

There are still very few occasions when the codes for *Octave* or MATLAB differ slightly. Thus it might be useful to have a command telling you whether you work with *Octave* or MATLAB. The command IsOctave() returns 1 if *Octave* is running and 0 otherwise. Copy the file below in directories where MATLAB/*Octave* will find it, e.g. for MATLAB in ˜/Documents/Matlab and for *Octave* in ˜/octave/site/.

```
                          ── IsOctave.m ──
function result = IsOctave()
% Returns true if this code is being executed by Octave.
% Returns false if this code is being executed by MATLAB,
% or any other MATLAB variant.
%
%    usage: result = isOctave()

persistent OctVerIsBuiltIn;
if (isempty(OctVerIsBuiltIn))
    OctVerIsBuiltIn = (exist('OCTAVE_VERSION', 'builtin') == 5);
    % exist returns 5 to indicate a built-in function.
end
result = octaveVersionIsBuiltIn;
```

```
% If OCTAVE_VERSION is a built-in function, then we must
% be in Octave. Since the result cannot change between function
% calls, it is cached in a persistent variable.  isOctave cannot
% be a persistent variable, because it is the return value of
% the function, so instead the persistent result must be cached
% in a separate variable.
end
```

1.1.7 Where and how to get help

There are different situations when help is useful and important

- You know the command and need to know more details. As an examine the command `plot()`.

  - Typing `help plot` will display information about the command `plot`. You will find a list of all possible arguments of this function.

  - Typing `doc plot` will put you in an on-line version of the *Octave* manual with the documentation on `plot()`. You can use this as a starting points to browse for similar commands. The command `doc` works with *Octave* only.

  - Typing `lookfor plot` will search in all of the on-line documentation and display a list of all the commands. Type `help lookfor` to find out more.

- Some commands have demos and example codes built in. Examine the command `quiver()`. With `example quiver` find a listing of working examples.

```
                                  ─── Octave ───
  example quiver
  -->
  quiver example 1:
   clf;
   [x,y] = meshgrid (1:2:20);
   h = quiver (x,y, sin (2*pi*x/10), sin (2*pi*y/10));
   title ('quiver plot')

  quiver example 2:
   clf;
   x = linspace (0, 3, 80);
   y = sin (2*pi*x);
   theta = 2*pi*x + pi/2;
   quiver (x, y, sin (theta)/10, cos (theta)/10, 0.4);
   axis equal tight;
   hold on; plot (x,y,'r'); hold off;
   title ('quiver() with scaled arrows');
```

By calling `demo quiver` the examples will be executed and you can examine code and results.

- Most of the *Octave* code is given as script file. You can look at the source. To locate the source code for the command `quiver` use `which`.

Octave
```
which quiver
-->
'quiver' is a function from the file
        /usr/local/share/octave/6.4.0/m/plot/draw/quiver.m
```

As it is Open Source code you can copy the file in your directory and modify the code perform the desired operation.

- If you only know the topic for which you need help, but not the exact command (yet), use the *Octave* manual.

  – Both are available on the net at http://www.gnu.org/software/octave/ and go to `Support`. You can browse in the HTML files or download the PDF file.

  – Both should be installed on your computer in HTML and PDF form.

    * Search your local disk for the file `octave.html`. It should be a directory and then the file `index.html` is the starting point into the HTML manual.
    * Search your local disk for the file `octave.pdf` with the PDF manual.

- The references given on page 2 might also be useful, and hopefully these notes too.

- For the *Octave* packages the site http://octave.sourceforge.io/ provides documentation.

1.1.8 Vectors and matrices

The basic data type in MATLAB (**MAT**rix **LAB**oratory) and *Octave* is a **matrix** of numbers.

- A matrix is enclosed in square brackets (`[ ]`) and is a rectangular set of numbers, real or complex.

- Different rows are separated by semicolons (`;`).

- Within each row the entries are separated by commas (`,`) or spaces.

Creation of vectors and matrices

Vectors are special matrices: a column vector is a $n \times 1$-matrix and a row vector a $1 \times n$-matrix. MATLAB/*Octave* does distinguish between row and column vectors, for beginners often a stumbling block.

There are different methods to create vectors:

- To create a row vector with known numbers we may just type them in, separated by commas or spaces.

```
x = [1 2 3 4 5]
```

To create the same vector as a column vector we may either use the transpose sign or separate the entries by columns.

```
x = [1 2 3 4 5]'     % create a column vector by tranposing
x = [1; 2; 3; 4; 5]  % create the column vector directly
```

- To create a matrix we use rows and columns, e.g. to create a 3×3 matrix with the numbers 1 through 9 as entries use:

```
A = [1 2 3; 4 5 6; 7 8 9]
```

- If the differences between subsequent values are know we can generate the vector more efficiently using colons (:). Examine the results of the elementary code below.

```
% x  = begin:step:end
x2 = 1:10        % all integer numbers from one to 10
x1 = 1:0.5:10    % all numbers from 1 to 10, with step 0.5
```

- With the command linspace() we can specify the first and last value, and the total number of points. To generate a vector with 30 points between 0 and 10 we may use

```
x = linspace(0,10,30)
```

With the command logspace(a,b,n) we generate n values between 10^a and 10^b, logarithmically spaced.

```
x = logspace(0,2,11)
```

- To create a matrix or vector to be filled with zeros or ones *Octave* provides special commands.

```
x0 = zeros(2,3)   % creates a 2 by 3 matrixfilled with 0
x1 = zeros(5,1)   % creates a column vector with 5 zeros
y1 = ones(10,6)   % creates a 10 by 6 matrix filled with 1
```

Octave has many built-in functions to generate vectors and matrices of special types. The code below first generates a row vector with 10 elements, all of the values are set to zero. Then the squares of the numbers 1 through 10 are filled in by a simple loop. Finally the result is displayed.

```
n = 10;
a = zeros(1,n);
for i = 1:n
  a(i) = i^2;
end%for
```

The result is generated more efficiently by using vectorized code, as shown below.

```
a = [1:10].^2
```

Vector operations

Addition and multiplication of matrices and vectors follows strictly (almost, except for broadcasting, see Section 1.1.9) the operational rules of matrix operations.

```
clear *
a = [1 2 3]  % create a row vector
b = [4;5;6]  % create a column vector

a+b        % not permitted, watch out for automatic broadcasting
a+b'       % permitted
a*a        % not permitted
a*a'       % permitted, leading to the scalar product
a'*a       % permitted, leading to a 3x3 matrix
a.*a       % permitted, leading to element wise multiplication
[a b']     % permitted, leading to a concatenation of the vectors
```

The code below will generate a plot of the function $y(x) = |\cos(x)|$ for $-10 \le x \le 10$.

```
clear
n = 1000;
x = linspace(-10,10,n);
for k = 1:n
  y(k) = abs(cos(x(k)));
end%for
plot(x,y);
```

This code is correct, but very inefficient. It does **not** use some of the best features of MATLAB and *Octave*. Many of the built-in functions apply directly to vectors. This is illustrated by the next implementation of the above calculations. A vector of 1000 numbers, ranging from -10 to +10 is generated, then the values of the cos-function is stored in the new vector y. The result is then element-wise multiplied with the sign of the cos-values and then plotted, leading to Figure 1.3.

```
clear
n = 1000;
x = linspace(-10,10,n);
y = cos(x);    s = sign(y);
plot(x,s.*y);
```

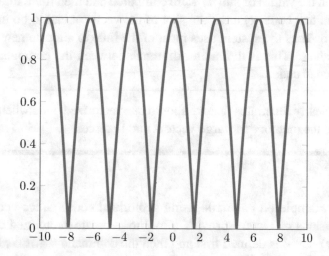

Figure 1.3: Graph of the function $|\cos(x)|$

1 Example : Speed of vectorized code

The three sections of code below compute the values of $\sin(n)$ for the arguments $n = 1, 2, 3, \ldots, 100000$. One might expect similar computation times.

```
──────────────────────── Octave ────────────────────────
clear; N = 100000;
tic();   for n = 1:N;   x(n) = sin(n); end%for
timer1 = toc()
tic();    x = zeros(N,1);   for n = 1:N;   x(n) = sin(n); end%for
timer2 = toc()
tic(); x = sin([1:N]);
timer3 = toc()
```

On a sample run (with an older version of *Octave*) we found

$$\text{timer1} = 48 \text{ sec} \qquad \text{timer2} = 2.4 \text{ sec} \qquad \text{timer3} = 0.013 \text{ sec}$$

and thus a drastic difference in performance. There are two major contributions to this effect:

- **vectorized code**

  In the third code segment the sin-function was called with a vector as argument and *Octave* could compute all values at once. The penalty for a function call had to be paid only once. This is the main difference between the computation time of the second and third code.

- **preallocation of vectors**[5]

  For the second and third code the resulting vector x was first created with the correct size and then the values of $\sin(n)$ were computed and then filled into the preallocated array. In the first code segment the size of the vector x had to be increased for each computation. Thus the system uses most of the time to allocate new vectors and then copy old values. This is the main difference between the computation time for the first and second code.

> This example clearly illustrates that we should use vectorized code whenever possible and preallocate the memory for large vectors and matrices.

Examine the example on page 104 using vectorized code. Since vectorization is important newer versions of *Octave* provide a tool to generate vectorized code. A function $F(x) = x \cdot \cos(x) \cdot e^{(x^2)}$ is defined first and then the command vectorize() is used to generate a vectorized version. More and applied examples of vectorized codes are shown in Section 1.5.

```
F = inline("x*cos(x)*exp(x^2)")
F(2)
Fv = vectorize(F)
Fv([2,3])
```

2 Example :   If the integral of the function in Figure 1.3 is to be computed, use the trapezoidal rule

$$\int_a^b f(x)\,dx \approx \sum_{k=1}^{n-1} \frac{f(x_k) + f(x_{k+1})}{2}\,(x_{k+1} - x_k)\,.$$

At first use a straightforward implementation of these formula, with a loop.

---

[5]Newer version of *Octave* use an improved memory allocation scheme and thus the first loop will be considerably faster.

```
n = 1000;   %% number of grid points
x = linspace(-10,10,n);
y = abs(cos(x));
plot(x,y);
intcgral = 0;
for k = 1:(n-1)
    integral = integral+0.5*(y(k)+y(k+1))*(x(k+1)-x(k));
end%for
integral
```

But this code does again **not** use some of the best features of MATLAB/*Octave*. The summation can be written as scalar product of two vectors

$$\langle \vec{x}, \vec{y} \rangle = \sum_{k=1}^{n} x_k\, y_k\ ,$$

or if the row vectors are regarded as $1 \times n$ matrices, as MATLAB does, use

$$\mathbf{x} \cdot \mathbf{y}' = \sum_{k=1}^{n} x_k\, y_k\ .$$

With the help of $dx_k = x_{k+1} - x_k$ the summation will run considerably faster.

```
y = cos(x);
y = sign(y).*y;

dx = diff(x);
ynew = (y(2:n)+y(1:n-1))/2;
nintegral = ynew*dx'
```

The built-in function `trapz` uses exactly the above idea to perform a numerical integration.

It is usually much faster to use the built-in vectorization than to use loops. Vectorization is one of the main speed advantages of MATLAB/*Octave* over other programs. It should be used whenever possible. Use `tic()` and `toc()` (see Section 1.1.10) to determine the time necessary to run through a section of code.

```
clear   % fast version
x = linspace(-10,10,10000);
tic();
for k = 1:10
    y = exp(sin(x.*x));
end%for
toc()
```

```
clear % slow version
x = linspace(-10,10,10000);
tic();
for k = 1:10
  for i = 1:10000
    y(i) = exp(sin(x(i)*x(i)));
  end%for
end%for
toc()
```

Since vectorization is important MATLAB and *Octave* provide support for this. Some of the basic operations (e.g. +-*/) can be performed element-wise on vectors or matrices, i.e. each entry will be computed separately. MATLAB/*Octave* will ignore the vector or matrix structure of the variable. Some books use the key word **point wise** operations instead of element wise operations. As a consequence MATLAB/*Octave* uses a preceding point to indicate element wise operations. As an example to compute $x(n) = n \cdot \sin(n)$ for $n = 1, 2, \ldots, 10$ we can use a loop

```
for n = 1:10
  x(n) = n*sin(n);
end%for
```

or the faster, vectorized code

```
n = 1:10;
x = n.*sin(n)
```

Matrices

Octave obviously has many commands for operations with matrices. Only very seldom loops have to be used for matrix operations.

```
clear      % clear all previously defined variables and functions
a = [1 2 3; 4 5 6; 7 8 10]
det(a)     % compute the determinant of the matrix
inv(a)     % compute the inverse matrix
a^2        % compute the square of the matrix, i.e. a*a
a.^2       % compute the square of each entry in the matrix
(a+a')/2   % compute the symmetric part of the matrix
a*inv(a)   % should yield the identity matrix
a.*inv(a)  % multiply each entry in the matrix with the
           % corresponding entry in the inverse matrix
```

Systems of linear equations

Obviously MATLAB should be capable of solving systems of linear equations. To solve a system $A\,\vec{x} = \vec{b}$ of 3 linear equations for 3 unknowns we best use the command x=A\b , i.e. we 'divide' the vector b from the left by the matrix A. Of course the inverse matrix could be used, but the computation is not done as efficiently and the results are not as reliable. As an example we consider the linear system

$$\begin{bmatrix} 1 & 2 & 3 \\ 4 & 5 & 6 \\ 7 & 8 & 10 \end{bmatrix} \cdot \vec{x} = \begin{pmatrix} 1 \\ 2 \\ 3 \end{pmatrix}$$

to be solved by

```
A = [1 2 3; 4 5 6; 7 8 10];
b = [1;2;3];
x = inv(A)*b
x = A\b
```

Computing the inverse matrix is rarely a reasonable way to solve a numerical problem. The other method is also more reliable as shown by the example using a matrix with zero determinant.

$$\begin{bmatrix} 1 & 2 & 3 \\ 4 & 5 & 6 \\ 7 & 8 & 9 \end{bmatrix} \cdot \vec{x} = \begin{pmatrix} 1 \\ 2 \\ 3 \end{pmatrix}$$

```
A = [1 2 3 ; 4 5 6; 7 8 9];
b = [1;2;3];
x1 = inv(A)*b;
control = A*x1
x2 = A\b
control2 = A*x2
```

MATLAB and *Octave* will show a warning, but then generate a result after all. Since this matrix A is not invertible the command null(A) will give more information about the solvability of the linear system.

One special 'feature' of *Octave* and MATLAB is that also systems with more equations than unknowns lead to a solution. The example considers 4 equations for 3 unknowns (A is a 4×3 matrix)

$$\begin{bmatrix} 1 & 2 & 3 \\ 4 & 5 & 6 \\ 7 & 8 & 10 \\ 1 & 2 & 4 \end{bmatrix} \cdot \vec{x} = \begin{pmatrix} 1 \\ 2 \\ 3 \\ 2 \end{pmatrix}$$

and is clearly **not** solvable, but consider the result for this system of over-determined equations:

```
clear
A = [1 2 3 ; 4 5 6; 7 8 10;1 2 4 ];
b = [1;2;3;2];
x = A\b
A*x
```

Octave and MATLAB return as solution vector x with residual vector r of smallest length, i.e. the best possible solution.

$$\text{Find vector } \vec{x} \text{ such that} \quad \|\vec{r}\| = \|\mathbf{A}\,\vec{x} - \vec{b}\| \quad \text{is minimal}$$

This can be a rather useful feature[6], but also create a problems if the user is not aware of what MATLAB or *Octave* are actually computing.

1.1.9   Broadcasting

In newer versions of *Octave* broadcasting is applied to some operations and then the computational rules for matrix operations are not strictly respected. As an example consider the subtraction

$$\begin{bmatrix} 1 & 2 & 3 \\ 4 & 5 & 6 \end{bmatrix} - \begin{bmatrix} 1 \\ 2 \end{bmatrix}$$

which is mathematical nonsense. With broadcasting *Octave* will automatically subtract the vector from each column of the matrix, but not without a warning.

```
A = [1 2 3; 4 5 6];    b = [1;2];
r1 = A-b
-->
r1 =    0   1   2
        2   3   4
```

Older versions of MATLAB do not know about automatic broadcasting. Since version of MATLAB (R2016b) seems to use broadcasting too. Both *Octave* and MATLAB have the broadcasting function bsxfun(), used below to achieve the same result.

```
r2 = bsxfun(@minus,A,b)
```

Use doc bsxfun to find out which operations can be broadcasted.

---

[6]This can be used to solve linear regression problems, see Section 3.2.

1.1.10   Timing of code and using a profiler

Timing with `tic()`, `toc()` **and** `cputime()`

In the above code we used `tic()` and `toc()` to determine the run-time of the loops. The resolution of this timer is not very good and it displays the actual time, and not the computation time used by the code. We may use a higher resolution timer based on the function `cputime()`. Examine the example below.

```
t0 = cputime();
x = sin([1:100000]);
timer = cputime()-t0
```

- The pair `tic()`, `toc()` is based on the wall clock, i.e. if you wait 10 seconds before typing `toc` those 10 seconds will be taken into account.

- `cputime()` is measuring the CPU time consumed by the current job. Thus just waiting will not increase `cputime()`.

- Some commands (e.g. `fft()`, `fft2()`) automatically use multithreaded libraries and `cputime()` will add up all the time consumed in the multiple threads, thus you might end up with more CPU time than wall time!

Using the profiler

Octave has a powerful profiler. It will analyze where your code is consuming time. Here is a simple example.

```
──────────────────────── Octave ────────────────────────
clear *
profile on  % turn the profiler on
n = 1000;
for jj = 1:100
  a = rand(n);   b = exp(cos(a));
end%for

T = profile('info');
profile off    % turn the profiler off
profshow(T)    % display the result
profexplore(T) % interactive exploration
```

and the results might look like

#	Function Attr	Time (s)	Time (%)	Calls
3	exp	1.530	42.13	100
2	cos	1.286	35.40	100
1	rand	0.816	22.47	100

4	profile	0.000	0.00	1
6	binary !=	0.000	0.00	1
5	nargin	0.000	0.00	1
7	__profiler_data__	0.000	0.00	1

Thus the above code called the function exp() 100 times and this consumed 1.5 sec of CPU time to evaluate the exp functions. This is not too bad, as each call actually computed a million[7] values of the exponential function. Thus it only took 0.015 μs to compute one value. Compare this to the result of the code below and act accordingly!

```
n = 1000;
for ii = 1:n
  for jj = 1:100
    a = rand(1);
    b = exp(cos(a));
  end%for
end%for
```

1.1.11   Debugging your code

There are different options to debug *Octave*/MATLAB code.

- In the editor window you can set a breakpoint. When running the code execution will stop at this point and you can use the command line to examine the current content variables, modify them, or continue with the execution of the program.

- To continue the running program type dbcont. Execution of the current function will continue. With dbquit the remaining part of the current function will not be executed, but MATLAB/*Octave* returns to the calling function.

- The GUI editor allows to single step through the code, step into sub-functions, or run the code to the next breakpoint.

- With the command dbstop you can set break point form the command line, e.g. dbstop myfunction 17 will set a breakpoint in the function myfunction() on line 17. With dbclear you can clear the breakpoint.

- You may also use the function keyboard() to interrupt running code on a selected line and the use the command line to examine the current values of the variables.

1.1.12   Command line, script files and function files

With *Octave* different methods can be used to write code and then test the results.

- Use the **command line** in the interface
  This is useful only for very small sections of code or for debugging.

---

[7] rand(n) generates a $n \times n$ matrix of random values. With $n = 1000$ this leads to a million values.

- Write a **script file**
  Longer pieces of code can be written with your favorite editor and stored in a file with the extension `.m`, e.g. `foobar.m`. Then the code can be run from the command line by typing `foobar`.

- Write a **function file**
  Functions to be used repeatedly can be written with an editor of your choice and then stored. A function can have one or multiple arguments and can also return one or multiple results.

Script files

Script files are used to write code to be run repeatedly, often with slight modifications. This is a common method to work with *Octave*.

If the code below is stored in the file `foobar.m`, then a circle with radius 3 in the complex plane will be generated by calling `foobar`.

foobar.m
```
t = linspace(0,2*pi,200);   % generate 200 points, equally spaced
z = 3*exp(i*t);             % compute the values in the complex plane
plot(z)                     % generate the plot
grid on; axis equal         % add a grid and equal scaling on both axis
title('Circle, radius 3')   % set a title
```

Observe that on the *Octave* command line you type `foobar` without the trailing extension `.m`. If working with the editor of the MATLAB or *Octave* GUI use the F5 key to save the file and run the code.

Function files

Function files can be used to define functions. As a first example we consider a statistical function to compute the mean value and the standard deviation of a vector of values. For a vector $\vec{x}$ with n values use

$$\text{mean} = \mu = \frac{1}{n}\sum_{k=1}^{n} x_k \quad \text{and} \quad \text{stdev}^2 = \sigma^2 = \frac{1}{n-1}\sum_{k=1}^{n}(x_k - \mu)^2 .$$

stat.m
```
function [mean,stdev] = stat(x)
% STAT Interesting statistics.
% This documentaion is displayed by 'help stat'
  n = length(x);
  mean = sum(x)/n;
  stdev = sqrt(sum((x-mean).^2)/(n-1));
```

This code has to be stored in a file `stat.m` and then can be used repeatedly, with one or two return arguments.

```
mymean = stat([1,2,3,4,5])
-->
mymean = 3
```

```
[mymean,mydev] = stat([1,2,3,4,5])
-->
mymean = 3
mydev = 1.5811
```

There are a few differences between MATLAB and *Octave* concerning script and function files.

- MATLAB[8] does not allow for definitions of functions within script files. Thus each function has to be given in a file of its own. This often leads to a large number of function files in a directory.

- In *Octave* the definition of a function may also be given within a script file and thus collections of functions in one file are possible. The end of the function has to be indicated by the keyword end or endfunction. MATLAB will not recognize the keyword endfunction.

1.1.13 Local and global variables, nested functions

In *Octave*/MATLAB the visibility of variables has to be taken into account. A variable declared in the workspace is not visible inside a function, unless you pass it as an argument.

```
                              —— Octave ——
a = 17     % set the value of a in the global workspace

function res = modify_a(x)
  % the variable a from the global workspace is not visible
  a = 2;   % here a is a new variable in the local context
  res = a*x;
endfunction

a2 = modify_a(a)
a          % will return the first value in the global context
-->
a  =  17
a2 =  34
a  =  17
```

---

[8]Starting with Version 2016b MATLAB allows definition of functions within script files too. The functions within a script file have to be at the end of the script.

If you really desire a variable to be visible inside a function (usually a bad idea) you can force Octave to do so by the keyword `global`, as illustrated by the code below.

```
                          ─────────── Octave ───────────
global a = 17     % set the value of a in the global workspace

function res = modify_a(x)
  global a  % the variable a is global variable
            % and its value is taken from the workspace
  res = a*x;
  a = 2;     % now the global variable is modified
endfunction

a2 = modify_a(a)
a
-->
a  =  17
a2 =  289
a  =  2
```

An important deviation from the above visibility of variable is given by nested functions. If a variable is declared in a function file, then it is visible inside all of the nested function in the same file.

```
                        ─────────── TestNest.m ───────────
function TestNest()
a = 17     % set the value of a in this function

function res = modify_a(x)
           % the variable a from the outer function is visible
  a = 2;   % the outer variable a is overwriten by 2
  res = a*x;
endfunction

a2 = modify_a(a)% display 34, the result of the function
a               % displays 2, modified by the function modify_a
endfunction
```

Since MATLAB allows nested functions, the above idea can be used to define many functions in one file. But observe that the functions defined within another function body will **not** be visible outside of the function. In addition nested functions can not be used inside program control statements e.g. `switch/case`, `if/else`, `while`, ...

1.1.14 Very elementary graphics

Find more information on graphics commands in Section 1.4. To generate a two–dimensional plot use the command `plot()` with the vector of the x and y values as arguments. The code below generates a plot of the function $v = \sin(x)$ with 21 values of x evenly distributed in the interval $0 \le x \le 9$. Find the result in Figure 1.4.

```
x = linspace(0,9,21);  y = sin(x);
plot(x,y)
```

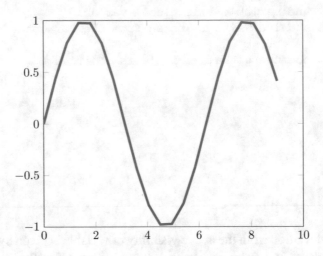

Figure 1.4: Elementary plot of a function

Multiple plots can be generated just as easily by calling plot() with more than one set of x and y values.

```
x = linspace(0,10,50);
plot(x,sin(x),  x,cos(x))
```

Considerably more information on graphics with MATLAB/*Octave* is given in Section 1.4, starting on page 86.

1.1.15 A breaking needle ploblem

The question

Assume that medical needles will penetrate skin if the applied force is larger than F_p and the same type of needles will break if the applied force is beyond a limiting breaking force F_b. The values of F_p are given by a normal distribution with mean value μ_p and standard deviation σ_p. The values of F_b are given by a normal distribution with mean value μ_b and standard deviation σ_b. To illustrate the method we work with hypothetical values for the parameters.

$$\mu_p = 2.0 \text{ [N]} \quad , \quad \sigma_p = 0.5 \text{ [N]} \quad , \quad \mu_b = 4.5 \text{ [N]} \quad , \quad \sigma_p = 0.635 \text{ [N]}$$

The theory and the resulting code

Since the normal distribution is given by the probability density function

$$p(x) = \frac{1}{\sigma \sqrt{2\pi}} \exp\left(-\frac{(x-\mu)^2}{2\sigma^2}\right)$$

we define a function in our script file to compute the value of this function for a vector of arguments.

```Octave
mup = 2.0; sigmap = 0.5; mub = 4.5; sigmab = 0.635;

% with Matlab put the function in a file normal_dist.m
function p = normal_dist(x,mu,sigma)
  p = 1/(sigma*sqrt(2*pi))*exp(-(x-mu).^2/(2*sigma*sigma));
endfunction
```

The above two distributions can then be visualized in Figure 1.5, generate by calling `plot()`.

```Octave
% generate values of forces from 0 to 8 N, stepsize df
df = 0.01; f = 0:df:8;
pp = normal_dist(f,mup,sigmap);
pb = normal_dist(f,mub,sigmab);
plot(f,pp,f,pb); grid on
```

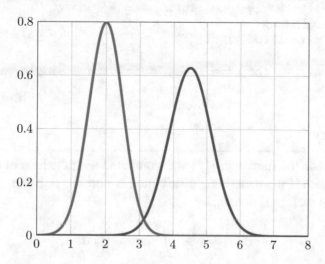

Figure 1.5: Probability of needles to penetrate, or to break

To be determined is the probability that a given needle will break before it penetrates the skin. To determine this value examine the following facts for very small values of Δf:

- With probability $p_p(f)\,\Delta f$ a needle penetrates at a force between f and $f + \Delta f$.

- For this needle to break the breaking force has to be smaller than f. This occurs with probability

$$p_2(f) = \int_{x=-\infty}^{f} p_b(x) \, dx$$

Since for our example find $p_b(x) \approx 0$ for $x < 0$ and thus

$$p_2(f) \approx \int_{x=0}^{f} p_b(x) \, dx$$

- The probability for a needle to break before penetrating at a force f is thus given by the probability density function.

$$p(f) = p_2(f) \cdot p_p(f)$$

and the total probability to fail is given by

$$P_{fail} = \int_{f=-\infty}^{\infty} p_2(f) \cdot p_p(f) \, df \approx \int_{f=0}^{8} p_2(f) \cdot p_p(f) \, df \ .$$

The above can be implemented, leading to Figure 1.6 of the probability distribution for failing and total probability of $P_{fail} \approx 0.00101 \approx 1/1000$.

```
%p2 = pb; p2(1) = 0;  % integration with a loop
%for k = 2:length(pb)
%      p2(k) = p2(k-1)+pb(k)*df;
%end%for
p2 = cumsum(pb);  % the same integration with a single command
pfail = pp.*p2;

plot(f,pfail);
trapz(f,pfail)
```

In the above code the function $p_2(f)$ was computed with the help of an integral, but is not the best approach if the underlying distribution is normal. A better approach is to use the **error function** erf(z), defined by

$$\text{erf}(z) = \frac{2}{\sqrt{\pi}} \int_0^z \exp(-t^2) \, dt \ .$$

Using the definition of the normal distribution[9]

$$p(x) = \frac{1}{\sigma\sqrt{2\pi}} \exp\left(-\frac{(x-\mu)^2}{2\sigma^2}\right)$$

---

[9]MATLAB and *Octave* have many and powerful commands to examine statistical questions of this type, see chapter 2.

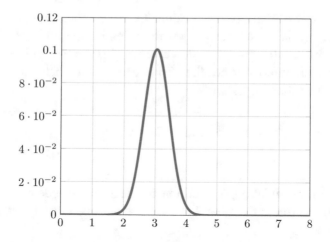

Figure 1.6: Probability distribution of needles to break before penetration

and basic calculus find

$$
\begin{aligned}
g(f) &= \int_{x=-\infty}^{f} p(x)\,dx = \int_{x=-\infty}^{\mu} p(x)\,dx + \int_{x=\mu}^{f} p(x)\,dx \\
&= \frac{1}{2} + \frac{1}{\sigma\sqrt{2\pi}} \int_{x=\mu}^{f} \exp\left(-\frac{(x-\mu)^2}{2\sigma^2}\right)\,dx
\end{aligned}
$$

$$
\text{substitution} \qquad t = \frac{x-\mu}{\sqrt{2}\,\sigma} \qquad , \qquad dx = \sqrt{2}\,\sigma\,dt
$$

$$
\begin{aligned}
&= \frac{1}{2} + \frac{\sqrt{2}\,\sigma}{\sigma\sqrt{2\pi}} \int_{t=0}^{\frac{f-\mu}{\sqrt{2}\,\sigma}} \exp\left(-t^2\right)\,dt \\
&= \frac{1}{2} \left(1 + \operatorname{erf}\left(\frac{f-\mu}{\sqrt{2}\,\sigma}\right)\right).
\end{aligned}
$$

Thus for the above problem we may equivalently write

$$
P_{fail} = \frac{1}{2} \int_{\infty}^{\infty} \left(1 + \operatorname{erf}\left(\frac{f-\mu_b}{\sqrt{2}\,\sigma_b}\right)\right) p_p(f)\,df .
$$

This eliminates the need for ignoring extreme values for the forces. In the above code the loop to compute the probability density function p_2 is replaced by a function call.

```
p2 = 0.5*(1+erf((f-mub)/(sqrt(2)*sigmab)));
```

A simple simulation

To verify the above computation we can run a simulation. The command `randn(1,NN)` creates NN random data with mean 0 and standard deviation 1. To obtain mean μ and standard deviation σ we multiply these values by σ and then add μ. With this type of simulated data for breaking and penetration forces count the number of pairs where the breaking force is smaller than the penetration force. This number of breaking needles is to be divided by the total number of needles. To understand the behavior of the comparison operator consider the following example.

```
res = [1 2 3 4 5 9] < [0 5 9 1 1 10]
-->
res =  0  1  1  0  0  1
```

Thus to examine, by a simulation, one million needles use the code below. The results will vary slightly from one run to the other, but should always be close to the above integration result.

```
NN = 1e6;    % one milion samples to be simulated
fpsimul = randn(NN,1)*sigmap+mup;
fbsimul = randn(NN,1)*sigmab+mub;
simulation = sum(fbsimul<fpsimul)/NN
-->
simulation = 9.6500e-04
```

A different approach to the problem is based on a different argument.

- With probability $p_b(f)\,\Delta f$ a needle does break at a force between f and $f + \Delta f$.

- For this needle to penetrate at a larger force find a probability

$$p_3(f) = \int_{x=f}^{\infty} p_p(x)\,dx \approx \int_{x=f}^{8} p_p(x)\,dx \ .$$

- The probability for a needle to break before penetration is thus

$$p(f) = p_3(f) \cdot p_b(f)$$

and the total probability to fail is given by

$$P_{fail} = \int_{f=-\infty}^{\infty} p_3(f) \cdot p_b(f)\,df \approx \int_{f=0}^{8} p_3(f) \cdot p_b(f)\,df \ .$$

The final result has to be the same as for the first approach.

```
p3 = pb; p3(length(pp)) = 0;
for k = fliplr(1:(length(pp)-1));
    p3(k) = p3(k+1) + pp(k)*df;
end%for
```

```
pfail = pb.*p3;
plot(f,pfail);
trapz(f,pfail)
-->
ans = 1.0108e-03
```

1.2 Programming with *Octave*

This short section cannot and will not cover all aspects of programing with MATLAB or *Octave*. You certainly have to consult the online documentation and other references, e.g. [Grif18] or [BiraBrei99]. The amount of available literature is enormous, free and nonfree. It is the goal of this section to get you started and point out some important aspects of MATLAB/*Octave*. It is assumed that you are familiar with a procedural programming language, e.g. C, C++, Java, ...

1.2.1 Displaying results and commenting code

If a line of code is not terminated by a semicolon (;) the results will be displayed in the command window with the name of the variable, followed be a new line. If you do not want to see the result, add the semicolon. With the command disp() display the result. One can use formatted output to add information, see Section 1.2.4.

```
x = exp(pi)
disp(x)
disp(sprintf('the value of exp(pi) is %g',x))
fprintf('the value of exp(pi) is %g\n',x)
-->
x =   23.141
23.141
the value of exp(pi) is 23.1407
the value of exp(pi) is 23.1407
```

An essential feature of good code is a readable documentation. For *Octave* you may use the characters % or # to start a comment, i.e. any text after those signs will not be examined by the *Octave* parser. In MATLAB only % can be used.

```
a = 1+2; % this is a comment in Octave and Matlab
b = 2-1; # this is a comment in Octave,
         # but will lead to an error in Matlab
```

There is no feature in MATLAB or *Octave* to comment out complete sections, you have to do line by line. Many editors provide functions to comment and uncomment sections.

1.2.2   Basic data types

MATLAB/*Octave* provides a few basic data types and a few methods to combine the basic data types. For a defined variable `var` *Octave* will display its data type with the command `typeinfo(var)`.

Numerical data types

The by far most important data type in *Octave* is a **double precision floating point number**. For most numerical operations this is the default data type and you **have to ask for** other data types. On systems that use the IEEE floating point format, values in the range of approximately 2.2251e-308 to 1.7977e+308 can be stored, and the relative precision is approximately 2.2204e-16, i.e you can expect at best 15 decimal digits to be correct. The exact values are given by the variables `realmin`, `realmax`, and `eps`, respectively. The information about your system can be obtained by the code below.

```
[realmin, realmax, eps]
-->
ans =    2.2251e-308   1.7977e+308   2.2204e-16
```

Based on floating point numbers we may built vectors and matrices with real or complex entries.[10] Complex numbers are described by their real and imaginary parts. All arithmetic operations and most mathematical functions can be applied to real or complex numbers.

```
a = 1.0+2i; b = 3*i;
[a+b, a*b, a/b]
[cos(a), exp(b), log(a+b)]
-->
ans = 1.00000 +5.00000i   -6.00000 +3.00000i    0.66667 -0.33333i
ans = 2.03272 -3.05190i   -0.98999 +0.14112i    1.62905 +1.37340i
```

Observe that the simplified syntax `2i` is equivalent to `2*i`.

Octave/MATLAB also provides the data type **single**, i.e. single precision floating point numbers. Its main advantage is to use less memory, the disadantage is a smaller range and resolution.

```
[realmin('single'), realmax('single'), eps('single')]
-->
ans =   1.1755e-38   3.4028e+38   1.1921e-07
```

You can convert between single and double variables with the commands `single()` and `double()`. The command `whos` will display all current variables, their size, memory foot print and their class.

---

[10]For many years a matrix of floating point numbers was the **only** data type in MATLAB.

```
clear *
a = rand(3);
aSingle = single(a);
whos
-->
Variables in the current scope:

  Attr Name        Size                   Bytes  Class
  ==== ====        ====                   =====  =====
       a           3x3                       72  double
       aSingle     3x3                       36  single

Total is 18 elements using 108 bytes
```

MATLAB/*Octave* has **integer data types** with fixed ranges. Find signed and unsigned integers with 8, 16, 32 or 64 bit resolution. In Table 1.2 find the types and their corresponding ranges.

type	min	max
int8	-128	+127
uint8	0	+255
int16	-32'768	+32'767
uint16	0	+65'535
int32	-2'147'483'648	+2'147'483'647
uint32	0	+4'294'967'295
int64	-9.2e+18	+9.2e+18
uint64	0	+1.8e+19

Table 1.2: Integer data types and their ranges

The basic arithmetic operations (+ - * /) are available for these types, with the usual results, as illustrated by an elementary example.

```
a = int16(100); b = int16(111);
[a+b, a-b, 3*b]
-->
ans =   211   -11   333
```

One has to watch out for the range of these types and the consequences on the results. The range used by uint8 is given by $0 \leq x \leq 255$

```
a = uint8(100); b = uint8(111);
[a+b, a-b, 3*b]
-->
ans = [211 0 255]
```

When applied to floating point numbers the commands int8(), int16(),... do not return
the integer part, but use rounding.

Integer data types with a prescribed resolution may be used to develop code for micro
controllers, as shown in Section 3.4, starting on page 293.

Observe that there are considerable differences in how the programming language C
and MATLAB/*Octave* handle integers.

- In *Octave* calculations are truncated to their range, e.g. the command int8(100)+
  30 leads to 127.

- Operations with different integer types are not allowed, i.e. using the command
  int8(10)+int16(70) will generate an error message.

- Operations with integer and floating types are allowed and lead to integer results,
  again truncated to their domain.

Characters

Individual letters can be given as characters, internally they are represented by integers
and conversions are possible, see Section 1.2.4. The internal representation leads to some
surprising results. You can subtract and add letters, or add numbers to letters, for the computations the ASCII codes are used.

```
char1 = 'a';   char2 = 'b'; char3 = 'A';
b_minus_a = char2-char1
a_minus_A = char1-char3
a97 = (char1==97)
-->
b_minus_a =  1
a_minus_A = 32
a97 =  1
```

1.2.3 Structured data types and arrays of matrices

Building on the basic data types *Octave* can work with a variety of structured data types:
vectors, matrices, strings, structures, cell arrays, lists.

Strings

Octave also works with strings, consisting of a sequence of characters, enclosed in either singe-quotes or double-quotes. With MATLAB only single quotes are allowed, thus one might consider using those exclusively. Internally strings are represented as vectors of single characters and thus they can be combined to longer strings.

```
                              Octave
name1 = 'John'   % a string in Octave and Matlab
name2 = "Joe"    % a string in Octave only
combined = [name1,' ',name2]
-->
name1 = John
name2 = Joe
combined = John Joe
```

One may also create a vector of strings, but it will be stored as a matrix of characters. As a consequence each string in the vector of strings will have the same length. Missing characters are replaced by spaces.[11]

```
                              Octave
combinedMat = [name1;name2]
size(combinedMat)
size(combinedMat(2,:))
-->
combinedMat=
John
Joe
ans = 2 4
ans = 1 4
```

Structures

One of the major disadvantages of matrices is, that all entries have to be of the same type, most often scalars. *Octave* supports structures, whose entries may be of different types. This feature is not yet used in many codes. Consider the trivial example below.

```
Customer1.name = 'John';
Customer1.age = 23;
Customer1.address = 'Empty Street'
-->
Customer1 =
{
  address = Empty Street
  age = 23
  name = John
}
```

---

[11]In MATLAB you will have to use the function `str2mat()`.

Lists and cell arrays

Instead of structures we may use cell arrays. To access and create cell arrays curly brackets (i.e. {}) have to be used. As illustration consider the example below.

```
c = {1, 'name', rand(2,2)}
-->
c =
{
  [1,1] =  1
  [1,2] = name
  [1,3] =
      0.17267    0.87506
      0.73041    0.85009
}
```

Each entry in a cell array may be used independently.

```
c{2}
-->
ans = name
```

```
m = c{3}
-->
m =   0.17267    0.87506
      0.73041    0.85009
```

or as a subset of the cell array.

```
c{2:3}
-->
ans = name
ans = 0.17267    0.87506
      0.73041    0.85009
```

Use cell arrays to construct multidimensional matrices. With the code below store the matrices

$$
\begin{bmatrix} 2 & 0 & 0 \\ 0 & 2 & 0 \\ 0 & 0 & 2 \end{bmatrix} \quad \text{and} \quad \begin{bmatrix} 1 & 1 & 1 \\ 1 & 1 & 1 \\ 1 & 1 & 1 \end{bmatrix}
$$

in two different layers of the $2 \times 3 \times 3$ matrix `mat3` and then compute

$$
\begin{bmatrix} 1 & 1 & 1 \\ 1 & 1 & 1 \\ 1 & 1 & 1 \end{bmatrix} - \begin{bmatrix} 2 & 0 & 0 \\ 0 & 2 & 0 \\ 0 & 0 & 2 \end{bmatrix}.
$$

```
mat3 = cell(2,1);
mat3{1} = 2*eye(3);
mat3{2} = ones(3);
mat3{2} - mat3{1}
-->
ans =
  -1   1   1
   1  -1   1
   1   1  -1
```

One has to observe that no standard computational rules for 3-d matrices are defined.

Cell arrays can be used as a counter in a loop.

```
disp("Loop over a cell array")
for i = {1,"two","three",4}
  i
end%for
```

Arrays of matrices or N-d matrices

With *Octave* we may also work with matrices of higher dimensions, i.e. arrays of matrices. The command below constructs 5 matrices of size 3×2 and fills it with random numbers. You may visualize this by stacking the 5 matrices on top of each other and thus obtain a 5 story building with one 3×2 matrix on each floor. Or consider it a matrix with three dimensions of size $3 \times 2 \times 5$. To access individual entries use three indices, e.g. A(2,1,4). As an example we compute for each position the average along the height, leading to a 3×2 matrix. The average is computed along the third dimension of the matrix.

```
A = rand(3,2,5);
Amean = mean(A,3)
-->
Amean =   0.79681   0.70946
          0.60815   0.48610
          0.38403   0.47336
```

To extract the "second floor" matrix you may use A2=A(:,:,2), but the result will be a 3 dimensional object of size $3 \times 2 \times 1$. To convert this into a classical 3×2 matrix use the command squeeze(). As an example multiply the transpose of the second floor with the fifth floor, leading to a 2×2 matrix.

```
squeeze(A(:,:,2))'*squeeze(A(:,:,5))
-->   1.06241   1.05124
      0.84722   1.03028
```

Arrays of matrices can be very convenient and many *Octave*/MATLAB commands are directly applicable to these N-d matrices, but not all commands.

1.2.4　Built-in functions

MATLAB/*Octave* provide a large number of built-in functions. Most of them can be applied to scalar arguments, real or complex. This may lead to a few surprises.

Functions with scalar arguments

- **trigonometric functions**: `sin()`, `cos()`, `tan()`, `tan()`, `asin()`, `acos()`, `atan()`, `atan2()`

  Observe that all trigonometric functions compute in radians, and not degrees. There are a few functions with a version using degree, e.g. `sind()`, `cosd()`. This author does not use those.

```
r1 = cos(pi)
r2 = acos(-1)
r3 = cos(i)
r4 = acos(-1.01)
r5 = acos(-1.01+i*1e-15)
r6 = tan(0.5)
r7 = atan(0.5)
r8 = atan2(-0.5,-1)
```

The function `atan2()` above is very useful to convert Cartesian coordinates to cylindrical coordinates. For given x and y we find the radius r and angle ϕ by solving

$$x = r \cos(\phi) \quad \text{and} \quad y = r \sin(\phi) \, .$$

Octave will compute the values by

```
r   = sqrt(x^2 + y^2)
phi = atan2(y,x)
```

Observe that using `atan((-y)/(-x))` will lead to the same result as `atan(y/x)`, while the calls `atan2(-y,-x)` and `atan2(y,x)` yield result that differ by $\pi = 180°$.

- **exponential functions**: `exp()`, `cosh()`, `sinh()`, `tanh()`, `pow2()`. For most of the exponential functions the corresponding inverse functions are also available: `log()`, `log2()`, `log10()`, `acosh()`, `asinh()`, `atanh()`. The only possible surprise is that the functions can be used with complex arguments. The code below verifies the Euler formula $e^{i\alpha} = \cos(\alpha) + i \sin(\alpha)$ for $\alpha = 0.2$.

```
alpha = 0.2;   [exp(i*alpha), cos(alpha), sin(alpha)]
-->
0.98007 + 0.19867i   0.98007 + 0.00000i   0.19867 + 0.00000i
```

- **generating random numbers**: for simulations it is often necessary to generate random numbers with a given probability distribution. The command rand(3) will generate a 3×3 matrix of random numbers, uniformly distributed in the interval $[0, 1]$.

```
r = rand(3)
-->
r =   0.694482   0.747556   0.266156
      0.609030   0.713823   0.054658
      0.461212   0.695820   0.769618
```

One often needs random numbers with a normal distribution, i.e. the command randn(). As an example we create a vector of 1000 random numbers, with mean 2 and standard deviation 0.5, then generate a histogram, see Figure 1.7.

```
N = 1000;
x = 2 + 0.5*randn(N,1);
hist(x)
```

Octave provides a few more distributions of random numbers.

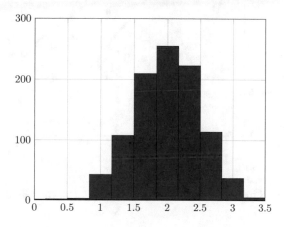

Figure 1.7: Histogram of random numbers with mean 2 and standard deviation 0.5

- rand() : uniformly distributed in $[0, 1]$.
- randi() : integer random numbers.

- randn() : normal distribution with mean 0 and variance 1.

- rande() : exponentially distributed.

- randp() : Poisson distribution.

- randg() : gamma distribution.

- **special functions**: many special functions are directly implemented in *Octave*, e.g. most of the Bessel functions bessel(), besselh(), besseli(), besselj(), besselk(), bessely(). One of the many applications of Bessel functions is radially symmetric vibrating drums, the zeros of the Bessel function $J_0(x)$ lead to the frequencies of the drum. The code below generates the plot of the function $f(x) = J_0(x)$ and its derivative $f'(x) = -J_1(x)$, find the result in Figure 1.8. There are many other special functions, e.g airy(), beta(), bincoeff(), erf(), erfc(), erfinv(), gamma(), legendre(). A good reference for special functions and their basic properties is [AbraSteg] or its modern version [DLMF15], freely accessible on the internet at DLMF.nist.gov.

```
x = 0:0.01:14;
plot(x, besselj(0,x),   x,-besselj(1,x))
legend('BesselJ0','derivative')
```

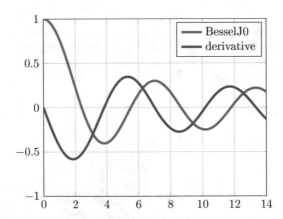

Figure 1.8: Graph of the Bessel function $J_0(x)$ and its derivative

Functions with matrix and vector arguments

A large number of numerical functions can be applied to vectors or matrices as arguments. The corresponding function will be computed for each of the values. As a simple example for vectors consider

```
x = [-1, 0, 1,3]
exp(x)
cos(x)
sqrt(x)
-->
x =   -1   0   1   3
exp(x) =      0.36788    1.00000    2.71828   20.08554
cos(x) =      0.54030    1.00000    0.54030   -0.98999
sqrt(x)=      0 + 1i     0+ 0i      1 + 0i    1.73205 + 0i
```

or for matrices

```
A = [ 0 pi 1; -1 -pi 2];
r1 = cos(A)
r2 = exp(A)
r3 = sqrt(A)
-->
r1 =
    1.00000   -1.00000    0.54030
    0.54030   -1.00000   -0.41615

r2 =
    1.000000    23.140693    2.718282
    0.367879     0.043214    7.389056

r3 =
    0.00000 + 0.00000i    1.77245 + 0.00000i    1.00000 + 0.00000i
    0.00000 + 1.00000i    0.00000 + 1.77245i    1.41421 + 0.00000i
```

3 Example : Element wise operations

Assume that for a vector of x values you want to compute $y = \sin(x)/x$. The obvious code

```
x = linspace(-1,1);
y1 = sin(x)/x
-->
y1 =   0.90164
```

produces certainly not the desired result. To compute $y_2 = \sin(x) \cdot x^2$ the code below yields not the expected vector either.

```
x  = linspace(-1,1);
y2 = sin(x)*x^2
-->
error: for A^b, A must be square
```

In both cases *Octave* (and MATLAB) use matrix operations, instead of applying the above operations to each of the components in the corresponding vector. To obtain the desired result the **dot operations** have to be used, as shown below.

```
x = linspace(-1,1);
y1 = sin(x)./x ;
y2 = sin(x).*x.^2;
plot(x,y1,x,y2)
```

Leaving out the dot (at first) is the most common syntax problem in MATLAB or *Octave*.

- x.*y will multiply each component of x with the corresponding component of y. Thus x and y have to be vectors or matrices of the same size. The result is of the same size.

- x./y will divide each component of x by the corresponding component of y. Thus x and y have to be vectors or matrices of the same size. The result is of the same size too.

- x.^2 will square each component of x and return the result as a vector or matrix of the same size.

◊

4 Example :   A few possible implementations of the norm function of a vector

$$\|\vec{x}\| = \sqrt{\sum_{i=1}^{n} x_i^2}$$

are given by

```
normX1 = sqrt(sum(x.^2))
normX2 = sqrt(sum(x.*x))
normX3 = sqrt(sum(x.*conj(x)))
```

All three codes return the correct result if the vector is real valued, i.e. $\vec{x} \in \mathbb{R}^n$. The last formula also generates the correct result for complex vectors $\vec{x} \in \mathbb{C}^n$.

5 Example : Matrix exponential

There are also functions that behave differently. The solution of the linear system of differential equations

$$\frac{d}{dt}\vec{x}(t) = \mathbf{A}\,\vec{x}(t) \quad \text{with} \quad \vec{x}(0) = \vec{x}_0$$

can formally be written as

$$\vec{x}(t) = \exp(t\,\mathbf{A})\,\vec{x}_0 \,.$$

Theoretically this matrix could be computed with a Taylor series

$$\exp(t\,\mathbf{A}) = \sum_{k=0}^{\infty}\frac{1}{k!}t^n\,\mathbf{A}^n = \mathbb{I} + t\,\mathbf{A} + \frac{1}{2}t^2\,\mathbf{A}^2 + \frac{1}{6}t^3\,\mathbf{A}^3 + \frac{1}{24}t^4\,\mathbf{A}^4 + \dots$$

Unfortunately this is neither fast nor reliable, as are many other simple ideas, most of them unreliable, see [MoleVanLoan03]. *Octave*/MATLAB can compute this matrix $\exp(t\,\mathbf{A})$ reliably with the command `expm()` (**matrix exponential**). As example consider

```
expm([1,2;3,4])
-->
     51.969     74.737
    112.105    164.074
```

Observe that the result is different from `exp([1,2;3,4])`. There also exists a similar function `logm()`, i.e. `A = logm(B)` will compute a matrix $\mathbf{A}$ such that $\mathbf{B} = \text{expm}(\mathbf{A})$. There is also a function `sqrtm()` to compute the square root of a matrix, illustrated below.

```
B  = [1,2;3,-4]
sB = sqrtm(B)
sB * sB
-->
B =   1    2
      3   -4

sB =    1.21218 + 0.31944i    0.40406 - 0.63888i
        0.60609 - 0.95831i    0.20203 + 1.91663i

sB*sB =    1.00000 - 0.00000i    2.00000 - 0.00000i
           3.00000 + 0.00000i   -4.00000 + 0.00000i
```

Observe that the result is different from `sqrt(B)`.   ◇

Formatted input and output functions

When reading or writing data one often has to insist on specific formats, e.g. the number of digits displayed. This can be done with the functions in Table 1.3 with the format templates listed in Table 1.4.

printf()	formatted output to stdout, not in MATLAB
sprintf()	formatted output to a string
fprintf()	formatted output to a stream (file) or stdout
sscanf()	formatted input from a string
fscanf()	formatted input from a stream (file)
disp()	display a string on the terminal

Table 1.3: Formatted output and input commands

As an example we display the number π in different formats.

Octave
```
printf("this is pi: %6.3f \n",pi)   % in Matlab use sprintf()
printf("this is pi: %10.3e \n",pi)
printf("this is pi: %3i \n",pi)
-->
this is pi:  3.142
this is pi:  3.142e+00
this is pi:    3
```

%i or %d	signed integer
%ni or %nd	signed integer in a field of length n, no leading zeros
%0ni or %0nd	signed integer in a field of length n, with leading zeros
%u	unsigned integer
%f	regular floating point number
%8.3f	in field of width 8 with precision 3
%e	floating point number with exponential notation
%10.5e	in field of width 10 with precision 5
%g	floating point number in normal or exponential format
%s	string, will stop at white spaces
%c	one or multiple characters

Table 1.4: Some output and input conversion templates

6 Example :   Generating numbered file names

Assume that your want to number your files with a three digit integer, e.g. data001.dat, data002.dat, data003.dat, ...This requires that you translate the integer number into a string, including the leading zeros. The name is composed of three sections.

- The first four characters are data.

- Then use formatted printing into a string (`sprintf()`) to generate the string containing the 3 digit number, including the leading zeros. Use the format `'%03d'`.

- The tail of the file name is given by its extension `.dat`. Since strings are vectors of characters, combine the three segments.

Here is one possibility to do so.

```
numbers = [ 0 1 2 3 115]; % numbers to be considered
for n = numbers
   filename = ['data',sprintf('%03d',n),'.dat']
end
-->
filename = data000.dat
filename = data001.dat
filename = data002.dat
filename = data003.dat
filename = data115.dat
```

If using the simpler format `'%d'`, then the length of the file names might change, see below.

```
numbers = [ 0 1 2 3 115]; % numbers to be considered
for n = numbers
   filename = ['data',sprintf('%d',n),'.dat']
end
-->
filename = data0.dat
filename = data1.dat
filename = data2.dat
filename = data3.dat
filename = data115.dat
```

In a real example you will obviously perform other operations in the loop too!   ◇

7 Example :   Formatted scanning

Formatted scanning has to be used when information is to be extracted from strings, as shown in this example. Within a string s a few digits of a number are displayed. First read the correct number of characters, then scan for the scalar number and finally store the remainder of the string. For subsequent calculation the scalar number `piApprox` can be used.

- The format string `%10c%e%s` consists of three contributions.

   - `%10c` : read a vector consisting of 10 characters
   - `%e` : read one floating point number
   - `%s` : read the remainder as a string

- The last parameter `"C"` indicates that we use a C style formatting, *Octave* only.

```
┌──────────────────────────── Octave ────────────────────────────┐
  s = 'pi equals 3.14 approximately';
  [head, piApprox, tail] = sscanf(s,'%10c%e%s','C')
  -->
  head = pi equals
  piApprox = 3.1400
  tail = approximately
└─────────────────────────────────────────────────────────────────┘
```

Codes of the above type have to be used when reading data from a file. MATLAB's version of scanf() behaves differently. The function textscan() serves as a replacement of sscanf(), as shown by the example.

```
┌──────────────────────────── Matlab ────────────────────────────┐
  A  =  textscan(s,'%10c%f%s')
  piApprox = A{2}
└─────────────────────────────────────────────────────────────────┘
```

Another possibility for problems of the above type is to use the function regexp().

◇

Conversion functions

If you need to translate one data type into another MATLAB/*Octave* provides a few functions.

- char() : will convert an integer to the corresponding character. The function can be applied to integers, vectors or matrices of integers. You may also use the function with cell arrays, see help char.

```
┌─────────────────────────────────────────────────────────────────┐
  c1 = char(65)
  c2 = char(65:90)
  c3 = char([65:75;97:107])
  -->
  c1 = A
  c2 = ABCDEFGHIJKLMNOPQRSTUVWXYZ
  c3 = ABCDEFGHIJK
       abcdefghijk
└─────────────────────────────────────────────────────────────────┘
```

- toascii() [12] : will convert a character to the corresponding ASCII code, an integer. The function can be applied to strings or vectors of strings.

```
┌──────────────────────────── Octave ────────────────────────────┐
  oneLetter = toascii('A')
  name = toascii('BFH-TI')
  mat = toascii(['1 2 3';'abcde'])
  -->
└─────────────────────────────────────────────────────────────────┘
```

---

[12]Currently MATLAB does not provide the function tosacii(), but you may obtain similar results by using the function int16().

```
oneLetter =   65
name =   66   70   72   45   84   73
mat  =   49   32   50   32   51
         97   98   99  100  101
```

- `int2str` : this function converts integers to strings, e.g.

```
s = int2str(4711)
-->
s = 4711
```

Observe that in the above code the variable s is of type string and not a number, e.g. you can not add 1 to s. The functions `num2str()` and `mat2str()` may be useful for similar tasks.

```
——————————————— Octave ———————————————
s2 = num2str(10*pi)
s3 = mat2str(rand(2),[4,3])
s4 = s3([1:4])
-->
s2 = 31.416
s3 = [0.6087,0.1361;0.6818,0.5794]
s4 = [0.6
```

- *Octave* provides a few more conversion functions: `str2double()`, `str2num()`, `hex2num()`, `num2hex()`, `num2cell()`,...

- To convert strings to numbers the formatted scanning and printing functions above can be used too, i.e see Section 1.2.4.

The above function can be used when reading data from a file or writing to a file, as examined in Section 1.2.8.

1.2.5   Working with source code

One of the big advantages of open source code projects is that you have access to the source and can thus examine and even modify it to meet your needs.[13]

- Locating the source code. With the command `which` find the location of the source code for a given function.

```
which logspace
-->  'logspace' is a function from the file
          /usr/local/share/octave/6.2.0/m/general/logspace.m
```

---

[13]One has to be careful when redistributing modified code. With MATLAB it is forbidden and with *Octave* you have to respect the GPL license.

Using the location of the source file you can copy it into a directory of yours and adapt the code. This is a very useful feature if a given function does almost what you need. Then take the source as a (usually) good starting point. Not all function are written with *Octave*, but C++ and FORTRAN are used too, e.g.

```
┌──────────────────────────────── Octave ────────────────────────────────┐
  which cosh
  -->
  'cosh' is a built-in function from the file
          libinterp/corefcn/mappers.cc
└─────────────────────────────────────────────────────────────────────────┘
```

- To just look at the code you can also use `type logspace` and *Octave* or MATLAB will display the source code of the function.

- The source code for many (*Octave* only) functions have built-in tests and you can call those with the command `test`.

```
┌──────────────────────────────── Octave ────────────────────────────────┐
  test logspace
  -->
  PASSES 6 out of 6 tests
└─────────────────────────────────────────────────────────────────────────┘
```

- The source for many (*Octave* only) functions have built in demos, as an example examine the result of the `demo delaunay`.

1.2.6 Loops and other control statements

Within MATLAB/*Octave* the standard loops and controls statements are available. We illustrate them with elementary examples[14].

Loops generated by `for`

The general form is given by

```
┌──────────────────────────────── Octave ────────────────────────────────┐
  for VAR = EXPRESSION
      BODY
  endfor
└─────────────────────────────────────────────────────────────────────────┘
```

To display a list of the squares of the integer numbers between 1 and 9, with a remark, use the following code.

─────────────────────────

[14]If MATLAB is used instead of *Octave*, then `endfor` has to be replaced by `end`, `endwhile` by `end` and a few more minor changes of the same type. This author considers the *Octave* version to be more readable. A possible workaround is to use a well placed comment sign, e.g `end%for` or `end%while`. I have adapted this notation in most places and it is working just fine.

```
—————————————————— Octave ——————————————————
for k = 1:9
  printf('the square of %i is given by %i \n', k, k^2)
endfor
```

It is not necessary to use subsequent numbering. As an example use all odd numbers from 1 through 20, or a given list of 3 numbers (4, 7 , 11).

```
—————————————————— Octave ——————————————————
for k = 1:2:20
  printf('the square of %i is given by %i \n', k, k^2)
endfor

for k = [4 7 11]
  printf('the square of %i is given by %i \n', k, k^2)
endfor
```

Loops generated by `while` or `until`

The general form is given by

```
—————————————————— Octave ——————————————————
while (CONDITION)
   BODY
endwhile
```

As sample code generate the first 10 numbers of the Fibonacci sequence.

```
—————————————————— Octave ——————————————————
fib = ones (1, 10);
i = 3;
while (i <= 10)
  fib (i) = fib (i-1) + fib (i-2);
  i++;
endwhile
fib
```

With the `while` command the condition is tested first. If the body of the loop has to be executed first, and then the test performed we may use the `until` command whose general form is

```
—————————————————— Octave ——————————————————
do
   BODY
until (CONDITION)
```

The famous Fibonacci sequence is defined by

$$c_i = c_{i-1} + c_{i-2} \quad \text{and} \quad c_0 = 0, \ c_1 = 1$$

and generated by

```
─────────────────────────── Octave ───────────────────────────
  fib = ones (1, 10);
  i = 2;
  do
    i++;
    fib (i) = fib (i-1) + fib (i-2);
  until (i == 10)
  fib
```

MATLAB does not provide a do--until loop. One may simulate this with the help of an extra flag.

```
─────────────────────────── Matlab ───────────────────────────
  flag = true;
  while flag
      BODY;
      flag = CONDITION;
  end%while
```

The if statements

If a body of code is to be used when a certain condition is satisfied we may use the if statement, whose general form is given by

```
─────────────────────────── Octave ───────────────────────────
  if (CONDITION)
     THEN-BODY
  endif
```

The code below will first generate a list of random integers between 0 and 100. Then from all numbers larger than 50 we will subtract 50.

```
  % generate 10 integer random numbers between 0 and 100
  vec = randi([0,100],1,10)
  for k = 1:10
     if (vec(k) >= 50)          % subtract 50, if larger than 50
        vec(k) += -50;          % Octave only
  %     vec(k) = vec(k)-50;     % Matlab and Octave
     end%if
  end%for
  vec
  -->
  vec =   46   71   18   26   58    3   80   69   92    8
  vec =   46   21   18   26    8    3   30   19   42    8
```

If 50 is either subtracted or added, depending on the size of the numbers, use the `else` statement whose general form is

```
                              Octave
  if (CONDITION)
      THEN-BODY
  else
      ELSE-BODY
  endif
```

From the random vector of number we subtract or add 50, depending on whether the number is smaller or larger that 50.

```
                              Octave
  vec = round(rand(1,10)*100)
  for k = 1:10
    if (vec(k) >= 50)
      vec(k) += -50;   % with Matlab use vec(k) = vec(k) - 50
    else
      vec(k) += +50;   % with Matlab use vec(k) = vec(k) + 50
    endif
  endfor
  vec
  -->
  vec =   54  36  39   3  16  92  63  61  58   5
  vec =    4  86  89  53  66  42  13  11   8  55
```

The third and most general form of the 'if' statement allows multiple decisions to be combined in a single statement. Its general form is given by

```
                              Octave
  if (CONDITION)
      THEN-BODY
    elseif (CONDITION)
      ELSEIF-BODY
    else
      ELSE-BODY
  endif
```

Jumping out of loops with `continue` and `break`

If the `continue` statement appears within the body of a loop (`for`, `until` or `while`) the rest of the body will be jumped over and the loop restarted with the next value. The example below prints only the even numbers of a selection of 10 random numbers between 0 and 100.

```
                              Octave
  vec = round(rand(1,10)*100)
  for x = vec
    if (rem(x,2) != 0)
```

```
   continue
  endif
  fprintf("%d ",x)
endfor
fprintf("\n") % generate a new line
-->
vec = 67   68   33    3   69   74   16   62   76   52
68 74 16 62 76 52
```

Observe that we do **not** leave the loop completely, but only ignore the remaining command for the current run through the loop. It is the `break` statement that will leave the loop completely. The code below will display the numbers, until encountering a value larger than 70. Then no further number will be displayed.

─────────────────────────── **Octave** ───────────────────────────
```
vec = round(rand(1,10)*100)
for x = vec
  if (x >= 70)
    break
  endif
  fprintf("%d ",x)
endfor
fprintf("\n") % generate a new line
-->
vec =   10    7   14   72    5   71   15   67   96    5
10 7 14
```

The `switch` statment

If a number of different cases have to be considered we may use multiple, nested `if` statements or the `switch` command.

─────────────────────────── **Octave** ───────────────────────────
```
switch EXPRESSION
  case LABEL
    COMMAND_LIST
  case LABEL
    COMMAND_LIST
    ...
  otherwise
        COMMAND_LIST
endswitch
```

A rather useless example of code is shown below. For a list of 10 random numbers the code prints a statement, depending on the remainder of a division by 5.

─────────────────────────── **Octave** ───────────────────────────
```
vec = round(rand(1,10)*100)
for k = 1:10
switch rem(vec(k),5)     % remainder of a division by 5
```

```
case (0)
  fprintf("%i is a multiple of 5\n",vec(k))
case (1)
  fprintf("A division of %i by of 5 leaves a remainder 1\n",vec(k))
case (3)
  fprintf("A division of %i by of 5 leaves a remainder 3\n",vec(k))
otherwise
  fprintf("A division of %i by of 5 leaves remainders 2 or 4\n",vec(k))
endswitch
endfor
```

1.2.7 Conditions and selecting elements

When selecting elements in a vector satisfying a given condition you have two options:

- Use the condition directly to obtain a vector of 0 and 1. A number 0 indicates that the condition is not satisfied. A number 1 indicates that the condition is satisfied.

- Use the command `find()` to obtain a list of the indices for which the condition is satisfied.

Both operations apply directly to vectors, which might have a large influence on computation time.

```
x = rand(1,10);    % create 10 random numbers with uniform distribution
ans1 = x < 0.5     % indicate the elements larger than 0.5
ans2 = find(x<0.5) % return indicies of elements satisfying condition
-->
ans1 =   0   1   1   1   0   1   0   1   0   1
ans2 =   2   3   4   6   8   10
```

8 Example : Selecting elements

We generate a large vector of random numbers with a normal distribution with mean value 0 and standard deviation 1. Then count the numbers between -1 and 1, expecting approximately 69% hits. The results very clearly illustrate the speed advantage of **vectorized code**. The difference in MATLAB is considerably smaller, caused by the built-in precompiler.

```
n = 1000000;
x = randn(n,1);    % create the random numbers

time0 = cputime(); % use a for loop with an if condition
counter = 0;
for i = 1:n
   if abs(x(i))<1
      counter = counter+1;
   end%if
```

```
end%for
percentage1 = counter/n
time1 = cputime-time0

time0 = cputime();
percentage2 = sum( abs(x)<1 )/n
time2 = cputime-time0
-->
percentage1 =  0.68374
time1 =   4.0528
percentage2 =  0.68374
time2 =   7.9490e-03
```

Assume you want to know the average values of the above random numbers, but only the ones larger than 1. To be able to use vectorized code proceed in three steps:

1. Use the command find() to generate the indices ind of the numbers satisfying the condition.

2. Generate a new vector with only those numbers, x(ind).

3. Use the command mean() to compute the average value.

```
ind = find(x>1);
result = mean(x(ind))
-->
result =  1.5249
```

You may also use an array of logical values to select elements. In the code below we generate a vector with numbers from 1 through 12. Then we select and display only the numbers between 5 and 8.7 . There is no need for the command find(). On some occasions this might be slightly faster.

```
x = 1:0.5:12;          % all numbers from 1 to 12 in steps of 0.5
ind = (x>=5)&(x<8.7)% flag the numbers between 5 and 8.7
x(ind)                 % the numbers between 5 and 8.7
-->
   0   0   0   0   0   0   0   0   1   1   1   1   1   1   1   1   0   0   0   0   0   0   0
   5.0000   5.5000   6.0000   6.5000   7.0000   7.5000   8.0000   8.5000
```

◇

1.2.8 Reading from and writing data to files

With *Octave* you have different options to read information from a file or write to a file:

- Loading and saving variables with `load()` and `save()`.

- Reading and writing delimited files with `dlmread()` and `dlmwrite()`.

- Read data from files with complicated structures of the data by scanning line by line.

`load()`	load *Octave* variables
`save()`	save *Octave* variables
`dlmread()`	read all data from a file
`dlmwrite()`	write data to a file
`textread()`	read data from a file
`strread()`	read data from a string
`fopen()`	open a stream (file)
`fclose()`	close a stream
`fgetl()`	read one line from the file
`sscanf()`	scan a string for formated data

Table 1.5: Reading and writing with files

Loading and saving *Octave* variables

With the command `save()` you can save some, or all, variables to a file. This allows for later loading of the information with the command `load()`. You can load and save information in different formats, including MATLAB formats. Use `help save` and `help load` for more information. In the example below a random matrix is created and saved to a file. Then all variables in *Octave* are cleared and the matrix is reloaded.

```
clear *
aMat = rand(2);       % create a random matrix
save data.mat aMat    % save this matrix to a file
clear                 % clear the valiables
aMat                  % try to access the matrix, should fail
load data.mat         % load the variable
aMat
-->
'aMat' undefined near line 5 column 1
aMat =    0.54174    0.83863
          0.17270    0.76162
```

The above commands work with files adhering to *Octave* and MATLAB standards only. Using options for the command one can read and write variables for many different versions on MATLAB and *Octave*. By using flags one can choose which format should be used, e.g. `save -v6 MyVariable.mat` will save the data on a MATLAB Version 6 specific format, which can be used by *Octave* and MATLAB.

When data is generated by other programs or instruments then the format is usually not in the above format and thus we need more flexible commands.

Delimited reading and writing, `dlmread()` **and** `dlmwrite()`

If your file contains data with known delimiters the command `dlmread()` is very handy. As an example consider the file

```
┌──────────────────────── SampleSimple.txt ────────────────────────┐

 1   1.2
 2   1.2
 3 1e-3
 4 -3E+3
 5 0

└───────────────────────────────────────────────────────────────────┘
```

to be read with a single command

```
┌───────────────────────────────────────────────────────────────────┐

 x = dlmread('SampleSimple.txt')
 -->
 x =
      1.0000e+00     1.2000e+00
      2.0000e+00     1.2000e+00
      3.0000e+00     1.0000e-03
      4.0000e+00    -3.0000e+03
      5.0000e+00     0.0000e+00

└───────────────────────────────────────────────────────────────────┘
```

It is also possible to read only selected columns and rows in a larger set of data. The delimiter used most often are spaces or commas, leading to CSV files, i.e. **C**omma **S**eparated **V**alues. Use `help dlmread` to find out how the delimiters (e.g. space, TAB, comma, ...) can be set. With `dlmwrite()` you can create files with data, to be read by other programs. These two commands replace `csvread()` and `csvwrite()`.

An application of the above approach is shown in Sections 3.6.2 and 3.2.13.

Using `textread()`

With the command `textread()` a file can be scanned line by line, and a format string can be provided. As example examine the following file with data in a `csv` format. The abbreviation `csv` stand for comma separated values.

```
┌──────────────────────── FuelConsumption.csv ────────────────────────┐
Toyota Auris Hybrid - Fuel Consumption
Date,km-reading,km driven,km driven, gas [l],cost [CHF],l/100km
,,calculated,manual,,,
01/17/15,10800,,,,,
02/20/15,11368,568,568,34.62,47.75,6.10
03/12/15,11987,619,619,34.56,51.50,5.58
04/01/15,12754,767,767,38.93,57.60,5.08
04/25/15,13506,752,752,38.26,56.60,5.09
...
└──────────────────────────────────────────────────────────────────────┘
```

This data is now used in the code `FuelConsumption.m`, leading to Figure 1.9. The essential function `textread()` takes a few arguments:

- The string `'FuelConsumption.csv'` is the name of the file with the data.

- The format string `'%s %f %f %f %f %f %f'` gives the format of the available data: first a string, then a sequence of 6 numbers. All numbers a read as data type double, even if integer would be possible. This is to avoid undesired type conversions in subsequent computations.

- The strings `'delimiter',','` indicate the the data is separated by commas.

- The last argument `'headerlines'`, 4 informs `textread` that the first 4 lines are header lines and should be ignored.

The first data entry is a string in a date format, e.g. '02/20/15'. This has to be converted to a number of days of using this car. This is performed with the help of `datenum()` and `datevec()` by the command `Day = datenum(datevec(Date));`. The first function `datevec()` converts this string into a serial number corresponding to this data, e.g. `[2015, 2, 20, 0, 0, 0]`. Then the command `datenum()` is used to convert this into a number of days since the start of the current year minus 50, e.g. `736015`. Using `Day = Day - Day(1)` arrive at the days of usage of this car by the current user. To finish the code two graphics are generated, see Figure 1.9,

- The fuel used for 100 km as function of the days of usage.

- A histogram of the fuel usage of this car.

```
┌──────────────────────── FuelConsumption.m ────────────────────────┐
% Fuel Consumption Auris 2018
[Date,Status_km,Distance,DistanceManual,FuelUsed,Price,Gasfor100km] = ...
        textread('FuelConsumption.csv','%s %f %f %f %f %f',...
              'delimiter',',','headerlines',4);
Day = datenum(datevec(Date));  Day = Day-Day(1);

FuelFor100km = 100*FuelUsed./Distance;
```

```
figure(1); plot( Day,FuelFor100km,'+')
           xlabel('Day'); ylabel('Fuel used [l/100km]')

figure(2); hist(FuelFor100km)
           xlabel('Fuel used [l/100km]')
```

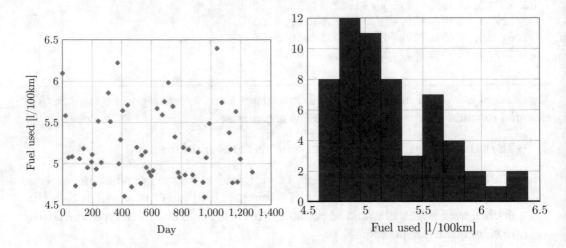

Figure 1.9: Fuel consumption for a hybrid car

Scanning a file, line by line

There are many files with data in a nonstandard format or with a few header lines. As an example consider the file `Sample.txt`.

― **Sample.txt** ―
```
this is a header line
the file was generated on Sept 25, 2007

a 4.03  5
b -5.8  4
c 1.0e3 3
d -4.7E-6 2
e 0 1
```

When reading information from a file (or another data stream) the following steps have to be taken:

- open the file for reading

- read the information, item by item or line by line

- scan the result to convert into the desired format

- close the file.

The commands in Table 1.5 are useful for this task and examine Section 1.2.4 for the scanning command `fscanf()`. The goal is to write code to read the information on the last lines of the file. The schema of the code is shown above. The formatted scanning has to be done carefully, since three different types of data are given on one line. The code bwlow wirks with *Octave* only, since MATLAB does not allow for scanning in C style.

- First read and ignore the three header lines.

- For each line read the string, then isolate the leading character with the format `'%1c'`. Then use `'%g'` to scan the floating point number, followed by `'%i'` for the trailing integer.

ReadSample.m

```
filename = 'Sample.txt';     % open the file for reading, in text mode
infile = fopen(filename,'rt');
c = blanks(20); x = zeros(1,20); n = x; % prealocate the memory

for k = 1:3                  % read and dump the three header lines
   tline = fgetl(infile);
end%for

k = 1;                       % initialize the data counter
tline = fgetl(infile);
while ischar(tline)
   % if tline not character, then we reached the end of the file
   % scan the string in the format: character float integer
   [ct,xt,nt] = sscanf(tline,"%1c%g%i","C");
   c(k) = ct; x(k) = xt;n(k) = nt; % store the data in the vectors
   tline = fgetl(infile);    % read the next line
   k = k+1;                  % increment the counter
end%while

fclose(infile);             % close the file
% use only the effectively read data
x = x(1:k-1); n = n(1:k-1);
c = c(1:k-1)
```

As a result we obtain the string c with content abcde, the vector x with the floating point numbers in the middle column of Sample.txt and the vector n with the numbers 5 through 1.

The above file can also be read by using command `textread()`, as shown below. This works with MATLAB and *Octave*.

```
[letters,num1,num2] = textread('Sample.txt','%s %f %u','headerlines',3)
-->
letters = {
  [1,1] = a
```

```
   [2,1] = b
   [3,1] = c
   [4,1] = d
   [5,1] = e }
 num1 =    4.0300e+00   -5.8000e+00    1.0000e+03   -4.7000e-06 0.0000e+00
 num2 =  5  4  3  2  1
```

Internally `textread()` is using `strread()`. Consult `help strread` to find out about the different formats supported.

Some applications of the above approach are shown in Sections 3.5, 3.7.1 and 3.8.1.

1.3 Solving Equations

In this subsection we show a few examples on how to solve different types of equations with the help of *Octave*. The examples are for instructional purposes only. We will examine:

- systems of linear equations

- zeros of polynomials

- zeros of single nonlinear functions

- zeros of systems of nonlinear functions

- optimization, maxima and minima.

Obviously the above list is by no means complete, it may serve as a starting point. There are many other types of very important problems to be solved that are ignored in this section:

- Ordinary differential equations: to be examined in Section 1.6, with a few examples.

- Numerical integration : to be examined in Section 3.1, with magnetic fields as application.

- Linear regression : to be examined carefully in Section 3.2, with real world, nontrivial examples.

- Nonlinear regression : to be examined carefully in Section 3.2.15, with real world, nontrivial examples.

- Fourier series, FFT : to be examined in Section 3.8, with a vibrating beam example.

1.3.1 Systems of linear equations

Since the main goal of MATLAB was to simplify matrix computations it should not come as a surprise that MATLAB and *Octave* provide many commands to work with linear systems. We illustrate some of the commands with elementary examples.

solving equations and optimization	
\	backslash operator to solve systems of linear equations
lu()	LU factorization of matrix, to solve linear systems
chol()	Cholesky factorization of matrix, to solve linear systems
roots()	find zeros of polynomials
fzero()	solve one nonlinear equation
fsolve()	solve nonlinear equations and systems
fsolveNewton()	Newtons algorithm, naive implementation
fmins()	mimimization, one or multiple variables
fminsearch()	mimimization, one or multiple variables
fminbnd()	constrained mimimization with respect to one variable
fminunc()	unconstrained mimimization, one or multiple variables

Table 1.6: Commands to solve equations and optimization

Using the backslash \ operator and lu() to solve linear equations

9 Example : A linear system with a unique solution

The linear system of three equations

$$
\begin{array}{rrrrrrr}
1\,x & + & 2\,y & + & 3\,z & = & 1 \\
4\,x & + & 5\,y & + & 6\,z & = & 2 \\
7\,x & + & 8\,y & + & 10\,z & = & 3
\end{array}
$$

should be rewritten using a matrix notation

$$
\begin{bmatrix} 1 & 2 & 3 \\ 4 & 5 & 6 \\ 7 & 8 & 10 \end{bmatrix} \begin{pmatrix} x \\ y \\ z \end{pmatrix} = \begin{pmatrix} 1 \\ 2 \\ 3 \end{pmatrix}.
$$

A linear system is solved by "dividing" by the matrix from the correct side

$$
\mathbf{A} \cdot \vec{x} = \vec{b} \qquad \Longleftrightarrow \qquad \vec{x} = \mathbf{A} \backslash \mathbf{A} \cdot \vec{x} = \mathbf{A} \backslash \vec{b}.
$$

In *Octave* and MATLAB this is implemented by

```
A = [1 2 3; 4 5 6; 7 8 10]; b = [1;2;3];   % create matrix and vector
x = A\b                % solve the system and display the result
-->
x =  -3.3333e-01
      6.6667e-01
      3.1713e-17
```

This confirms the exact solution $(\frac{-1}{3}, \frac{2}{3}, 0)^T$. This approach is clearly better than computing the inverse matrix $\mathbf{A}^{-1}$ and then computing $\mathbf{A}^{-1}\vec{b}$. For performance and stability reasons it is (almost) never a good idea to compute the inverse matrix.

If many linear systems have to be solved with the same matrix $\mathbf{A}$, then one shall not use the operator \ many times. One may either use a matrix for the right hand side or use the LU factorization presented in the two examples below. If you want to solve the above system and also

$$
\begin{aligned}
1x &+& 2y &+& 3z &=& -1 \\
4x &+& 5y &+& 6z &=& 0 \\
7x &+& 8y &+& 10z &=& 11
\end{aligned}
$$

then use the same matrix as above, but replace the vector with a 3×2 matrix.

```
b = [1, -1;
     2,  0;
     3, 11];
x = A\b
-->
x =  -3.3333e-01    1.1667e+01
      6.6667e-01   -2.1333e+01
      3.1713e-17    1.0000e+01
```

Octave uses the Gauss algorithm with partial pivoting to solve the system. This can be written as a matrix factorization.

$$\mathbf{L} \cdot \mathbf{U} = \mathbf{P} \cdot \mathbf{A}$$

- $\mathbf{L}$ is a lower triangular matrix

- $\mathbf{U}$ is an upper triangular matrix

- $\mathbf{P}$ is a permutation matrix

```
[L,U,P]  = lu(A)
-->
L = 1.00000    0.00000    0.00000
    0.14286    1.00000    0.00000
    0.57143    0.50000    1.00000

U = 7.00000    8.00000    10.00000
    0.00000    0.85714     1.57143
    0.00000    0.00000    -0.50000

P =    0    0    1
       1    0    0
       0    1    0
```

Then use

$$\mathbf{A}\vec{x} = \vec{b} \qquad \Longleftrightarrow \qquad \mathbf{L}\,\mathbf{U}\vec{x} = \mathbf{PA}\,\vec{x} = \mathbf{P}\vec{b} \qquad \Longleftrightarrow \qquad \begin{cases} \mathbf{L}\vec{y} & = & \mathbf{P}\vec{b} \\ \mathbf{U}\vec{x} & = & \vec{y} \end{cases}.$$

Instead of solving $\mathbf{A}\,\vec{x} = \vec{b}$ directly, first solve the lower triangular system $\mathbf{L}\,\vec{y} = \mathbf{P}\,\vec{b}$ and then the upper triangular system $\mathbf{U}\,\vec{x} = \vec{y}$.

```
x =   U\(L\(P*b))
-->
x =   -3.3333e-01
       6.6667e-01
       3.1713e-17
```

If many more linear system with different right hand sides $\vec{b}$ have to be solved, only the last step has to be repeated. Thus computing the LU factorization is equivalent to determining the inverse matrix, but with better numerical stability. $\diamond$

10 Example : Solving linear systems is a n^3 process

According to results you have seen in your class on linear algebra the computational effort to solve linear systems is proportional to n^3, the number of equations and unknowns. This can be verified with the help of a simulation.

- First generate a list of sizes n of matrices to be examined.

- For each value of n generate a random matrix of size $n \times n$ and add a diagonal matrix to assure that the system is uniquely solvable.

- Measure the CPU time it takes to solve the linear system.

Then generate a plot of the CPU time a function of the size n. We expect

$$\text{CPU} \approx c\,n^3$$
$$\log(\text{CPU}) \approx \log(c) + 3\,\log(n) .$$

On a doubly logarithmic scale expect a straight line with slope 3. This is confirmed by the code below and the resulting Figure 1.10[15].

```
                          LinearSystemSpeed.m
nlist = floor(logspace(3.4,4,10));
timer = zeros(size(nlist));
for k = 1:length(nlist)
  n = nlist(k);
  A = rand(n)-0.5 + n*eye(n);
```

---

[15]Most of the time for this simulation is used up for generating the random numbers. But only the solving time is measured by calling cputime().

```
   f = rand(n,1);
   t0 = cputime();
   x = A\f;
   timer(k) = cputime() - t0;
end%for

MFlops = 1/3*nlist.^3./timer/1e6

figure(1); plot(nlist,timer,'+-')
           xlabel('n, size of system'); ylabel('time [s]');  grid on

figure(2); plot(log10(nlist),log10(timer),'+-')
           xlabel('log(n)'); ylabel('log(time)');  grid on
```

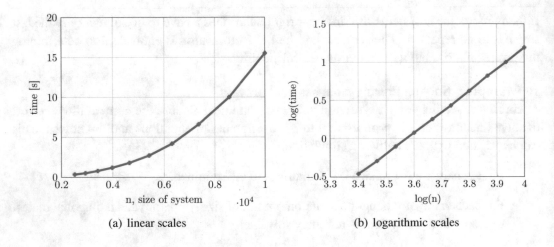

(a) linear scales   (b) logarithmic scales

Figure 1.10: Performance of the linear system solver

One has to be a bit careful when choosing large values of n. A matrix of size $n \times n$ needs to store n^2 real numbers, requiring approximately $8\,n^2$ bits of memory. For $n = 1024$ this leads to 8 MB of memory, but for $n = 10'240$ 800 MB of memory are required.   ◊

Linear systems without solution, over- and under-determined systems

Even for systems $\mathbf{A}\,\vec{x} = \vec{b}$ without unique solution the backslash operator will lead to a result.

- Even if a linear system is over-determined and has no solution, *Octave* and MATLAB will give an answer. In this case the result will be the solution of a linear regression problem, see Section 3.2. The answer $\vec{x}$ is determined such that the norm of the residual vector $\vec{r} = \mathbf{A}\,\vec{x} - \vec{b}$ is minimal.

- If the system is under-determined and has infinitely many solutions, the backslash operator \ will give one answer. It will return the solution $-\vec{x}$ with minimal norm. You have to use linear algebra to be able to generate all solutions.

11 Example : A linear system without solution

This will be even more useful if there are no unique solutions. The system

$$
\begin{aligned}
1x &+& 2y &+& 3z &=& 1 \\
4x &+& 5y &+& 6z &=& 2 \\
7x &+& 8y &+& 9z &=& 4
\end{aligned}
$$

does **not** have a unique solution. We find

```
A = [1 2 3; 4 5 6; 7 8 9]; b = [1;2;4];
[L,U,P] = lu(A)
-->
L =    1.00000    0.00000    0.00000
       0.14286    1.00000    0.00000
       0.57143    0.50000    1.00000

U =    7.00000    8.00000    9.00000
       0.00000    0.85714    1.71429
       0.00000    0.00000   -0.00000

P =  0    0    1
     1    0    0
     0    1    0
```

Then solving $\mathbf{L}\,\vec{y} = \mathbf{P}\,\vec{b}$ leads to

```
y = L\(P*b)
-->
y =    4.00000
       0.42857
       0.50000
```

and thus the system $\mathbf{U}\,\vec{x} = \vec{y}$ turns into

$$
\begin{aligned}
7\,x_1 &+& 8\,x_2 &+& 9\,x_3 &=& 4 \\
&& 0.85714\,x_2 &+& 1.71429\,x_3 &=& 0.42857 \\
&& && 0\,x_3 &=& -0.5
\end{aligned}
$$

Obviously the last equation does **not** have a solution. Using the inverse matrix to solve this system with `inv(A)*b` will return a solution, accompanied by a warning message.

```
┌─────────────────────────── Matlab ───────────────────────────┐
  xInv = inv(A)*b
  -->
  Warning: Matrix is close to singular or badly scaled.
          Results may be inaccurate. RCOND = 2.202823e-18.

  xInv = 3.1525e+15
        -6.3050e+15
         3.1525e+15
└───────────────────────────────────────────────────────────────┘
```

The backslash operator \ leads to a similar warning.

```
┌─────────────────────────── Matlab ───────────────────────────┐
  xBack = A\b
  -->
  Warning: Matrix is close to singular or badly scaled.
          Results may be inaccurate. RCOND = 2.202823e-18.

  xBack = 3.1525e+15
         -6.3050e+15
          3.1525e+15
└───────────────────────────────────────────────────────────────┘
```

The results by *Octave* are slightly different, but not reliable either. Keep this example in mind and **do not ignore warning messages**.

Since the determinant of the above 3×3 matrix $\mathbf{A}$ vanishes, the linear system has only solutions for vectors $\vec{b}$ of a special form. Consult you linear algebra book for details. The system has (nonunique) solutions only if the right hand side $\vec{b}$ of the equation is in the range of the matrix $\mathbf{A}$. Obtain a basis for this space with the command orth().

```
┌───────────────────────────────────────────────────────────────┐
  RangeSpace = orth(A)
  -->
  RangeSpace =   0.21484  -0.88723
                 0.52059  -0.24964
                 0.82634   0.38794
└───────────────────────────────────────────────────────────────┘
```

This implies that for vectors

$$\vec{b} = \lambda_1 \begin{pmatrix} 0.21484 \\ 0.52059 \\ 0.82634 \end{pmatrix} + \lambda_2 \begin{pmatrix} -0.88723 \\ -0.24964 \\ 0.38794 \end{pmatrix}$$

the system has a solution $\vec{x}$ of $\mathbf{A}\,\vec{x} = \vec{b}$. This solution is not unique, but we can add a multiple of a vector in the null-space or kernel of the matrix $\mathbf{A}$. The command null() computes a basis for the null-space.

```
ns = null(A)
-->
ns =   -0.40825
       +0.81650
       -0.40825
```

The result implies that for vectors $\vec{c} = \alpha\,(1, -2, 1)^T$ we have $\mathbf{A}\vec{c} = \vec{0}$ and any vector $\vec{x} + \mu\,\vec{c}$ is another solution of $\mathbf{A}\,\vec{x} = \vec{b}$. ◇

12 Example : Solving an over-determined system of linear equations
With the system of four equations

$$
\begin{aligned}
1\,x_1 + 2\,x_2 &= 1 \\
4\,x_1 + 5\,x_2 &= 2 \\
9\,x_1 + 8\,x_2 &= 4 \\
3\,x_1 + 6\,x_2 &= 0
\end{aligned}
$$

we only have two unknowns. The system is **over–determined**. Surprisingly *Octave* and MATLAB give a solution without any warning.

```
A = [1 2; 4 5; 9 8; 3 6]; b = [1;2;3;0];
x = A\b
-->
x =    0.48610
      -0.14297
```

A quick test shows that this is **not** a solution.

```
A*x - b
-->
 -0.79984
 -0.77045
  0.23114
  0.00048
```

At first sight this might be surprising and can lead to problems for uninformed users. In fact *Octave* and MATLAB solve an optimization problem. The resulting vector x minimizes the norm of the residual $\vec{r} = \mathbf{A}\,\vec{x} - \vec{b}$. Thus we might say that *Octave* returns the best possible solution. For many applications this is the desired solution, e.g. for linear regression problems (see Section 3.2). Internally *Octave* is using a QR factorization to solve this problem, i.e the matrix is factored in the form $\mathbf{A} = \mathbf{Q}\,\mathbf{R}$. Some details are spelled out in Section 3.2.7. ◇

13 Example : Solving an under-determined system of linear equations

The linear system of 2 equations

$$
\begin{aligned}
1\,x \;+\; 2\,y \;+\; 3\,z \;&=\; 7 \\
4\,x \;+\; 5\,y \;+\; 6\,z \;&=\; -5
\end{aligned}
$$

must have infinitely many solutions. You find a description of all solutions by finding one particular solution $\vec{x}_p \in \mathbb{R}^3$ and the vector $\vec{n} \in \mathbb{R}^2$ generating the null space. Then all solutions are of the form $\vec{x} = \vec{x}_p + \lambda\,\vec{n}$, where $\lambda \in \mathbb{R}$. *Octave* can generate those vectors.

```
A = [1 2 3; 4 5 6]; b = [7; -5];
xp = A\b
n = null(A)
-->
xp =   -8.8333
       -1.3333
        6.1667

n =     0.40825
       -0.81650
        0.40825
```

The particular solution found by *Octave* is the one with the smallest possible norm, as can be verified by the orthogonality $\langle \vec{x}_p\,,\,\vec{n}\rangle = 0$.   ◇

Commands to solve special linear systems

For matrices with special properties *Octave* can take advantage of these properties and find the solution with better reliability, or faster.

14 Example : Cholesky factorization for symmetric, positive definite matrices

If the matrix $\mathbf{A}$ is known to be symmetric and positive definite we can use a more efficient and reliable algorithm, based on the Cholesky factorization of the matrix.

$$
\mathbf{A} = \mathbf{R}^T \cdot \mathbf{R}
$$

where $\mathbf{R}$ is an upper triangular matrix. A linear system of equations can then be solved as a sequence of systems with triangular matrix.

$$
\mathbf{A}\,\vec{x} = \vec{b} \qquad \Longleftrightarrow \qquad \mathbf{R}^T\,\mathbf{R}\,\vec{x} = \vec{b} \qquad \Longleftrightarrow \qquad
\begin{cases}
\mathbf{R}^T\,\vec{y} \;=\; \vec{b} \\
\mathbf{R}\,\vec{x} \;=\; \vec{y}
\end{cases}
$$

To examine the system

$$
\begin{aligned}
3\,x \;+\; 0\,y \;+\; 1\,z \;&=\; 1 \\
0\,x \;+\; 3\,y \;+\; 2\,z \;&=\; 2 \\
1\,x \;+\; 2\,y \;+\; 9\,z \;&=\; 3
\end{aligned}
$$

we use the code below, verifying that we have the same solution with both solution methods.

```
A = [3,0,1; 0,3,2; 1,2,9];
R = chol(A)
b = [1;2;3];
x1 = A\b;
xChol = R\(R'\b)
MaxError = max(abs(x1-xChol))
-->
R =    1.73205    0.00000    0.57735
       0.00000    1.73205    1.15470
       0.00000    0.00000    2.70801

xChol =    0.27273
           0.54545
           0.18182

MaxError = 0
```

A second output argument of `chol()` indicates whether the matrix was positive definite or not. Use `help chol` to find out more. ◇

15 Example : Sparse matrices

There are many applications where the matrix **A** consists mostly of zero entries and thus the standard algorithms will waste a lot of effort dealing with zeros. To avoid this **sparse matrices** were introduced. As an example we consider the $n \times n$ matrix **A** given by

$$\mathbf{A} = \frac{1}{(n+1)^2} \begin{bmatrix} 2 & -1 & & & & \\ -1 & 2 & -1 & & & \\ & -1 & 2 & -1 & & \\ & & \ddots & \ddots & \ddots & \\ & & & -1 & 2 & -1 \\ & & & & -1 & 2 \end{bmatrix}$$

This type of matrix appears very often when using the finite difference method to solve differential equations, e.g. heat equations.

Using the command `spdiags()` we create a sparse matrix where *Octave* only stores the nonzero values and their position in the matrix.

```
n = 10;
A = spdiags([-ones(n,1),2*ones(n,1),-ones(n,1)],[ -1,0,1],n,n)/(n+1)^2
-->
A = Compressed Column Sparse (rows = 10, cols = 10, nnz = 28 [28%])
   (1, 1) ->  0.016529
   (2, 1) -> -0.0082645
```

```
(1, 2) -> -0.0082645
(2, 2) ->  0.016529
(3, 2) -> -0.0082645
(2, 3) -> -0.0082645
(3, 3) ->  0.016529
...
```

Then a system of linear equations can be solved as before, but *Octave* will automatically take advantage of the sparseness of the matrix.

```
b = ones(n,1);
x = A\b;    % solve the sparse system
x'          % diplay as row vector
-->
605.0  1089.0  1452.0  1694.0  1815.0  1815.0  1694.0 1452.0  1089.0
605.0
```

By changing the size n of the matrix you can solve a large number of linear equations, e.g. $n = 100'000$. Working with full matrices you would run out of memory.   ◇

16 Example : Sparse Cholesky factorization
The matrix in the previous example is symmetric and positive definite, thus we may use the Cholesky factorization $\mathbf{A} = \mathbf{R}^T\mathbf{R}$. MATLAB/*Octave* return $\mathbf{R}$ as a sparse matrix.

```
R = chol(A);
xChol = R\(R' \b)
```

If the matrix $\mathbf{A}$ is known to be sparse we can call the function chol() with three output arguments and find a sparsity preserving permutation matrix $\mathbf{Q}$ such that

$$\mathbf{Q}^T \cdot \mathbf{A} \cdot \mathbf{Q} = \mathbf{R}^T \cdot \mathbf{R}$$

The permutation matrix $\mathbf{Q}$ is best returned as a permutation vector. This allows to save large amounts of memory for some applications. To solve the system we have to take the permutation matrices into account. Use the fact that $\mathbf{Q}^{-1} = \mathbf{Q}^T$ to examine the system $\mathbf{A}\,\vec{x} = \vec{b}$ with the help of

$$\mathbf{Q}^T\mathbf{A}\,\mathbf{Q}\mathbf{Q}^T\vec{x} = \mathbf{Q}^T\vec{b} \quad \Longleftrightarrow \quad \mathbf{R}^T\,\mathbf{R}\mathbf{Q}^T\vec{x} = \mathbf{Q}^T\vec{b} \quad \Longleftrightarrow \quad \begin{cases} \mathbf{R}^T\,\vec{y} &=& \mathbf{Q}^T\vec{b} \\ \mathbf{R}\,\mathbf{Q}^T\vec{x} &=& \vec{y} \end{cases}$$

and thus

$$\vec{x} = \mathbf{Q}\mathbf{R}^{-1}(\mathbf{R}^T)^{-1}\mathbf{Q}^T\vec{b}\,.$$

This is implemented by

```
[R, P, Q] = chol(A);
x3 = Q*(R\(R' \ (Q' *b)));
```

This code is longer than just x = A\b, but there are cases where it is considerably faster and saves memory. ◇

1.3.2 Zeros of polynomials

Real or complex polynomials of degree n have exactly n zeros, maybe complex and maybe with higher multiplicity. Thus MATLAB/*Octave* provides a special command to determine those zeros, often called roots, of polynomials. To determine the zeros of of the cubic polynomial

$$p(z) = 1 + 2\,z + 3\,z^3$$

use the command

```
roots([3 0 2 1])
-->
    0.20116 + 0.88773i
    0.20116 - 0.88773i
   -0.40232 + 0.00000i
```

Thus there is one real root at $x \approx -0.4$ and two complex conjugate roots. This is confirmed by the graph of the polynomial $p(x)$ in Figure 1.11.

```
x = -1:0.01:2;           % choose values between -1 and 2
y = polyval([3 0 2 1],x) % evaluate the polynomial at those points
plot(x,y)                % generate the plot
xlabel('x'); ylabel('y = 1+2*x+3*x^3')
```

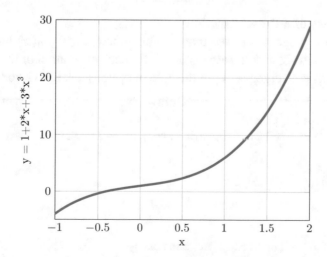

Figure 1.11: Graph of the polynomial $1 + 2\,x + 3\,x^3$

1.3.3   Nonlinear equations

Not all equations are linear equations or zeros of polynomials in one variable. Many applications lead to nonlinear equations or systems of equations. Solving nonlinear equations, or systems, can be very difficult. *Octave*/MATLAB provide a few algorithms to help solving nonlinear equations.

Solving a single nonlinear equation

With the commands `fzero()` or `fsolve()` can solve single equations or systems of equations. Most algorithms require a good initial guess for a solution of $f(x) = 0$. To determine the obvious solution $x = \pi$ of $\sin(x) = 0$ use

```
x0 = fsolve(@(x)sin(x),3)
-->
x0 =   3.1416
```

The command `fzero()` is very reliable at solving a single equation for one unknown. In particular if you can bracket the solution, i.e. you have one value with $f(a) < 0$ and another with $f(b) > 0$, then there is a solution in the interval between a and b.

```
format long
x0 = fzero(@(x)sin(x),[3,3.2])
difference = x0-pi
-->
x0 =   3.14159265358979
difference =    -1.33226762955019e-15
```

17 Example :   Zero of a Bessel function

Examine Figure 1.8 (page 44) to see that the function $f(x) = J_0(x)$ has a zero close to $x_0 = 6$. Since the value of x and y are displayed for each evaluation of the function observe the iterations and its convergence. With the help of options ask for 12 correct digits.

```
                              ── Octave ──
x0 = 6.0;   clear options
options.TolFun = 1e-12;    options.TolX   = 1e-12;

function y = f(x)
  y = besselj(0,x);
  [x y]
endfunction

[x, fval, info] = fsolve(@f,x0,options)
-->
    6.00000    0.15065
    6.00004    0.15066
    5.99996    0.15064
```

```
5.455533  -0.022077
5.455566  -0.022065
5.455500  -0.022088
5.5198e+00  -9.8715e-05
5.5201e+00  -9.3923e-07
5.5201e+00  -8.9600e-09
5.5201e+00  -8.5477e-11
5.5201e+00  -8.1499e-13

x =  5.5201
fval = -8.1499e-13
info =  1
```

The algorithm seems to lead to a linear convergence, i.e. the number of correct digits increases at a constant rate. 11 iterations were necessary to arrive at the desired accuracy. The return value of `info` contains information on whether the solution converged or not. Find information on the interpretation of the values of info by `help fsolve`. ◇

18 Example : Using a user provided Jacobian
The algorithm used in `fsolve()` uses a Newton iteration

$$x_{n+1} = x_n - \frac{f(x_n)}{f'(x_n)}$$

to determine an approximate solution. This algorithm should converge quadratically, i.e. the number of correct digits is (approximately) doubled at each step. The documentation of *Octave* states that one can pass a function for the derivative to the command `fsolve()` and then this derivative will be used for the iteration. This excellent feature might be very efficient, in particular for systems of equations. One should observe fewer evaluations[16] of the function f, provided the initial value is close enough to the true solution.

```
─────────────────── Octave ───────────────────
x0 = 6.0  % choose the starting value

function [y,dy] = f2(x)
  y  = besselj(0,x);      dy = -besselj(1,x);
  % display value and derivative for each evaluation of the function
  disp([x,y])
endfunction

clear options            % set the options for fsolve
options.TolFun = 1e-12;
options.TolX   = 1e-12;
options.Jacobian = 'on';
options.Updating = 'off';
```

---

[16]Without any further measures the function will be called twice for each step, which is not necessary. Read `help fsolve` to learn how to avoid this double evaluation.

```
[x2, fval, info] = fsolve(@f2,x0,options)
-->
    6.00000        0.15065
    6.00000        0.15065
    5.455533      -0.022077
    5.455533      -0.022077
    5.5198e+00    -9.8715e-05
    5.5198e+00    -9.8715e-05
    5.5201e+00    -2.5912e-09
    5.5201e+00    -2.5912e-09
    5.5201e+00     3.2470e-16
```

Observe that the number of correct digits is doubled for each iteration. Only 5 iterations were necessary to arrive at the desired accuracy.

19 Example : **Options of** `fsolve()` [17]
The function `fsolve()` allows to set a few options.

- `TolX` : the tolerance in x values, The default value is $1.5 \cdot 10^{-8}$.

- `FunX` : the tolerance in the function values, The default value is $1.5 \cdot 10^{-8}$.

- `MaxIter` : maximal number of iterations to be used. The default value is 400.

- `Jacobian` : a user supplied derivative may be used.

- `Updating` : if set to "off" a Newton algorithm is used.

The default values of those options can be found in the source file for the function `fsolve()`. Type `which fsolve` on the *Octave* command promt to find the exact location of the source file and then examine the file with your favourite editor. As an example we compute the first zero of $\sin(x)$ without and with options.

```
──────────────────────── Octave ────────────────────────
clear options
res1 = fsolve(@(x)sin(x),3)-pi
options.TolFun = 1e-15;   options.TolX    = 1e-15
res2 = fsolve(@(x)sin(x),3,options)-pi
-->
res1 = -2.8937e-10
options =  scalar structure containing the fields:
    TolFun =     1.0000e-15
    TolX    =     1.0000e-15
res2 =   4.4409e-16
```

---

[17]This is specific for *Octave*. The options for MATLAB are slightly different.

The command `fsolve()` is rather powerful, it can also examine over-determined systems of equations and may be used for nonlinear regression problems, see Section 3.2.15.

Solving systems of nonlinear equations

The command `fsolve()` can also be used to solve systems of equations. Use some geometry (intersection of ellipses) to convince yourself that the system

$$
\begin{aligned}
x^2 + 4\,y^2 - 1 &= 0 \\
4\,x^4 + y^2 - 1 &= 0
\end{aligned}
$$

must have a solution close to $x \approx 1$ and $y \approx 1$. This solution can be found by `fsolve()`.

```octave
                              ─── Octave ───
  x0 = [1;1];   % choose the starting value

  function y = f(x)   % define the system of equations
    y = [x(1)^2 + 4*x(2)^2-1;
         4*x(1)^4 + x(2)^2-1];
  endfunction

  [x,fval,info] = fsolve(@f,x0)   % determine one of the possible solutions
  -->
  x =   0.68219
        0.36559
  fval =  -1.3447e-07
           1.1496e-07
  info =  1
```

Implementing Newton's Algorithm

As a first example of an extended function we develop code for Newton's algorithm to solve systems of equations.

The main tool to solve a system of nonlinear equations of the form $\vec{f}(\vec{x}) = \vec{0}$ is Newton's algorithm. For a well chosen starting vector $\vec{x}_0$ apply the iteration

$$
\vec{x}_{n+1} = \vec{x}_n - (DF(\vec{x}_n))^{-1}\vec{f}(\vec{x}_n)
$$

where the Jacobian matrix of partial derivatives is given by

$$
DF(\vec{x}) =
\begin{bmatrix}
\frac{\partial f_1(\vec{x})}{\partial x_1} & \frac{\partial f_1(\vec{x})}{\partial x_2} & \cdots & \frac{\partial f_1(\vec{x})}{\partial x_n} \\
\frac{\partial f_2(\vec{x})}{\partial x_1} & \frac{\partial f_2(\vec{x})}{\partial x_2} & & \frac{\partial f_2(\vec{x})}{\partial x_n} \\
\vdots & & \ddots & \vdots \\
\frac{\partial f_n(\vec{x})}{\partial x_1} & \frac{\partial f_n(\vec{x})}{\partial x_2} & \cdots & \frac{\partial f_n(\vec{x})}{\partial x_n}
\end{bmatrix}
$$

Consult your calculus lecture notes for information on the algorithm. Below the algorithm is implement in *Octave* and MATLAB. The code will by no means replace the *Octave* function `fsolve.m` or the MATLAB function `fsolve()` in the Optimization toolbox. Both have more options and will examine over-determined systems too (see Section 3.2.15).

All the code segments below have to be in a file `fsolveNewton.m`, together with a copyright statement. Then the code can be used to solve the system

$$
\begin{aligned}
x^2 + 4\,y^2 - 1 &= 0 \\
4\,x^4 + y^2 - 1 &= 0
\end{aligned}
$$

by calling

Octave
```
x0 = [1;1];  % choose the starting value
% for matlab put the functions in separate files
function y = f(x)   % define the system of equations
   y = [x(1)^2 + 4*x(2)^2-1;
        4*x(1)^4 + x(2)^2-1];
endfunction

function y = dfdx(x)   % Jacobian
   y = [2*x(1),  8*x(2);
        16*x(1)^3, 2*x(2)^2];
endfunction
% use a finite difference Jacobian
[x,iter] = fsolveNewton('f',x0,1e-4*[1;1])
[x,iter] = fsolveNewton('f',x0,1e-6)           % higher accuracy
[x,iter] = fsolveNewton('f',x0,1e-6,'dfdx')    % user provided Jacobian
```

This example was already solved in the previous Section 1.3.3.

The lines below show the source code for the function `fsolveNewton()`, including the documentation. The built–in function `fsolve()` is a more sophisticated implementation of the same algorithm.

• Define the function name and give the basic documentation.

---
fsolveNewton.m
```
function [x,iter] = fsolveNewton(f,x0,tolx,dfdx)
% [x,iter] = fsolveNewton(f,x0,tolx,dfdx)
%
% use Newtons method to solve a system of equations f(x)=0,
% the number of equations and unknowns have to match
% the Jacobian matrix is either given by the function dfdx or
% determined with finite difference computations
%
% input parameters:
% f     string with the function name
%       function has to return a column vector
```

```
% x0    starting value
% tolx  allowed tolerances,
%          a vector with the maximal tolerance for each component
%          if tolx is a scalar, use for each of the components
% dfdx  string with function name to compute the Jacobian
%
% output parameters:
% x      vector with the approximate solution
% iter number of required iterations
%          if iter>20 then the algorithm did not converge
```

- Verify the input arguments and set up the starting point for the loop to come.

```
if ((nargin < 3)|(nargin>=5))
  help fsolveNewton
  error('fsolveNewton');
end%if
maxit = 20;      % maximal number of iterations
x = x0;
iter = 0;
if isscalar(tolx) tolx = tolx*ones(size(x0)); end%if
dx = 100*abs(tolx);
f0 = feval(f,x);
m = length(f0);    n = length(x);
if (n ~= m)
  error('number of equations not equal number of unknown')
end%if
if (n ~= length(dx))
  error('tolerance not correctly specified')
end%if
```

- Start the loop. Compute the Jacobian matrix, either by calling the provided function or by using a finite difference approximation.

```
jac = zeros(m,n); % reserve memory for the Jacobian
done = false;      % Matlab has no 'do until'
while done
  if nargin==4     % use the provided Jacobian
    jac = feval(dfdx,x);
  else  % use a finite difference approximation for Jacobian
    dx = dx/100;
    for jj= 1:n
      xn = x; xn(jj) = xn(jj)+dx(jj);
      jac(:,jj) = ((feval(f,xn)-f0)/dx(jj));
    end%for
  end%if
```

- Apply a Newton step and close the loop.

```
dx = jac\f0;
x = x - dx;
iter = iter+1;
f0 = feval(f,x);
if ((iter>=maxit)|(abs(dx)<tolx))
    done = true;
end%if
end%while
```

To estimate the derivatives $\frac{\partial f(x)}{\partial x}$ the above codes uses finite difference approximations of the form

$$\frac{\partial f(x)}{\partial x} \approx \frac{f(x+h) - f(x)}{h} \quad \text{for } h \text{ small enough.}$$

1.3.4 Optimization

In this section some commands from the optimization package at SourceForge[18] are presented. Thus have a quick look at the package, resp. its documentation. If the package is not installed on your system (hint: `pkg list`) you will have to download, install and load the package, using the instructions in Section 1.1.2 on page 7.

20 Example : The function $f(x) = \sin(2x)$ has a local minimum at $x_{min} = \frac{3\pi}{4} \approx 2.3562$. For the command `fmins()` provide an initial guess and then obtain an approximate answer.

```
─────────────────────────── Octave ───────────────────────────
xMin = fmins(@(x)sin(2*x), 2.5)
-->
xMin =   2.3560
```

If we want better accuracy of the solution we have to choose the correct options. Find the documentation with `help fmins`. In the example below we ask `fmins()` to show intermediate results and work with better accuracy, which is obtained for the final result. The intermediate results show the Nelder–Mead simplex algorithm at work.

```
─────────────────────────── Octave ───────────────────────────
options = [1, 1e-5];
xMin = fmins(@(x)sin(2*x),2.5,options)
-->
f(x0) = 9.5892e-01
Iter.  1,  how = initial    nf =   2,  f = 9.5892e-01  (0.0%)
Iter.  2,  how = shrink,    nf =   5,  f = 9.5892e-01  (0.0%)
Iter.  3,  how = shrink,    nf =   8,  f = 9.5892e-01  (0.0%)
```

---

[18]In MATLAB with the Optimization toolbox installed you may use the function `fminsearch()` instead of `fmins()`. In *Octave* both are available.

```
Iter.    4,    how = shrink,      nf =   11,   f = 9.5892e-01   (0.0%)
Iter.    5,    how = contract,    nf =   13,   f = 9.9969e-01   (4.3%)
Iter.    6,    how = shrink,      nf =   16,   f = 9.9969e-01   (0.0%)
Iter.    7,    how = shrink,      nf =   19,   f = 9.9969e-01   (0.0%)
Iter.    8,    how = contract,    nf =   21,   f = 9.9990e-01   (0.0%)
Iter.    9,    how = contract,    nf =   23,   f = 9.9999e-01   (0.0%)
Iter.   10,    how = contract,    nf =   25,   f = 9.9999e-01   (0.0%)
Iter.   11,    how - contract,    nf =   27,   f = 1.0000e+00   (0.0%)
Iter.   12,    how = shrink,      nf =   30,   f = 1.0000e+00   (0.0%)
Iter.   13,    how = shrink,      nf =   33,   f = 1.0000e+00   (0.0%)
Iter.   14,    how = contract,    nf =   35,   f = 1.0000e+00   (0.0%)
Iter.   15,    how = shrink,      nf =   38,   f = 1.0000e+00   (0.0%)
Iter.   16,    how = contract,    nf =   40,   f = 1.0000e+00   (0.0%)
Iter.   17,    how = shrink,      nf =   43,   f = 1.0000e+00   (0.0%)
Iter.   18,    how = shrink,      nf =   46,   f = 1.0000e+00   (0.0%)
Simplex size 8.0951e-06 <= 1.0000e-05...quitting
xMin = 2.3562
```

If the function depends on one variable only the command `fminbnd()` can be used too. It uses a different algorithm and is usually more efficient for functions of one variable.

```
xMin = fminbnd(@(x)sin(2*x), 0,pi)
-->
xMin =   2.3562
```

21 Example : Optimization with respect to multiple variables

The algorithms in `fmins()` and `fminsearch()` can also optimize functions of multiple variables. Both are using a simplex method and thus do not require derivatives of the function to be optimized. Instead of a maximum of

$$f(x,y) = -2\,x^2 - 3\,x\,y - 2\,y^2 + 5\,x + 2\,y$$

search for a minumum of $-f(x,y)$. Examine the graph of $f(x,y)$ in Figure 1.12.

```
─────────────────────────── Octave ───────────────────────────
[xx,yy] = meshgrid( [-1:0.1:4],[-2:0.1:2]);

function res = f(x,y)
   res = -2*x.^2 -3*x.*y-2*y.^2 + 5*x+2*y;
endfunction

surfc(xx,yy,f(xx,yy));    xlabel('x'); ylabel('y');
```

Using the graph we conclude that there is a maximum not too far away from $(x,y) \approx$ $(1.5,\,0)$. We use `fmins()` with the function $-f$ and the above starting values.

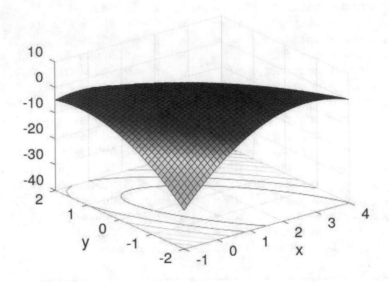

Figure 1.12: Graph of a function $h = f(x, y)$, with contour lines

```
┌──────────────────────────── Octave ────────────────────────────┐
  xMin = fmins(@(x)-f(x(1),x(2)),[1.5,0])
  -->
  xMin = 1.9996  -1.0005
└─────────────────────────────────────────────────────────────────┘
```

An exact computation will find the exact position of the maximum at $(x, y) = (2, -1)$. To obtain better accuracy use options, e.g.

```
┌──────────────────────────── Octave ────────────────────────────┐
  xMin = fmins(@(x)-f(x(1),x(2)),[1.5,0], [0, 1e-10])
  -->
  xMin =  2.0000  -1.0000
└─────────────────────────────────────────────────────────────────┘
```

$\diamondsuit$

22 Example : Hardware implementation of division

For an efficient hardware implementation of division of floating point numbers one needs a good approximation of the function $1/x$ on the interval $[\frac{1}{2}, 1]$, e.g. by a polynomial of degree 2. The maximal error should be minimal, i.e. seek the minimum of

$$f(a, b, c) = \max_{0.5 \leq x \leq 1} |\frac{1}{x} - a x^2 - b x - c|$$

As a starting guess for the parabola use the straight line through the points $(0.5, 2)$ and $(1, 1)$. Another option would be to generate the Chebyshev approximating polynomial of degree 2, see Section 3.4.6 on page 312. To improve accuracy and reliability, repeat the call of `fmins()` until the result stabilizes.

```octave
x = linspace(0.5,1,1001);
format long  % display (too) many digits
function res = toMin(p,x)
  res = max(abs(1./x - polyval(p,x)));
endfunction

pOptim = fmins(@(p)toMin(p,x),[0,-2,3],[0, 1e-15])
pOptim = fmins(@(p)toMin(p,x),pOptim,[0, 1e-15])
-->
pOptim =   2.74516598909596  -6.05887449195935   4.32842712578459
pOptim =   2.74516598880275  -6.05887449156216   4.32842712566662
```

The result can be verified visually by Figure 1.13. This is an excellent approximation of $1/x$ by a second order polynomial.

```octave
yOptim = polyval(pOptim,x);
figure(1); plot(x,1./x,x,yOptim);
           xlabel('x'); ylabel('1/x and polynomial approx.')
figure(2); plot(x,yOptim-1./x);
           xlabel('x'); ylabel('difference')
```

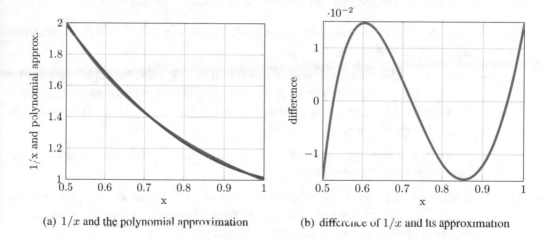

(a) $1/x$ and the polynomial approximation   (b) difference of $1/x$ and its approximation

Figure 1.13: $1/x$ and a polynomial approximation of degree two

$\diamond$

Linear and nonlinear regression problems can also be considered as minimization problem. For this special type of problem there are better algorithms, see Chapter 3.2 and Section 3.2.15 for nonlinear regression problems.

1.4   Basic Graphics

For the graphics commands there are only some small differences between `MATLAB` and
Octave. In addition there ar1e many more options and possibilities than it is possible to
illustrate in the given time and space for these notes.

- *Octave* and `MATLAB` will by default open up a graphics window on screen to display
  graphics.

- After a graph is displayed you can change its appearance with many commands:
  choose different axis, put on labels or a title, add text, ...

- If you generate a new picture, the old one will be overwritten, You can change this
  behavior with `hold` or `figure()`.

- You can open up multiple graphics windows, using the command `figure()`.

- Within a graphics window you can use the mouse to zoom, move or rotate the picture.

- With the command `print()` you can write the current figure into a file, choosing
  from many different formats.

- When starting up *Octave* you can choose which graphics toolkit to use.

  - `qt` : this toolkit is used with *Octave* 4.0.1 or newer and allows for interactive
    modifications of the figure.

  - `gnuplot` : this was the default. *Octave* will use *Gnuplot* as graphics engine to
    generate the graphics on screen or in files. This is a time tested method, but not
    very efficient for large 3D graphics. Use `gnuplot_binary` to find out which
    binary is actually used.

  - `fltk` : this is a graphics engine using OpenGL. It will use the specialized
    hardware on the graphics card.

  - Switching forth and back between the two toolkits within one *Octave* session
    is possible. Close all graphics windows with `close all` and then change the
    graphics toolkit, e.g. with the command below.

```
                                         Octave
    graphics_toolkit          % display the current graphics toolkit
    graphics_toolkit fltk     % switch to the fltk toolkit
```

- With `MATLAB` there is only one graphics toolkit, thus there is nothing to choose.

In this section a few examples will be shown, but much more is possible.

1.4.1   2-D plots

The basic `plot()` **command**

Graphs of known functions are easy to generate with MATLAB or *Octave*, as shown with the code below and the resulting Figure 1.14(a).

```
x = 0:0.1:7; y1 = sin(x);
plot(x,y1)
```

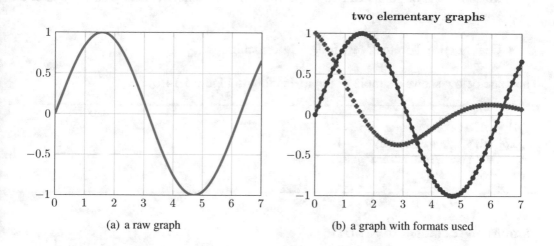

(a) a raw graph   (b) a graph with formats used

Figure 1.14: Elementary graphs of functions

MATLAB and *Octave* will essentially plot a number of points and you can choose how to connect the points.

- The command `plot()` can only[19] display points and straight line connections between them. Thus you have to compute the coordinates of the points first and then generate the graphics.

  - The command `plot()` with one argument only will use the numbering of the given values for the horizontal coordinate and the values for the vertical coordinates,

  - The command `plot()` with two arguments requires the values of the x coordinates as the first argument and the y components as the second argument.

  - A third argument of `plot()` may be a format string, specifying how `plot()` has to display and connect the individual points. There are many options:

    * Choose a lines style: – lines, . dots, ˆ impulses, L steps.

---

[19]We knowingly ignore the commands `ezplot()`, `fplot()` and its friends. They are of very limited use and you are better off using the *Octave*/MATLAB way to generate figures: first compute the data, then display the data.

          * Choose the color by a letter: k (black), r (red), g (green), b (blue), c (cyan), m (magenta), w (white).

          * Use +, *, o or x in combination with the lines style to choose the point style.

      – In *Octave* you can set the text for the legend with `';key text;'`.

      – In *Octave* and MATLAB you can set the text for the legend with the command `legend('key text')`.

- Multiple sets of points can be displayed with one single call to `plot()`. List the arguments sequentially.

- Use `help plot` to access further information.

The code below shows a simple example, leading to Figure 1.14(b).

```
y2 = cos(x).*exp(-x/3);
plot(x,y1,'-*r',x,y2,'+b')
title('two elementary graphs');
grid on
```

Options and additions to `plot()`

The result of the basic command `plot()` may be modified by a number of options and parameters, most of which are shown in Table 1.7. The code below is using some of these options with the result shown in Figure 1.15(a).

- To generate a PDF (Portable Document Format) use `print -dpdf` (MATLAB and *Octave*). This will lead to a full PDF page. You will have to crop the large margins of the graphics if you only need the picture and the are many tools achieve this, e.g. the command `pdfcrop` to be used outside of MATLAB.. With *Octave* you may use `print -dpdfwrite` to generated the cropped graphics directly. This file is suitable to be included in LaTeX documents when using PDF as graphics format, e.g. by `pdflatex`.

- An encapsulated Postscript file `graph3.eps`, containing the graphics, will be created in the current directory. The command in the code below generates a level 2 encapsulated file, using a tight bounding box and a different font. This file is suitable to be included in LaTeX documents when using EPS as graohics format, e.g. by `latex`.

- For LibreOffice or Word documents the PNG format is useful. It is important to generate bitmap files with the correct size and not rescale them with the word processor! In the example below an image of size 600 by 400 is generated.

plot commands	
`plot()`	basic command to plot one or multiple functions
`semilogx()`	same as `plot()` but with logarithmic horizontal scale
`semilogy()`	same as `plot()` but with logarithmic vertical scale
`loglog()`	same as `plot()` but with double logarithmic scales
`hist()`	generate and plot a histogram
`bar()`	generate a bar chart
`plotyy()`	generate a plot with 2 independent y axes
options and settings	
`figure()`	choose the display window on the screen
`title()`	set a title for the graphic
`xlabel()`	specify a label for the horizontal axis
`ylabel()`	specify a label for the vertical axis
`zlabel()`	specify a label for the third axis
`text()`	put a text at a given position in the graph
`legend()`	puts a legend on the plot or turns them on/off
`grid`	turn grid on (`grid on`) or off (`grid off`)
`axis()`	choose the viewing area, use `axis()` to reset
`xlim()`	choose the limits on the x axis, similar for y and z axis
`hold`	toggle the hold state of the current graphic
`colorbar()`	add a colorbar to the graphic
`subplot()`	create one of the figures in a multiplot
`print()`	save the current figure in a file
`clf`	clear the current figure
`graphics_toolkit`	choose the graphics engine to be used (*Octave*)

Table 1.7: Generating 2D plots

- Observe that with MATLAB a command `print('myPic.png')` will print directly to the printer of your system, and not write the graphics to a file. If you want the file specify the device explicitly, e.g. `print('myPic.png','-dpng')`.

- The are many more formats available, see `help print` or examine Section 1.4.2.

Octave

```
clf
x = -4:0.1:4;
y = (1+x.^2).*exp(3*x);
semilogy(x,y);
text(-3,2000,'Text in Graph');
title('logarithmic scale in vertical direction')
xlabel('Distance [m]'); ylabel('Temp [K]')
print('graph3.eps','-depsc2','-FTimes-Roman:20','-tight')
% Matlab is slightly different
print('graph3.pdf','-dpdfwrite')   % Octave only
print('graph3.png','-S600,400')    % in Matlab slightly different
%% to obtain a 4 by 3 inches picture with a resolution of 200dpi use
%% the code below. The resulting PNG file will be a 800x600 picture
% set(gcf,'PaperUnits','inches','PaperPosition',[0 0 4 3])
% print('graph3.png','-dpng','-r200')
```

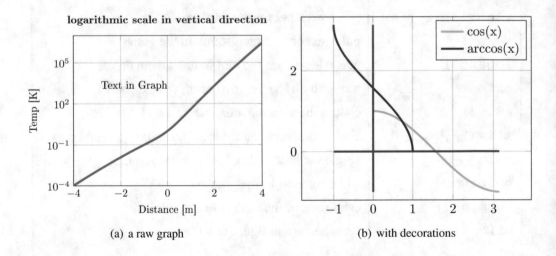

(a) a raw graph   (b) with decorations

Figure 1.15: Graphs with text and legends

```
clf % clear the current figure
x = linspace(0,pi,50);   y = cos(x);
plot(x,y,'g',y,x,'r');
axis([-1.2 pi+0.2 -1.2 pi+0.2],'equal')
hold on
```

```
plot([-1 pi],[0 0],'b',[0 0],[-1 pi],'b');  % show coordinate axis
grid on
hold off
legend('show')
legend('cos(x)','arccos(x)','location','northeast')
legend('boxon')
```

With `subplot()` one can generate multiple graphs in one figure. The code on the last two lines generates a title over all subfigures.

```
x = linspace(-2*pi,2*pi,200);
clf
axis('normal');
axis()               % leave the scaling up to Octave/Matlab
subplot(2,2,1); plot(x,sin(x));   subplot(2,2,2); plot(x,cos(x));
subplot(2,2,3); plot(x,sinh(x)); subplot(2,2,4); plot(x,cosh(x));
ha = axes('Position',[0 0 1 1],'Xlim',[0 1],'Ylim',[0 1],...
    'Box','off','Visible','off','Units','normalized','clipping','off');
text(0.3, 0.95,'Titel Over all Subplots')
```

Find another example in Figure 1.28 on page 111.

Interactive manipulations

With MATLAB and new versions of *Octave* you can manipulate properties of graphics interactively.

- You can zoom in and out.

- Rotate a 3D graph. Works with 2D graphs too!!

- You can add text at any position in the picture.

- Label and rescale the axis.

- You can turn the grid on and off.

  - With Matlab choose the small arrow in the menu line of the graphics to edit the plot. Then click in the graph and use the right mouse button to obtain the menu with the options to choose.
  - With *Octave* click on the graph and either hit the key "G" (Gnuplot toolkit) or use the menue item in the top line.

Grid lines and tick marks

One can modify the tick marks and the grid lines in a plot by setting properties of the axis object. With the code below we generate Figure 1.16(a).

- With the command `set()` the properties of the current axis are modified. The function `gca()` returns a handle to the current axis.

- On the x-axis the tick marks are shown at multiples of $\pi/2$ and labeled accordingly.

- On the y–axis the labels are unevenly spaced, but more grid lines are shown.

```
x = linspace(0,10);
figure(1); plot(x,sin(x))
set(gca(),'XTick',[0:pi/2:3*pi],...
    'XTickLabel',{'0','\pi/2','\pi','3\pi/2','2\pi','5\pi/2','3\pi'})
set(gca(),'XGrid','on')
set(gca(),'YTick',[-0.5,0,0.4,1])
set(gca(),'YMinorGrid','on');
```

An axis object has many properties to be modified by `set()`. To examine the list of all those use the command `get(gca())`. The list will be rather long!

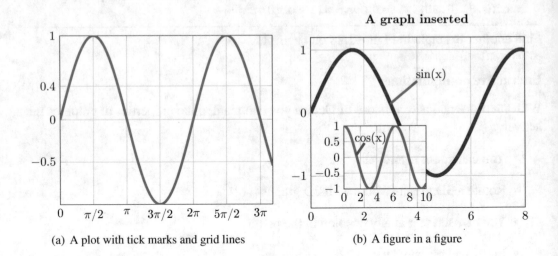

(a) A plot with tick marks and grid lines   (b) A figure in a figure

Figure 1.16: Tick marks, grid lines and a figure in a figure

A figure in a figure

Octave/Matlab allow for many more tricks with pictures, e.g. you can generate a figure in a figure, as shown in Figure 1.16(b).

```
x = linspace(0,10);   f1 = sin(x);   f2 = cos(x);% generate the data
figure(1); clf
plot(x,f1,'r','linewidth',3)
axis([0 8 -1.5 1.2])
```

```
set(gca,'fontsize',16);        grid on
line([3 4],[sin(3) 0.5]);      text(3.8,0.6,'sin(x)','fontsize', 16)
title('A graph inserted')

axes('position',[0.25 0.20 0.3 0.3])            % set the new frame
set(gca, 'xlim' ,[3,10],'ylim',[-1.1 1.1])
axis([0 10 -1.2 1.2])
plot(x,f2,'linewidth',2); grid on
line([1.5 2.5], [cos(1.5) 0.45]);  text(0.8,0.6,'cos(x)')
```

When MATLAB/*Octave* generate a graphics it has many default options set. You can access those through the command get(). Find more information on this in section 15.3 *Graphics Data Structure* of the *Octave* manual

Size of graphics when printing

When printing or generating a PostScript or PDF file the size is specified by the figure's 'papersize' property and the position on the page by 'paperposition'. In the code below we generate a figure that is wider than usual, leading[20] to Figure 1.17. With the command gca() we can get a handle to the current axis, and then modify some of the properties. To find out more, generate an arbitrary graphics and then examine the result of get(gca()). In the above example we set the markers on the x–axis at specific locations and used the special symbol π.

```
x = linspace(0,10); y1 = sin(x); y2 = cos(x);

h = figure(1); clf; plot(x,y1,x,y2)
               legend('sin(x)','cos(x)'); xlabel('x')
               set(h,'paperunits','centimeters')
               set(h,'papersize',[15,6])
               set(h,'paperposition',[0,0,16,6])
               set(gca(),'xtick',[0,pi,2*pi, 3*pi],...
                  'xticklabel',{'0','\pi','2 \pi','3 \pi'})

print -dpdfwrite SizeAndTick.pdf
print -depsc     SizeAndTick.eps
```

1.4.2 Printing figures to files

The GUI of *Octave* and MATLAB allow you to save a figure in different formats. In many applications it is more convenient to generate the file with the figure by a command. Here you can choose and modify many different aspects of a figure.

---

[20]Currently the MATLAB version looks nicer than the *Octave* version.

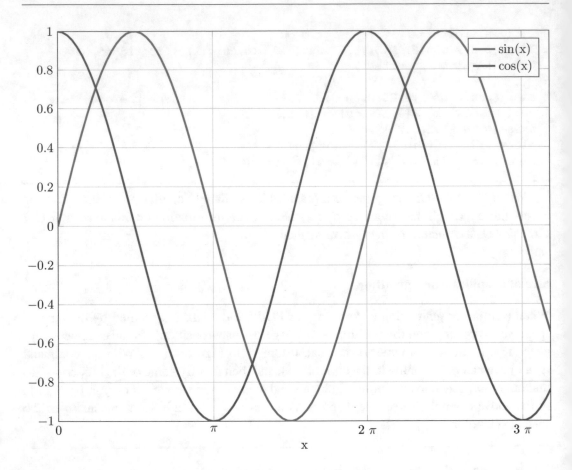

Figure 1.17: A graphics of specific size and with special tick marks

Generating files in different formats

In the above section we already mentioned that on-screen figures can be written to files, for them to be included in your favorite text processing tool or to be used with LaTeX. The basic command to be used is `print()`. It takes the name of the file to be generated and other options as parameters. The `print()` command has many options, consult the documentation with `help print` or `doc print` and Table 1.8 shows a few options.

   Octave and MATLAB usually choose the file format by looking at the given extension of the given file name. If no device is given explicitly MATLAB will send the file to the printer, while *Octave* writes to the given file.

```
print MyFigure.png % Matlab sends to printer, Octave to file or printer
print -dpng MyFigure.png  % Matlab and Octave write to the file
```

There are different types of files that might be useful. Here only 4 formats will be commented: `eps`, `png`, `pdf` and `fig`.

print	the basic command to write a picture in a file
options	
-d...	choose the device to be used
-depsc	colored EPS (Encapsulated PostScript)
-dpng	PNG (Portable Network Graphics)
-dpdf	PDF (Portable Document Format), full page
-dpdfwrite	PDF, with bounding box (Octave only)

Table 1.8: the `print()` command and its options

EPS : Encapsulated PostScript files can be used with LaTeX or other good text processing tools. Many printers, resp. their drivers, can print these files directly to paper.

– EPS files may be generated by either one of the lines below. The file `graph.eps` will contain the image.

```Octave
print('graph.eps','-depsc','-FTimes-Roman:20','-tight')
print -depsc -FTimes-Roman:20 -tight graph.eps
```

– These figures can be rescaled without problems and the fonts for the included characters remain intact.

– For photographs the created files might be unnecessarily large.

– LibreOffice/OpenOffice/Word can not handle Postscript pictures

PNG : Portable Network Graphics files are a good format for bitmaps, e.g. photos.

– The files may be generated by either of the lines below. The image will have a resolution of 600 columns and 400 lines. The file `graph.png` will contain the image.

```Octave
print('graph3.png','-S600,400')
print '-S600,400' graph3.png
```

– These figures **can not be rescaled** without serious loss of image quality. Thus you have to generate it in the correct size and the above option to specify the resolution of the image is essential.

– The codes in MATLAB are slightly different. To generate a picure of dimension 6×4 cm with a resolution of 100 dpi use

```
┌──────────────────────────── Matlab ────────────────────────────┐
│ set(gcf,'PaperUnits','centimeters','PaperPosition',[0 0 6 4])   │
│ print -dpng graph3.png -r100                                    │
└─────────────────────────────────────────────────────────────────┘
```

- LATEX and LibreOffice/OpenOffice/Word can handle PNG files.

PDF : Portable Document Format can be used in LATEXor to directly print the figure. PDF can be used instead of EPS.

- PDF files may be generated by either one of the lines below. The file `graph.pdf` will contain the image.

```
┌──────────────────────────── Octave ────────────────────────────┐
│ print('graph.pdf')              % full page, Matlab/Octave      │
│ print('graph.pdf','-dpdfwrite') % with bounding box, Octave     │
└─────────────────────────────────────────────────────────────────┘
```

Use the second line if you need to include the generated picture into another document, as the first line will always generate a full page. Use system tools the remove excess margins, e.g. `pdfcrop`.

- For photographs the created files might be unnecessarily large.
- LibreOffice/OpenOffice/Word can not handle PDF pictures

fig : If you want to apply further modifications to your figure the `fig` format might be handy. You can then use the Unix program `xfig` to modify your picture, e.g. change fonts, colors, thickness of lines, and many more. Then save the figure in the desired format for the final usage. To generate a colored figure use to option `-color`, e.g. by calling `print -color graph.fig`.

Converting an image to different formats

On occasion it is necessary to convert between different formats. Most Unix systems provide powerful tools for this task.

ImageMagick is a cross platform, open source software suite for displaying, converting, and editing raster image files. It can read and write over 100 image file formats. It can not only convert, but also apply many operations to images: resize, rescale, rotate, ... To convert a file `graph.gif` to the PNG format you may type `convert graph.gif graph.png` on a Unix command line. If you want to achieve identical results from within *Octave* you have to use the system command.

```
┌─────────────────────────────────────────────────────────────────┐
│ system('convert graph.gif graph.png')                           │
└─────────────────────────────────────────────────────────────────┘
```

For these lecture notes I had to convert many EPS figures into the PDF format. For this I used the tool `epstopdf`, e.g. type `epstopdf graph.eps` on a Unix command line to generate the PDF file. This can be done within *Octave* by using the `system()` command, as shown above.

Since the command often has to be applied to all *.eps files in a sub-directory I created a shell command to convert all images at once. This shell script only works on Unix systems!

```
─────────────────────────── doEPStoPDF.sh ───────────────────────────
#!/bin/bash
for file in *.eps
do
    epstopdf $file
done
```

Using `printFigureToPdf.m` to generate PDF files

The script file `printFigureToPdf.m` uses features of MATLAB and *Octave* to generate a PDF with a desired size. Observe that the size of the fonts are adapted accordingly[21]. If you have to fiddle with the exact form of an output, have a closer look at the source of this command. It uses a few `set()` commands to generate the desired size and then calls to standard commands. It might just give you the necessary hints.

```
─────────────────────────── Octave ───────────────────────────
% script file to test the PDF output of printFigureToPdf()
x = linspace(0,10);
plot(x,sin(x),x,cos(x))
legend('sin(x)','cos(x)')
xlabel('x')
% size given in inches, with a 10% border
printFigureToPdf('Sin_Cos.pdf',[6,4],'inches',[0.1,0.1,0.1,0.1])
% size given in inches again (problems with cm),
% with a 10% border, except at the top only 5%
printFigureToPdf('Sin_Cos_cm.pdf',[4,3],...
                 'inches',[0.1,0.1,0.1,0.05])
```

Using the TikZ package to save graphics

There is an excellent set of scripts `matlab2tikz` to convert MATLAB/*Octave* figures to the tikz format, which can then be used by LATEX. Find it on the MATLAB Central web site or the current version at https://github.com/matlab2tikz/matlab2tikz. Most of the figures in these notes were generated by `matlab2tikz()` and a subsequent call of `pdflatex`. With *Octave* the print device `tikzstandalone` is similar.

1.4.3 Generating histograms

The command `hist()` will create a histogram of the values in the given vector. The following code and the resulting Figure 1.18 illustrate that the values around ± 1 are more likely to show up as results of the sin–function on $-5\pi \leq x \leq 5\pi$. The interval of all occurring values ($[-1, 1]$) is divided up into subintervals of equal length. Then the command

---

[21]With version 6.2.0 *Octave* the legends are not handled correctly.

hist(y,20) counts the number of values in each of the 20 subintervals and displays the
result as height of the column. This leads to Figure 1.18(a). The histogram in Figure 1.18(b)
is normalized, such that the sum of all heights equals 1. Thus we can read the probability
for the values to fall into one of the bins. The values of the centers are in the vector center
and the corresponding heights are stored in height and thus available for further computations. The resulting graph can also be generated by bar(center,height) or with
plot(center,height). Observe that the scaling has to be left to MATLAB/Octave by
a call of axis() without arguments. The codes for Octave and MATLAB differ slightly
and are shown below.

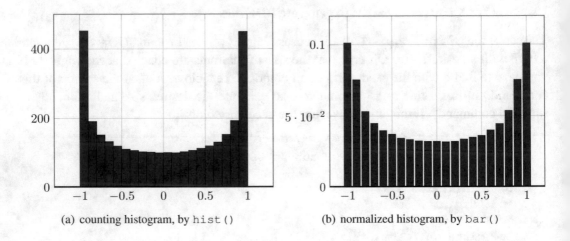

(a) counting histogram, by hist()   (b) normalized histogram, by bar()

Figure 1.18: Histogram of the values of the sin–function

```
x = -5*pi:0.01:5*pi;    y = sin(x);
figure(1); hist(y,20)
          axis([-1.3 1.3]);          % Octave
          % axis([-1.3 1.3 0 500]); % for Matlab use this line

[height,center] = hist(y,-1:0.1:1,1)
height = height/sum(height);
figure(2); bar(center,height);
          axis([-1.2 1.2]);          % Octave
          % axis([-1.1 1.1 0 0.12]) % for Matlab use this line
```

1.4.4 Generating 3-D graphics

With Octave and MATLAB three dimensional plots can also be generated. A list of some of
the commands is shown in Table 1.9.

plot commands	
`plot3()`	to plot a curve in space
`meshgrid()`	generate a mesh for a surface plot
`surf()`	generate a surface plot on a mesh
`surfc()`	generate a surface plot and contour lines
`mesh()`	generate a mesh plot on a mesh
`meshc()`	generate a mesh plot and contour lines
`contour()`	graph the contour lines of a surface
`contourf()`	graph the contour lines with colored patches
`quiver()`	generate a vector field
options and settings	
`view()`	set the viewing angles for 3d–plots
`caxis()`	choose the colormap and color scaling for the surfaces

Table 1.9: Generating 3D plots

Curves in space: `plot3()`

To examine a curve in space $\mathbb{R}^3$ use the command `plot3()`. One may also plot multiple curves and choose styles, just as for the command `plot()`. The only difference is that you have to provide the data for three components. The code below leads to Figure 1.19.

```
t = 0:0.1:5*pi;
x = cos(t);   y - 2*sin(t);   z = t/(2*pi);
plot3(x,y,z)
grid on
view(25,45);
```

A surface plot in space: `meshgrid()`, `surf()` **and** `mesh()`

If a surface of the type $z = f(x,y)$ in space is to be plotted then one has to apply a few steps.

- Choose the values for x and y.

- Generate matrices with values for the coordinates at each point on a mesh with the help of `meshgrid()`.

- Compute the values of the height at those points with the help of the given function $z = f(x,y)$.

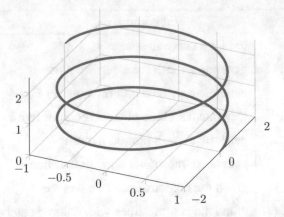

Figure 1.19: A spiral curve in space

- Generate the graphics with surf() or mesh() and choose a good view point and scaling.

To better understand the effect of the command meshgrid one may examine the result of the small grid generated by the code below.

```
[xx,yy] = meshgrid(1:6,-1:3)
-->
xx =
   1   2   3   4   5   6
   1   2   3   4   5   6
   1   2   3   4   5   6
   1   2   3   4   5   6
   1   2   3   4   5   6

yy =
  -1  -1  -1  -1  -1  -1
   0   0   0   0   0   0
   1   1   1   1   1   1
   2   2   2   2   2   2
   3   3   3   3   3   3
```

The commands mesh(),meshc(),meshz(), surf(), surfc() and surface() allow to visualize surfaces in the space $\mathbb{R}^3$.

To examine the surface generated by the function

$$z = f(x,y) = e^{-x^2-y^2} \quad \text{for} \quad -2 < x < 2 \quad \text{and} \quad -1 < y < 3$$

use the codes below to create Figure 1.20(a).

```
x = -2:0.1:2;   y = -1:0.1:3;   [xx,yy] = meshgrid(x,y);
zz = exp(-xx.^2-yy.^2);
figure(1); mesh(xx,yy,zz)
           grid on;    view(120,40)
           xlabel('x'); ylabel('y'); zlabel('height');
```

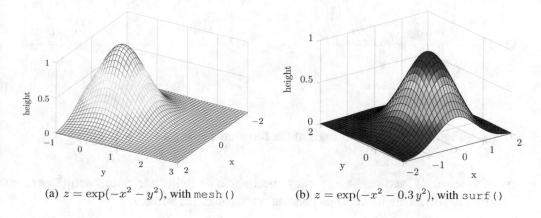

(a) $z = \exp(-x^2 - y^2)$, with `mesh()`   (b) $z = \exp(-x^2 - 0.3y^2)$, with `surf()`

Figure 1.20: Graphs of two functions, as sufaces in $\mathbb{R}^3$

Instead of a mesh, as shown in Figure 1.20(a) create fully colored patches with the command `surf()`. Find the result in Figure 1.20(b).

```
x = -2:0.1:2;   y = -1:0.1:2;   [xx,yy] = meshgrid(x,y);
zz = exp(-xx.^2-0.3*yy.^2);
figure(3); surf(xx,yy,exp(-xx.^2-yy.^2))
           xlabel('x'); ylabel('y');zlabel('height')
```

To draw contour lines of a given graph use `contour()` with the result if Figure 1.21(a).

```
──────────────────────────── Octave ────────────────────────────
x = -2:0.1:2;   y = -1:0.1:2;   [xx,yy] = meshgrid(x,y);
zz = exp(-xx.^2-0.3*yy.^2);

figure(2); contour(xx,yy,zz,15)
           xlabel('x'); ylabel('y');        axis equal
```

For the contours we can also insist that the levels shown are between 0 and 1, with steps of 0.1 and the levels are labeled, as shown in Figure 1.21(b).

```
cc = [0:0.1:1]                       % select the desired levels
figure(4); [C,h] = contour(xx,yy,zz,cc);  % compute and display
           clabel(C,h,cc,'FontSize',10);  % the level curves
           xlabel('x'); ylabel('y');       axis equal
```

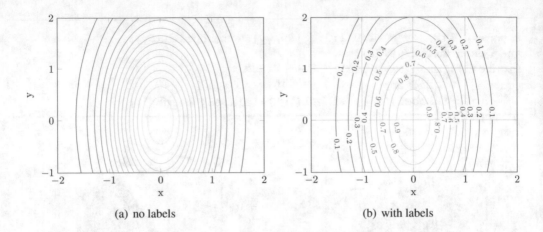

(a) no labels   (b) with labels

Figure 1.21: Contour lines of the function $z = \exp(-x^2 - 0.3\,y^2)$

- With mesh() and meshc() only the lines connecting the points are drawn, no surface patches are used. Thus you can see through the surface.

- With surf() and surfc() the rectangles connecting the lines are filled with colored patches and thus you can not see through the surface.

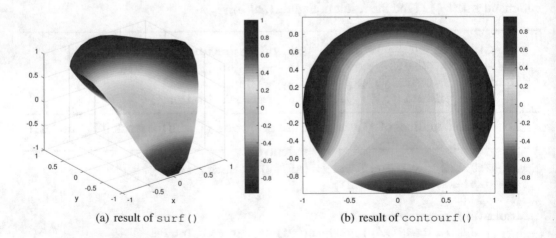

(a) result of surf()   (b) result of contourf()

Figure 1.22: Surface and contour plots, without visible lines

With contourf() not only the lines are drawn, but also colored patches. With an option you can not display the contour lines at all, but only the collored patches. In addition a colorbar can be displayed with the shading is based on an interpolation. Find the result in Figure 1.22.

```
alpha = linspace(0,2*pi,31)'; radius = linspace(0,1,31);
x = cos(alpha)*radius; y = sin(alpha)*radius;
z = x.^2 + y.^3;
figure(1); H1 = surf(x,y,z); xlabel('x'); ylabel('y')
           colormap(jet());
           shading interp
           set(H1,'EdgeColor','none');
           colorbar

figure(2); colormap(jet());
           [C,H2] = contourf(x,y,z,50);
           colorbar
           shading interp
           set(H2,'LineStyle','none');
```

The above structure of commands allows to examine rather involved surfaces. Any surface parameterized over a rectangle can be displayed. As an example we consider the torus in Figure 1.23. The torus is parameterized by two angles u and v.

```
u1 = linspace(0,2*pi,51); v1 = linspace(pi/4,2*pi-pi/4,21);
[u,v] = meshgrid(u1,v1);
r0 = 4; r1 = 1;
x = cos(u).*(r0+r1*cos(v)); y = sin(u).*(r0+r1*cos(v));
z = r1*sin(v);
mesh(x,y,z)
```

1.4.5 Generating vector fields

Octave has commands to display vector fields. As an example we consider the planar vector field

$$\vec{F}(\vec{x}) = \begin{pmatrix} y \\ -x \end{pmatrix}$$

on the domain $-1 \leq x, y \leq 1.5$. Thus at a point $(x, y) \in \mathbb{R}^2$ we attach the vector $(y, -x)$. To display this vector field we have to generate a set of points $(x_i, y_i) \in \mathbb{R}^2$ at which the vectors are to be plotted. In this example the horizontal component of the vector field is given by $+y$ and the vertical component by $-x$. To obtain a good result the length of the vectors have to be scaled. The effect of the scaling factor might depend on the version of *Octave*/MATLAB used! Then the vector field is generated, find the result in Figure 1.24.

```
xvec = -1:0.2:1.5; yvec = -1:0.2:1.5;
[x,y] = meshgrid(xvec,yvec);
x = x(:); y = y(:);        % convert the matrix into a column vector.
Vx = y;  Vy = -x;          % define the vector field
```

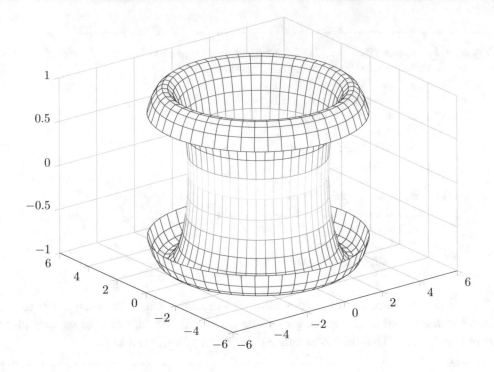

Figure 1.23: A general 3D surface

```
scale =  2;                    % scaling
quiver(x,y,Vx,Vy,scale)        % display the vector field
```

23 Example : MATLAB/*Octave* provide the command `peaks()` to generate a a nice, reasonably complicated surface. If you want to find out about the function used in `peaks.m`, examine its source code. With the help of `gradient` and `quiver` we can also generate the gradient vector field belonging to this function.

```
[xx,yy,zz] = peaks(20);
figure(1); meshc(xx,yy,zz);
figure(2); contour(xx,yy,zz)
figure(3); [Dx,Dy] = gradient(zz,yy(2)-yy(1)); quiver(xx,yy,Dx,Dy)
```

◇

24 Example :   As a second example consider visualizing the magnetic field generated by two vertical conductors. According to Ampère's law the field strength is given by $B = \frac{\mu_0 I}{2\pi r}$ where I is the current and r the horizontal distance from the conductor. The direction of the vector is tangential to a circle with the center at the wire and follows the right hand rule.

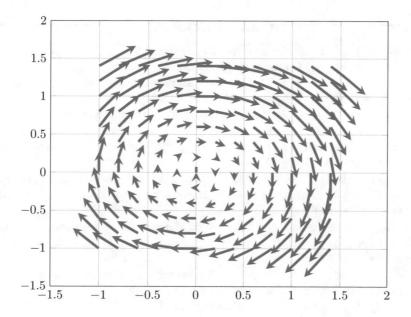

Figure 1.24: A vector field

This can be written in the form

$$\vec{B}(x, y) = \frac{\mu_0 I}{2\pi} \frac{1}{x^2 + y^2} \begin{pmatrix} -y \\ x \end{pmatrix}.$$

With this information write code to generate the vector field.

- First set the desired parameters. Put the the value of $\frac{\mu_0 I}{2\pi}$ into a constant `cI1`, resp. `cI2`.

- Generate the points where the vector field is to be computed by `linspace()` and `meshgrid()`.

- Compute the distance from the wires with a vectorized call. Then use `find()` to set the values of the distance to NaN for points too close to the wire.

- Compute the two components of the vector field and use `quiver()` to generate Figure 1.25.

- Since most of the vectors are very short it is difficult to detect the direction in Figure 1.25(a). Thus modify the vectors such that they all have length 1 and then generate Figure 1.26(a). In addition one can remove the heads of the vectors by `set(h,'maxheadsize',0)`.

MagneticField.m

```
Lx = 2; Ly = +1.5 ;   % domain to be examined
cI1 = 1; cI2 = +1 ;   % the two currents
D = 1;                % half the distance of the two conductors
Nx = 35; Ny = 25;     % number of grid points
Dmin = 0.1;           % minimal distance from conductors

x = linspace(-Lx,Lx,Nx); y = linspace(-Ly,+Ly,Ny);% generate grid
[xx,yy] = meshgrid(x,y);
Dist1 = sqrt((xx-D).^2+ yy.^2); % distance of from the first wire
remove1 = find(Dist1<Dmin);     % remove points too close to wire
Dist1(remove1) = NaN;
Dist2 = sqrt((xx+D).^2+ yy.^2); % distance from the second wire
remove2 = find(Dist2<Dmin);     % remove points too close to wire
Dist2(remove2) = NaN;
                                % compute the vector field
Vy = +cI1*(xx-D)./Dist1.^2 + cI2*(xx+D)./Dist2.^2;
Vx = -cI1*(yy)./Dist1.^2   - cI2*(yy)./Dist2.^2;


figure(1); h = quiver(xx,yy,Vx,Vy,2);
           xlabel('x'); ylabel('y'); axis([-Lx,Lx,-Ly,Ly]);
           axis equal
norms = sqrt(Vx.^2 + Vy.^2);
Vxn = Vx./norms; Vyn = Vy./norms;
figure(2); h = quiver(xx,yy,Vxn,Vyn);
           set(h,'maxheadsize', 0)
           xlabel('x'); ylabel('y'); axis([-Lx,Lx,-Ly,Ly]);
           axis equal
```

In Figures 1.25(a), 1.25(b) and 1.26(a) both currents are positive. In Figures 1.25(c), 1.25(d) and 1.26(b) find the similar results with currents of opposite sign. Observe that between the two conductors the behavior of the magnetic field is drastically different. Use the commands `stream2()` or `streamline()` to generate stream lines, generating Figure 1.26.

1.5   Basic Image Processing

For the graphics commands there are some noticeable differences between MATLAB and *Octave*. Thus it is necessary to consult the corresponding help files to find the documentation. In addition there are many more options and possibilities than it is possible to illustrate in the given time and space for these notes.

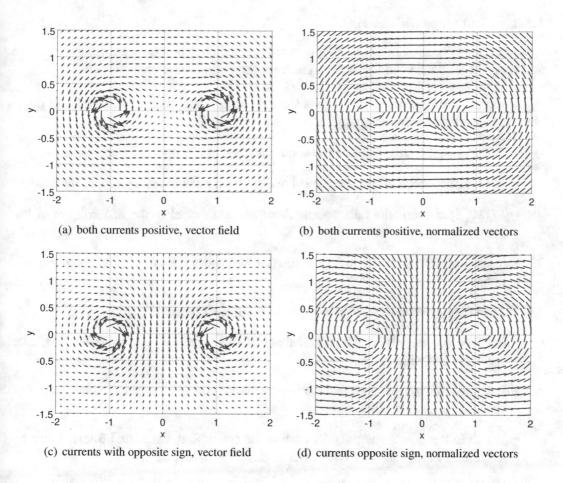

(a) both currents positive, vector field

(b) both currents positive, normalized vectors

(c) currents with opposite sign, vector field

(d) currents opposite sign, normalized vectors

Figure 1.25: Magnetic fields with either both currents positive or opposite sign

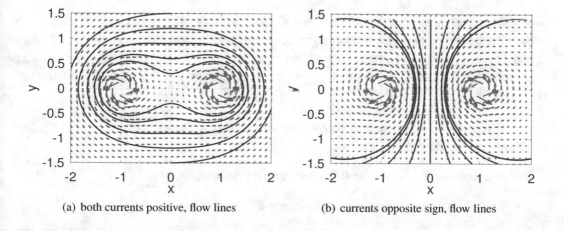

(a) both currents positive, flow lines

(b) currents opposite sign, flow lines

Figure 1.26: Flow lines of magnetic fields

1.5.1   First steps with images

Most of the information in this section is based on the image processing toolbox of *Octave* and you will need to install the toolbox and have access to its documentation.

Mathematically speaking an image is a matrix of numbers, or matrix of triples of numbers. If the image is represented by the $n \times m$ matrix M, then

- $M(1, 1)$ contains the information about the pixel in the top left corner of the image.

- $M(1, 10)$ contains the information about the tenth pixel in the top row of the image.

- $M(10, 1)$ contains the information about the tenth pixel in the first column of the image.

- the vector $M(:, 1)$ contains the information about all pixels in the first column of the image.

The information about each pixel can be given in different forms:

- For BW images each pixel is represented by 0 or 1, since the only two colors available are black and white.

- For grayscale images each pixel is represented by the level of gray, which can be of type `uint8`, `uint16` or `double`.

- For RGB images the intensity for each of the colors Red, Green and Blue is given by a number, which can be of type `uint8`, `uint16` or `double`.

- For **indexed** images each pixel is represented by the number of its color, i.e. an integer. Then you need the **colormap** with the translation of the number of the color to the actual color, usually given by RGB codes.

Indexed images require less memory and *Octave* uses the command `colormap()` to switch between the many colormaps (`autumn()`, `bone()`, `cool()`, `copper()`, `flag()`, `gray()`, `hot()`, `hsv()`, `jet()`, `ocean()`, `pink()`, `prism()`, `rainbow()`, `spring` `summer()`, `white()`, `winter()`, `contrast()`, `gpmap40()`). You can create your own colormap.

It is sometimes necessary to convert from one image format to another and *Octave* provides the commands:

- `rgb2gray()`: convert an RGB image to a gray scale image.

- `gray2ind()`: convert gray scale image to an index image

- `ind2gray()`: convert an indexed image to a gray scale image

- `ind2rgb()`: convert an indexed image to an RGB image

- There are more possible conversions and also the commands to detect of what type an image is. Consult the manuals.

- With the command `imformats()` generate a list of all the image formats available on your platform.

Tables 1.10 and 1.11 show a rather **incomplete** selection of commands related to image processing. It is essential to consult the available documentation. *Octave* and MATLAB provide basic commands and data structures to implement image processing operations efficiently. There are also many more resources on image processing with *Octave* and MATLAB on the internet.

`imshow()`	display the image
`image()`	display a matrix as image
`imagesc()`	scale image and display
`imfinfo()`	obtain information about an image in a file
`imread()`	load an image from a file
`imwrite()`	write an image to a file
`imformats()`	list all the image formats available
`colormap()`	return or set the colormap
`rgb2gray()`	convert RGB to gray scale image
`rgb2ind()`	convert an RGB image to an indexed image
`ind2gray()`	convert indexed image to gray scale image
`gray2ind()`	convert a gray scale image to an indexed image

Table 1.10: Commands for image loading, writing and converting

25 Example : RGB, grayscale

A picture `WallaceGromit.png` in the PNG format (portable network graphics) is loaded into *Octave* and `imfinfo()` displays information on the image. It is an image of size 724×666 and each pixel (picture element) consists of three colors (RGB), encoded by an integer between 0 and 255. Thus the "matrix" `im` in the code below is of size $724 \times 622 \times 3$ and the image is shown in Figure 1.27(a).

```
imfinfo('WallaceGromit.png')          % show information on the file
im = imread('WallaceGromit.png');     % load the file
size(im)
figure(1); imshow(im)                 % display the original picture
```

`imresize()`	change the size of an image
`imrotate()`	rotate an image matrix
`fspecial()`	create filters for image processing
`imfilter()`	apply an image filter
`imsmooth()`	smooth an image with different algorithms
`imshear()`	shear an image
`edge()`	use a selection of edge detection algorithms
`conv(), conv2()`	convolution, one and two dimensional
`fft2(), ifft2()`	2D Fast Fourier Transforms

Table 1.11: Commands for elementary image processing

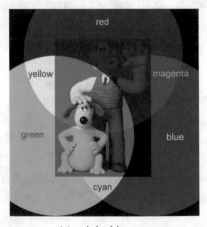

(a) original image

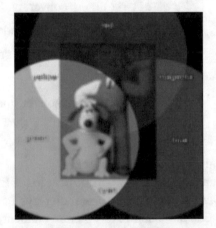

(b) after an averaging filter

Figure 1.27: Wallace & Gromit, original and with an averaging filter applied. Image courtesy of Aardman Animations LTD.

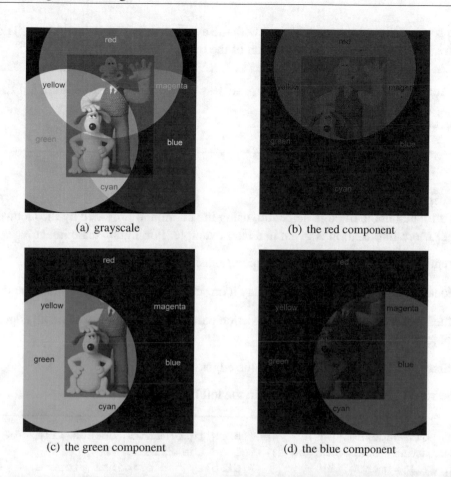

Figure 1.28: Wallace & Gromit, as grayscale and R, G and B images

The image can be converted to a grayscale image with the help of rgb2gray() and then displayed. Each color in RGB can be displayed independently, where a white spot indicates a high intensity of this color. Find the results in Figure 1.28.

```
imGray = rgb2gray(im);
imR = im(:,:,1); imG = im(:,:,2); imB = im(:,:,3);
redmap = zeros(256,3);            % fill the colormaps with zeros
greenmap = redmap; bluemap = redmap;
redmap(:,1) = linspace(0,1,256); % put numbers in the first column
greenmap(:,2) = linspace(0,1,256); bluemap(:,3) = linspace(0,1,256);
figure(2); subplot(2,2,1); imshow(imGray)
           subplot(2,2,2); imshow(imR)
           subplot(2,2,3); imshow(imG)
           subplot(2,2,4); imshow(imB)
```

The image package in *Octave* and the Image Processing Toolbox in MATLAB contain many and powerful commands for image processing. As an elementary example consider a filter to replace the values at each pixel with the average values of its neighbors. The result in Figure 1.27(b) is a smeared out version of the original image 1.27(a).

```
pkg load image   %% Octave only, in Matlab use Image Processing Toolbox
F = fspecial('average', 12);
imFilter = imfilter(im, F);
figure(3); imshow(imFilter)
```

26 Example : Edge Detection

This is a first example of edge detection, using the command provided by the Octave Forge package. More information is given in a later example. For a gray-scale image

1. read some information about the picture, using the command `imfinfo()`

2. load the image in *Octave* and display it on screen, using `imread()` and `imshow()`.

3. then use one of the many edge detection parameters to hopefully find all edges of the objects displayed.

4. finally we display the image with the edges only.

Find the result of the commands below in the left half of Figure 1.29.

```
imfinfo('shapessm.jpg')       % show information on the file
im = imread('shapessm.jpg');  % load the file
figure(1); imshow(im)         % display the original picture
edgeim = edge(im,'Canny');    % run one of the possible edge detections
figure(2); imshow(edgeim)     % display the picture with edges only
```

Figure 1.29: A grayscale and BW picture and edge detection. Find the original picture at [Kove20].

Since the sections in the original picture are either very dark, or very bright, we may convert the gray scale image into a BW (black and white) picture and try another edge detection. This might get rid of some artifacts. Find the result of the commands below in the right half of Figure 1.29. To not obtain black blobs on paper the roles of black an white are inverted by displaying `1-imbw` instead of the BW image.

```
imbw = im2bw(im,0.5); % convert to a black/white picture (0 and 1 only)
figure(3); imshow(1-imbw)
edgeimbw = bwmorph(imbw,'remove'); % run one  edge detections
figure(4); imshow(1-edgeimbw)      % display the picture, edges only
```

27 Example : Filtering with FFT

The idea presented in this example is also used for the JPEG compression. Find a very readable description in [Stew13, §9].

Using FFT write an image as sum of periodic signals with different frequencies. Then we can filter high or low frequencies. As an example examine the image in Figure 1.31. First load the image, convert it to a grayscale image and display the result.

```
imfinfo('Lenna.jpg') % read and convert to a grayscale image
imG = rgb2gray(imread('Lenna.jpg')); imG = rgb2gray(im);
figure(1); imshow(imG)    % display the result
```

Then apply the two dimensional FFT with the help of the command fft2() and choose the number of frequencies to keep by n=40. This corresponds to a perfect low-pass filter. Due to the symmetries in the FFT of real valued signals and images we have to keep the lowest n frequencies **and** the highest $n-1$ frequencies and thus 4 blocks of the FFT of the image are copied into the FFT of the filtered image. This algorithm is visualized in Figure 1.30. The codes is using block operations instead of loops, for speed reasons.

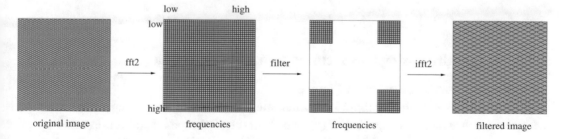

Figure 1.30: Apply a low pass filter to an image, based on FFT

```
imFFT = fft2(im2double(imG));    % convert to floating numbers, apply FFT
n = 40                           % number of frequencies to keep
[nx,ny] = size(imFFT)            % size of the image, and FFT
imFilter = zeros(nx,ny);         % zero matrix
imFilter(1:n+1,1:n+1)      = imFFT(1:n+1,1:n+1);       % top left
imFilter(1:n+1,ny-n+1:ny)  = imFFT(1:n+1,ny-n+1:ny);   % top right
imFilter(nx-n+1:nx,1:n+1)  = imFFT(nx-n+1:nx,1:n+1);   % bottom left
imFilter(nx-n+1:nx,ny-n+1:ny) = imFFT(nx-n+1:nx,ny-n+1:ny); % right
```

Finally apply the inverse FFT by `ifft2()` and keep the real part only. Then display the filtered image, as shown in Figure 1.31(b).

```
┌─────────────────────────── Octave ───────────────────────────┐
 newIm = real(ifft2(imFilter));        % apply inverse FFT
 figure(2); imshow(newIm)              % display the filtered image
 imwrite(newIm,'LennaFiltered.png');   % save the filtered image
└──────────────────────────────────────────────────────────────┘
```

(a) original image   (b) with low a pass filter

Figure 1.31: Original image of Lenna, and with a lowpass filter by FFT. This image is taken from http://www.lenna.org and one of the most often use pictures for image processing.

1.5.2 Image processing and vectorization, edge detection

28 Example : Adding Noise to a Picture
The main purpose of this example is to illustrate the power of vectorized code. We start with the Lena picture in Figure 1.31(a). Each pixel of the picture is represented by an integer value between 0 and 255. We add some noise to this picture by adding a random number, generated by a normal distribution with average 0 and a standard deviation given by `NoiseAmp = 20`. The code below applies this idea with nested loops.

```
┌─────────────────────────── Octave ───────────────────────────┐
 im  = imread('Lenna.jpg');
 imG = rgb2gray(im);                   % convert to a grayscale image
 figure(1); imshow(imG)                % display the result
 [Nlines,Ncols] = size(imG);           % compute the size of the picture
 NoiseAmp = 20;                        % amplitude of noise
 newIm = imG;                          % copy the image
 t0 = cputime();
 for lin = 1:Nlines                    % loop over al rows
   for col = 1:Ncols                   % loop over al columns
```

```
     newIm(lin,col) = newIm(lin,col) + NoiseAmp*randn(1);% add noise
   end%for
 end%for
 timingLoop = cputime()-t0
 figure(2); imshow(newIm)
```

A sample run took 2 seconds and produced the expected, noisy result. The same algorithm can be applied using vectorized code, getting rid of the loops. Generate a matrix of random numbers with one command (NoiseAmp*randn(NLines,Ncols)) and add it to the original matrix.

```
──────────────────────────── Octave ────────────────────────────
 t0 = cputime();
 newIm2 = uint8(single(imG) + NoiseAmp*randn(Nlines,Ncols));
 timingVectorized = cputime()-t0
 figure(3); imshow(newIm2)
```

This code only used 0.017 sec of CPU time, i.e. the code is 120 times faster. Using MATLAB the speed difference is not as drastic, but still relevant. $\diamond$

29 Example : Edge detection
In many applications the first step of image processing is edge detection, e.g. to identify objects. As example consider Figure 1.32 and the goal is to mark the edges of the different objects.

Figure 1.32: A few objects, for edge detection

• The basic idea
The basic idea is illustrated by analyzing one line in the picture. The size of the

picture is 800×1200 and we pick a horizontal line, cutting through the shadow below the pocket knife, through the scissors and the Pritt stick. The darker sections correspond to lower values in Figure 1.33(a).

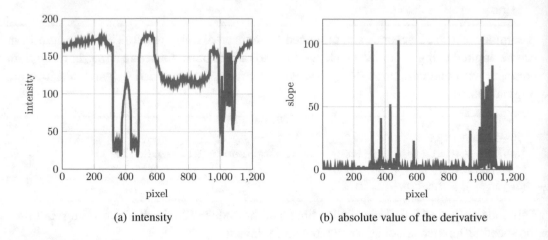

(a) intensity   (b) absolute value of the derivative

Figure 1.33: Intensity along a horizontal line through the objects

The idea of an edge detection is to look for steep slopes in the graph in Figure 1.33(a). Based on the definition of the derivative examine

$$f'(x) \approx \frac{f(x+h) - f(x-h)}{2\,h} \ .$$

For the discrete values $u(k)$ of the intensity define

$$\text{edgeLine}(k) = +u(k+1) - u(k-1) \quad \text{for} \quad 2 \le k \le n-1$$

or with *Octave*/MATLAB

```
imRaw = rgb2gray(imread('Edgetest2.png'));
figure(1); imshow(imRaw);
          hold('on')
          plot([1,1200],[400,400],"r");
test_line = double(imRaw(400,:)); % pick one line to be examined
                                  % convert to data type double
figure(2); plot(test_line)
          xlabel('pixel'); ylabel('intensity')
n = length(test_line);
edgeLine = test_line(3:n)-test_line(1:n-2);

figure(3); plot(abs(edgeLine))
          xlabel('pixel'); ylabel('slope')
```

This leads to Figure 1.33(b). Now select a cutoff level, e.g. 40, to decide where an edge is visible. This will point towards a few points sitting on an vertical edge in Figure 1.32, which is visually confirmed by scanning along the red horizontal line in that figure.

The above implementation uses a `for...end` loop, which should be replaced by a vector operation, mainly for speed reasons.

```
edgeline = [0 , -test_line(1:n-2) + test_line(3:n) , 0];
```

The above can also be considered as a convolution[22] of the vector of the intensity values along the line to be examined with the vector.

−1	0	1

• **Detecting horizontal edges, with the Sobel filter**
The above idea has to be carried over to the full image. Replace the vector by the 3×3 matrix

−1	0	+1
−2	0	+2
−1	0	+1

This corresponds to a weighted average of the slopes in three lines. To apply the procedure to the matrix

$$
\mathbf{A} =
\begin{bmatrix}
1 & 11 & 11 & 1 & 1 & 1 \\
2 & 12 & 12 & 2 & 2 & 2 \\
3 & 13 & 13 & 3 & 3 & 3 \\
4 & 14 & 14 & 3 & 3 & 3 \\
5 & 15 & 15 & 3 & 3 & 3 \\
6 & 16 & 16 & 3 & 3 & 3
\end{bmatrix}
$$

proceed as follows:

1. Put the central element of the 3×3 filter matrix over one entry in the matrix $\mathbf{A}$.

2. Multiply and add the overlapping numbers.

3. Put the result in the new, filtered matrix.

Examine a few examples:

---

[22]Check you math lecture notes for the definition of convolution, either with Laplace or Fourier transforms.

– In the second row and second column obtain

$$b_{2,2} = \left\{ \begin{array}{lllll} +(-1)\cdot 1 & + & (0)\cdot 11 & + & (+1)\cdot 11 \\ +(-2)\cdot 2 & + & (0)\cdot 12 & + & (+2)\cdot 12 \\ +(-1)\cdot 3 & + & (0)\cdot 13 & + & (+1)\cdot 13 \end{array} \right\} = 30$$

This indicates an edge with positive slope, i.e. an increasing intensity.

– In the second row and fourth column obtain

$$b_{2,4} = \left\{ \begin{array}{lllll} +(-1)\cdot 11 & + & (0)\cdot 1 & + & (+1)\cdot 1 \\ +(-2)\cdot 12 & + & (0)\cdot 2 & + & (+2)\cdot 2 \\ +(-1)\cdot 13 & + & (0)\cdot 3 & + & (+1)\cdot 3 \end{array} \right\} = -30$$

This indicates an edge with negative slope, i.e. a decreasing intensity.

– In the second row and fifth column obtain

$$b_{2,5} = \left\{ \begin{array}{lllll} +(-1)\cdot 1 & + & (0)\cdot 1 & + & (+1)\cdot 1 \\ +(-2)\cdot 2 & + & (0)\cdot 2 & + & (+2)\cdot 2 \\ +(-1)\cdot 3 & + & (0)\cdot 3 & + & (+1)\cdot 3 \end{array} \right\} = 0$$

This indicates no edge.

This should be implemented directly by matrix operations and applied it to the given image. Observe the code without loops, leading to fast computations.

```
[nx,ny] = size(imRaw);      % size of picture
ix = 2:nx-1; iy =  2:ny-1; % indices, unshifted
imRaw = double(imRaw);
Gx = -1*imRaw(ix-1,iy-1) + 1*imRaw(ix+1,iy-1)...
     -2*imRaw(ix-1,iy  ) + 2*imRaw(ix+1,iy  )...
     -1*imRaw(ix-1,iy+1) + 1*imRaw(ix+1,iy+1);

edgeLevel = 100;  % choose the detection level
imshow(1-(abs(Gx)>edgeLevel))
```

Find the result on in Figure 1.34(a). Observe that only horizontal edges are marked. The obvious vertical edges in the original image are not detected.

- **Detecting vertical edges**
  For vertical edges apply a similar procedure, but use the matrix

+1	+2	+1
0	0	0
−1	−2	−1

and the code

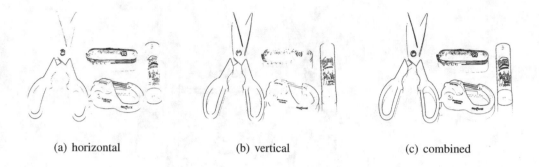

| (a) horizontal | (b) vertical | (c) combined |

Figure 1.34: Sobel edge detection

```
Gy = +1*imRaw(ix-1,iy-1) + 2*imRaw(ix,iy-1) + 1*imRaw(ix+1,iy-1)...
     -1*imRaw(ix-1,iy+1) - 2*imRaw(ix,iy+1) - 1*imRaw(ix+1,iy+1);
```

leading to the result in Figure 1.34(b).

- **Combining horizontal and vertical edges**
  Now combine the two basic detection algorithms, leading to Figure 1.34(c). Finally all edges are visible.

```
imshow(1-((sqrt(Gx.^2+Gy.^2))>cdgeLevel))
```

- **The function** `edge()`
  The image package has a built in function `edge()`, which can be used to apply the Sobel edge detection algorithm, leading to Figure 1.35. The function `edge()` allows to use a few other filters for edge detection, e.g. Sobel, Prewitt[23], Roberts, Canny, ... All those algorithms are using the above idea, with different filter matrices. Examine the result of `help edge` or the source code of `edge.m` to find out more.

```
imSobel = edge(uint8(imRaw),'Sobel');
imshow(1-imSobel)
```

---

[23]For the Prewitt filter use the matrices

$$\begin{bmatrix} +1 & 0 & -1 \\ +1 & 0 & -1 \\ +1 & 0 & -1 \end{bmatrix} \quad \text{and} \quad \begin{bmatrix} +1 & +1 & +1 \\ 0 & 0 & 0 \\ -1 & -1 & -1 \end{bmatrix}.$$

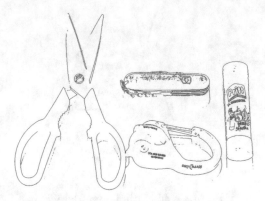

Figure 1.35: Result of a Sobel edge detection with the command `edge`

30 Example : Observations on edge detection
The above idea can be modified in many different ways.

- Use other filters, e.g. the matrices

+3	0	−3
+10	0	−10
+3	0	−3

and

+3	+10	+3
0	0	0
−3	−10	−3

.

Since the above filter operations can be written as convolution , use the command `conv2()` in *Octave* to apply the edge detection filters. Particular attention has to be paid to the behavior of the convolution at the boundary of the images, consult the documentation on `conv()` and `conv2()`. This leads to very efficient code and Figure 1.36(a) as result.

```
imRaw = rgb2gray(imread('Edgetest.png'));
Sx = single([ 3 0 -3 ; 10 0 -10 ; 3 0 -3 ]);
Sy = single([ 3 10 3 ; 0 0 0 ; -3 -10 -3 ]);
imRawEdge = sqrt(conv2(imRaw,Sx).^2 + conv2(imRaw,Sy).^2);
figure(2); imshow(1-(imRawEdge>250))
```

- The above fails miserably if noise is added, as can be seen in Figure 1.36(b). Our eye and brain can still see the real edges, but the code does not.

```
[n,m] = size(imRaw);
imNoise = imRaw + 10*randn(n,m);
figure(3); imshow(imNoiseEdge)
```

```
imNoiseEdge = sqrt(conv2(imNoise,Sx).^2 + conv2(imNoise,Sy).^2);
figure(4); imshow(1-(imNoiseEdge>250))
```

- The effect of noise can sometimes be controlled by an averaging filter. The value at one pixel is replaced by a weighted average of points close to this pixel. To average over a 5×5 section we can (many other options are possible) use the matrix

$$
\mathbf{A} = \frac{1}{m}
\begin{bmatrix}
1 & 3 & 6 & 3 & 1 \\
3 & 8 & 10 & 8 & 3 \\
6 & 10 & 12 & 10 & 6 \\
3 & 8 & 10 & 8 & 3 \\
1 & 3 & 6 & 3 & 1
\end{bmatrix}
$$

where m is chosen such that the sum of all values in $\mathbf{A}$ equals 1. This assures that the average intensity of the image is not modified. Use a convolution of the noisy image matrix with this averaging matrix to generate a new picture. Now we can detect the edges in a noisy image, as seen in Figure 1.36(c).

```
AvgMat = [1  3  6  3 1;  3  8 10  8 3;    6 10 12 10 6;
          3  8 10  8 3;  1  3  6  3 1];
AvgMat = AvgMat/sum(AvgMat(:));
% apply the averaging filter
imNoiseAvg = conv2(imNoiseEdge,AvgMat,'same');
% run the two edge detection filters and add
imNoiseAvgEdge = sqrt(conv2(imNoiseAvg,Sx).^2 +...
                      conv2(imNoiseAvg,Sy).^2);
% invert the intensity and display
figure(5);  imshow(1-(imNoiseAvgEdge>250))
```

(a) without noise   (b) with noise   (c) with noise and averaging filter

Figure 1.36: Edge detection, using a different filter

$\Diamond$

31 Example :   Filters can also be applied directly with the help of the commands `fspecial()` and `imfilter()`. As an example try to detect edges in the flower picture. Examine and explain the result!

```
FlowersEdge.m
im = imread ('SvenMieke.png');
imH = imfilter(im,fspecial("Sobel") ,'replicate');
imV = imfilter(im,fspecial("Sobel")','replicate');
figure(11); subplot(1,3,1); imshow(im);
            subplot(1,3,2); imshow(imH);
            subplot(1,3,3); imshow(imV);
```

$\diamond$

1.6   Ordinary Differential Equations

From your class on Engineering Mathematics or Ordinary Differential Equations you should have some basic knowledge and suitable examples for fixed step size algorithms, e.g. Euler, Heun and Runge-Kutta. Thus we concentrate on the usage of the *Octave* commands to solve differential equations. It is assumed that you are familiar with the theoretical aspects of ODEs (Ordinary Differential Equations).

For a given, smooth function $f(x, t)$ and given initial time t_0 and initial values x_0 the initial value problem

$$\frac{d}{dt} x(t) = f(t, x(t)) \quad \text{with} \quad x(t_0) = x_0$$

has exactly one solution, a function $x(t)$. One can attempt to find a solution, by analytical or numerical methods. *Octave* and `MATLAB` use numerical methods to determine approximations to solutions of differential equations. In this section we present a few basic ideas:

1. With *Octave* or `MATLAB` use one of the `ode??()` commands to solve a single ODE or a system. A logistic equation and the the Volterra–Lotka model are used as typical examples. Due to current events (2020-2022) a simple model for the spreading of a virus is briefly examined.

2. To start out only the command `ode45()` will be used. In Section 1.6.2 a few more of the available algorithms in *Octave*/`MATLAB` are documented.

3. Examine how to use `C++` code within *Octave* to improve the speed.

4. Examine how to perform further calculations with solutions of ODEs. We examine the period of a Volterra–Lotka solution.

5. To close the section a code from your author's lecture notes is examined.

Introduction

The simplest form of an ODE (Ordinary Differential Equation) is

$$\frac{d}{dt} u(t) = f(t, u(t))$$

for a given function $f : [t_{init}, t_{end}] \times \mathbb{R} \longrightarrow \mathbb{R}$. An additional initial time t_0 and an initial value $u_0 = u(t_0)$ lead to an IVP (Initial Value Problem). A solution has to be a function, satisfying the differential equation and the initial condition. One can show that for differentiable functions f any IVP has a unique solution on an interval $a < t_0 < b$. The final time $t < b$ might be smaller that $+\infty$, since the solution could blow up in finite time.

32 Example : The logistic differential equation
The behavior of the size of a population $0 \le p(t)$ with a limited nutrition supply can be modeled by the logistic differential equation

$$\frac{d}{dt} p(t) = (\alpha - p(t)) \, p(t) \, .$$

In this case the differential equation with $f(p) = (\alpha - p) \, p$ is autonomous, i.e. f does not explicitly depend on the time t. In Figure 1.37 three solutions for $\alpha = 2$ with different initial values are shown. The vector field is generated by displaying many (rescaled) vectors $(1, f(p))$ attached at points (t, p). The function $p(t)$ being a solution of the ordinary differential equation is equivalent to the slope of the curve to coincide with the directions of the vector field. Thus vector fields can be useful to understand the qualitative behavior of solutions of ODEs. Figure 1.37 is generated by the code below.

```
                              Logistic.m
 t_max = 5; p_max = 3;
 [t,p] = meshgrid(linspace(0, t_max,20),linspace(0,p_max,20));
 v1 = ones(size(t)); v2 = p.*(2-p);
 figure(1);  quiver(t,p,v1,v2)
             xlabel('time t'); ylabel('population p')
             axis([0 t_max, 0 p_max])
 [t1,p1] = ode45(@(t,p)p.*(2-p),linspace(0,t_max,50),0.4);
 [t2,p2] = ode45(@(t,p)p.*(2-p),linspace(0,t_max,50),3.0);
 [t3,p3] = ode45(@(t,p)p.*(2-p),linspace(0,t_max,50),0.01);
 hold on; plot(t1,p1,'r',t2,p2,'r', t3,p3,'r'), hold off
```

Systems of ordinary differential equations

The above idea can be applied to a system of ODEs. As example consider the famous predator–pray model by Volterra–Lotka[24].

---

[24]Proposed by Alfred J. Lotka in 1910 and Vito Volterra in 1926.

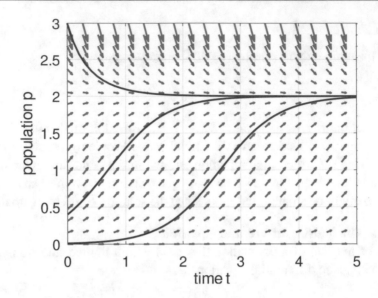

Figure 1.37: Vector field and three solutions for a logistic equation

33 Example : Volterra–Lotka predator–prey model
Consider two different species with the size of their population given by $x(t)$ and $y(t)$.

$$x(t) \quad \text{population size of pray at time } t$$
$$y(t) \quad \text{population size of predator at time } t$$

The predators y (e.g. sharks) are feeding of the pray x (e.g. small fish). The food supply for the pray is limited by the environment. The behavior of these two populations can be described by a system of two first order differential equations.

$$\frac{d}{dt}\,x(t) \;=\; (c_1 - c_2\,y(t))\,x(t)$$
$$\frac{d}{dt}\,y(t) \;=\; (c_3\,x(t) - c_4)\,y(t)$$

where c_i are positive constants. This function can be implemented in MATLAB/*Octave* as a function file `VolterraLotka.m`.

VolterraLotka.m
```
function res = VolterraLotka(t,x)
  c1 = 1; c2 = 2; c3 = 1; c4 = 1;
  res = [(c1-c2*x(2))*x(1);
         (c3*x(1)-c4)*x(2)];
end%function
```

With the help of the above function generate information about the solutions of this system of ODEs.

- Generate the data for the vector field and then use the command `quiver` to display the vector field, shown in Figure 1.38(b).

- Use `ode45()` to generate numerical solutions with initial values $(x(0), y(0)) = (2, 1)$ and for 100 times $0 \le t_i \le 15$. Display the result in Figure 1.38.

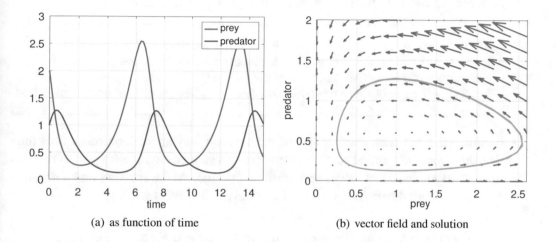

(a) as function of time   (b) vector field and solution

Figure 1.38: One solution and the vector field for the Volterra-Lotka problem

```
─────────────────────── VolterraLotkaField.m ───────────────────────
x = 0:0.2:2.6;   % define the x values to be examined
y = 0:0.2:2.0;   % define the y values to be examined
n = length(x);   m = length(y);
Vx = zeros(n,m); Vy = Vx; % create zero vectors for the vector field
for i = 1:n
  for j = 1:m
    v = VolterraLotka(0,[x(i),y(j)]); % compute the vector
    Vx(i,j) = v(1); Vy(i,j) = v(2);
  end%for
end%for

t = linspace(0,15,100); [t,XY] = ode45(@VolterraLotka,t,[2;1]);

figure(1); plot(t,XY)
           xlabel('time'); legend('prey','predator');
           axis([0,15,0,3]); grid on
figure(2); quiver(x,y,Vx',Vy',2); hold on
           plot(XY(:,1),XY(:,2));
           axis([min(x),max(x),min(y),max(y)]);
           grid on; xlabel('prey'); ylabel('predator'); hold off
```

◇

34 Example : Determine the period of a Volterra-Lotka solution

Using Figure 1.38(a) one might guess that the solution of the Volterra-Lotka equation is periodic with a period $6.5 \leq T \leq 7$. Use *Octave* to confirm this proposition.

- First choose the initial values and solve the system of differential equations using `ode45()`. Use the inital time 0 and generate values of the solution at 1000 times between 6.5 and 7.0 . We require an absolute and relative tolerance of 10^{-10}.

```
x0 = 2; y0 = 1;                  % initial values
t = [0,linspace(6,7,1000)];     % examine times between 6 and 7
ode_opt = odeset('AbsTol',1e-10,'RelTol',1e-10);
[t,XY] = ode45(@VolterraLotka,t,[x0,y0],ode_opt); % solve the ODE
```

- Examine the first component (prey) of the solution only and determine at what time its value crosses the initial value. For this detect a sign change of $x(t) - x(0)$ by multiplying two subsequent values and examine the sign. At the crossing find a negative sign. As a result determine in which time interval the period will be.

```
y = XY(2:end,1)-x0; t = t(2:end);   % examine the first component
plot(t,y);                           % visual test for zero
grid on                              % detect sign changes
s = sign((XY(2:end,1)-x0).*(XY(1:end-1,1)-x0));
pos = find(s<0);                     % position of sign change
```

- To increase **accuracy** use a linear interpolation. If a function $f(t)$ crosses zero between a and b, replace the actual function by a straight line and search the zero of this linear interpolation. By solving

$$f(a + \Delta t) \approx g(\Delta t) = f(a) + \frac{f(b) - f(a)}{b - a} \Delta t = 0$$

find

$$\Delta t = \frac{-f(a)}{f(b) - f(a)} (b - a) .$$

Using this idea solve the Volterra-Lotka equation again with initial time given by the time just before the first component of the solution crosses its initial value again. Thus we only have to solve for a very short time interval.

```
% use linear interpolation to determine the partial time step
dt = (t(pos)-t(pos-1))*y(pos-1)/(y(pos-1)-y(pos))
T = t(pos-1)+dt% estimate the period, then compute the value at T
[tnew,XYnew] = ode45(@VolterraLotka,[T-dt,T],XY(pos,:));
XYnew(end,:)-[x0,y0] % difference to initial values
-->
```

```
dt  =  1.6494e-04
T   =  6.9411
ans =  1.3768e-07  -6.8739e-08
```

The numerical result confirms that **both** components are periodic with a period of $T \approx 6.9411$.

35 Example : The SIR model for the spreding of a pandemic

A pandemic might start with a few infected individuals, but the infection can spread very quickly. One of the mathematical models for this is the SIR model (Susceptible, Infected, Recovered). Find a description at https://www.maa.org/press/periodicals/loci/joma/the-sir-model-for-spread-of-disease-the-differential-equation-model and a YouTube video on the SIR model at https://www.youtube.com/watch?v=NKMHhm2Zbkw.

This (overly) simple model for the spreading of a virus uses three time dependent variables:

$$
\begin{aligned}
S(t) &= \text{the susceptible fraction of the population} \\
I(t) &= \text{the infected fraction of the population} \\
R(t) &= \text{the recovered fraction of the population}
\end{aligned}
$$

and $S(t) + I(t) + R(t) = 1$ implies

$$\frac{d\,S(t)}{dt} + \frac{d\,I(t)}{dt} + \frac{d\,R(t)}{dt} = 0 \,.$$

Assuming there are N individuals in the population, use two parameters to describe the spreading of the virus.

b = number of contacts per day of an infected individual.

  $b\,N\,I(t)$ new attempted infections, but only the fraction $S(t)$ is susceptible.

k = fraction of infected individuals that will recover in one day.

  $k\,N\,I(t)$ individuals will recover in one day.

Interpretation of the parameters:

- The value $\frac{1}{k}$ can be considered as number of days an individual can spread the virus to new patients.

- Every day an infected individual will make contact to b other individuals, possibly infecting them with the virus. Only the fraction $0 < S(t) < 1$ is susceptible, thus $b\,S(t)$ will actually be infected by this individual.

- During his sick period an individual will thus infect $\frac{b}{k}\,S(t)$ newly infected patients.

- As we have a total of $N\,I(t)$ sick individuals we will observe $b\,N\,I(t)\,S(t)$ new infection every day.

This leads to a system of ODEs for the three ratios $S(t)$, $I(t)$ and $R(t)$.

$$\frac{d}{dt}S(t) = -b\,S(t)\,I(t)$$

$$\frac{d}{dt}I(t) = -\frac{d\,S(t)}{dt} - \frac{d\,R(t)}{dt} = +b\,S(t)\,I(t) - k\,I(t)$$

$$\frac{d}{dt}R(t) = +k\,I(t)$$

Rewrite this as an ODE for $I(t)$ and $R(t)$, using $S(t) = 1 - I(t) - R(t)$.

$$\frac{d}{dt}I(t) = +b\,(1 - I(t) - R(t))\,I(t) - k\,I(t) = (+b\,(1 - I(t) - R(t) - k)\,I(t)$$

$$= (+b - k - b\,I(t) - b\,R(t))\,I(t)$$

$$\frac{d}{dt}R(t) = +k\,I(t)$$

This system of ODE is solved numerically using e.g `ode45()`. Find the results of the code below in Figure 1.39. The additional code in the file `SIR_Model.m` generates the vector fields in Figure 1.40.

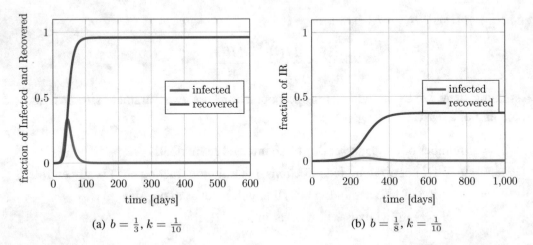

(a) $b = \frac{1}{3}, k = \frac{1}{10}$ (b) $b = \frac{1}{8}, k = \frac{1}{10}$

Figure 1.39: SIR model, with infection rate b and recovery rate k. The positive effect of a small infection rate b is obvious.

```
                          SIR_Model.m
IO = 1e-4;  S0 = 1 - I0;  R0 = 0; % the initial values
b = 1/3;  k = 1/10;               % the model parameters
%% b = 1/8;                       % use a smaller infection rate
```

```
%% for MATLAB comment out this function and put it a file SIR.m
function res = SIR(t,I,R,b,k) %%x = IR
  res = [(+b-k-b*I-b*R).*I;  k*I];
end%function

[t,IR] = ode45(@(t,x)SIR(t,x(1),x(2),b,k),linspace(0,600,601),[I0,R0]);

figure(1); plot(t,IR)
           xlabel('time [days]');
           ylabel('fraction of Infected and Recovered')
           ylim([-0.05 1.05])
           legend('infected','recovered', 'location','east')
```

◇

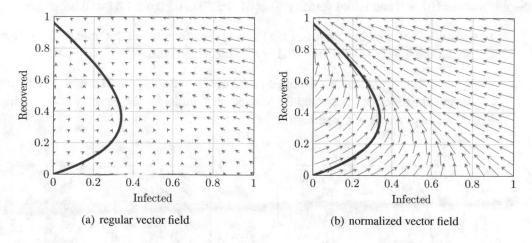

(a) regular vector field   (b) normalized vector field

Figure 1.40: SIR model vector field with $b = \frac{1}{3}$ and $k = \frac{1}{10}$

Converting an ODE of higher order to a system of order 1

Ordinary differential equation of higher order can be converted to systems of order 1. Thus most numerical algorithm are applicable to system of order 1, but not to higher order ODEs. The method is illustrated by the example of a pendulum equation.

36 Example : ODE for a damped pendulum
The equation

$$\ddot{x}(t) + \alpha\,\dot{x}(t) + k\,x(t) = f(t)$$

describes a mass attached to a spring with an additional damping term $\alpha\,\dot{x}(t)$. Introducing the new variables $y_1(t) = x(t)$ and $y_2(t) = \dot{x}(t)$ leads to

$$\frac{d}{dt}\,y_1(t) = \dot{x}(t) = y_2(t)$$

and

$$\frac{d}{dt}\, y_2(t) = \frac{d}{dt}\, \dot{x}(t) = \ddot{x}(t) = f(t) - \alpha\, \dot{x}(t) - k\, x(t) = f(t) - \alpha\, y_2(t)(t) - k\, y_1(t)\;.$$

This can be written as a system of first order equations

$$\frac{d}{dt} \left(\begin{array}{c} y_1(t) \\ y_2(t) \end{array} \right) = \left(\begin{array}{c} \dot{y}_1(t) \\ \dot{y}_2(t) \end{array} \right) = \left(\begin{array}{c} y_2(t) \\ f(t) - k\, y_1(t) - \alpha\, y_2(t)(t) \end{array} \right)$$

or

$$\frac{d}{dt}\, \vec{y}(t) = \vec{F}(\vec{y}(t))\;.$$

With the help of a function file the problem can be solved with computations very similar to the above Volterra–Lotka example. The code below will compute a solution with the initial displacement $x(0) = 0$ and initial velocity $\frac{d}{dt}\, x(0) = 1$. Then Figure 1.41 will be generated.

- In Figure 1.41(a) find the graphs of $x(t)$ and $v(t)$ as function of the time t. The effect of the damping term $-\alpha\, v(t) = -0.1\, v(t)$ is clearly visible.

- In Figure 1.41(b) find the vector field and the computed solution. The horizontal axis represents the displacement x and the vertical axis indicates the velocity $v = \dot{x}$. This is the **phase portrait** of the second order ODE.

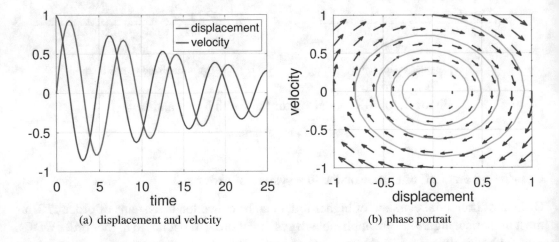

(a) displacement and velocity   (b) phase portrait

Figure 1.41: Vector field and a solution for a spring-mass problem

—————————————————————— **Spring_Model.m** ——————————————————————
```
y = -1:0.2:1;   v = -1:0.2:1;   n = length(y);  m = length(v);
Vx = zeros(n,m); Vy = Vx; % create zero vectors for the vector field
function ydot = Spring(y)
  ydot = zeros(size(y));
  k = 1; al = 0.1;
```

```
  ydot(1) = y(2);
  ydot(2) = -k*y(1)-al*y(2);
end%function
for i = 1:n
  for j = 1:m
    z = Spring([y(i),v(j)]);         % compute the vector
    Vx(i,j) = z(1); Vy(i,j) = z(2); % store the components
  end%for
end%for

t = linspace(0,25,100);    [t,XY] = ode45(@(t,y)Spring(y),t,[0;1]);

figure(1); plot(t,XY)
           xlabel('time'); legend('displacement','velocity')
           axis(); grid on
figure(2); plot(XY(:,1),XY(:,2),'g'); % plot solution, phase portrait
           axis([min(y),max(y),min(v),max(v)]);
           hold on
           quiver(y,v,Vx',Vy','b');
           xlabel('displacement'); ylabel('velocity');
           grid on; hold off
```

◇

1.6.1 Using C++ code to speed up computations

There are two reasons to use C/C++ code within *Octave* or MATLAB:

- Speed: if your code can not be vectorized you might gain a lott of speed by incorporating C/C++ code.

- Hardware: you might want to acces special hardware through a library provided by the hardware producer.

There are two way to incorporate C or C++ code into *Octave* or MATLAB:

- Use OCT files with *Octave* only. This approach has better performance, since all structures are based on *Octave*.

- Use MEX files with *Octave* or MATLAB. This approach allows to share code between MATLAB and *Octave*, but has lower performance. To quote the *Octave* manual: "In particular, to support the manner in which variables are passed to mex functions there are a significant number of additional copies of memory blocks when calling or returning from a mex-file function."

C++ code in *Octave*, using OCT files

When solving a differential equation $\dot{x}(t) = \vec{F}(\vec{x}(t))$ numerically the function $\vec{F}$ will have to be called many times. Thus we have to look out for fast computations. *Octave* has a good interface for C++ code to be integrated into the *Octave* environment. As an example we rewrite the *Octave* function file `VolterraLotka.m` as C++ code. Observe that within *Octave* the indexing of arrays starts with 1, but in C and C++ the first index is 0. Thus the components $x(1)$ and $x(2)$ in *Octave* now become $x(0)$ and $x(1)$ in C++.

```
──────────────────────── VolterraLotkaC.cc ────────────────────────
#include <octave/oct.h>
DEFUN_DLD (VolterraLotkaC, args, ,
  "Function for a Volterra Lotka model"){
  ColumnVector dx (2);
  ColumnVector t (args(0).vector_value ());
  ColumnVector x (args(1).vector_value ());
  double c1 = 1.0, c2 = 2.0, c3 = 1.0, c4 = 1.0;
  dx(0) = (c1-c2*x(1))*x(0);
  dx(1) = (c3*x(0)-c4)*x(1);
  return octave_value (dx);}
```

Then launch the *Octave* compiler in a shell (or within the *Octave* environment) in the current directory by the command `mkoctfile VolterraLotkaC.cc`. A compiled version `VolterraLotkaC.oct` will be created and used by *Octave*. To use the compiled code type the command `VolterraLotkaC()` in *Octave*, if this file is in the current directory or visible in `path`. To compare the results generated by the function file and the compiled code.

```
──────────────────────────── Octave ────────────────────────────
[VolterraLotka(0,[2.22,1.12]), VolterraLotkaC(0,[2.22,1.12])]
-->
  -2.7528  -2.7528
   1.3664   1.3664
```

The main advantage of C++ code is speed. With the code below one can compare the performance of the script file with the C++ code.

```
──────────────────────────── Octave ────────────────────────────
t = linspace(0,5000,100);
ode_opt = odeset('AbsTol',1e-10,'RelTol',1e-10);
t0 = cputime();
[t1,XY1] = ode45(@VolterraLotka,t,[2,1],ode_opt);
TimeScript = cputime()-t0

t0 = cputime();
[t2,XY2] = ode45(@VolterraLotkaC,t,[2,1],ode_opt);
TimeCPP = cputime()-t0

ratio = TimeScript/TimeCPP
```

On my current (2021) test system the C++ ran 1.4 times faster than the function file. The difference for more complicated examples can be considerably larger.

Other examples are provided with the *Octave* distribution, e.g. `oregonator.cc`.

1.6.2 ODE solvers in MATLAB/*Octave*

Most of the ODE solver in *Octave* and MATLAB follow a very similar syntax. Thus it is very easy to switch the solvers and find the one most suitable for your problem. In the next subsection four of the available algorithms will be described with more details.

- Octave 6.2.0: `ode15i`, `ode15s`, `ode23`, `ode23s`, `ode45`.

- Matlab R2019a: `ode113`, `ode15i`, `ode15s`, `ode23`, `ode23s`, `ode23t`, `ode23tb`, `ode45`.

For most questions one of the four algorithms mentioned below will lead to reliable results at a reasonable computational cost.

`ode45` : an implementation of a Runge–Kutta (4,5) formula, the Dormand–Prince method of order 4 and 5. This algorithm works well on most non–stiff problems and is a good choice as a starting algorithm.

`ode23` : an implementation of an explicit Runge–Kutta (2,3) method, the explicit Bogacki–Shampine method of order 3. It might be more efficient than `ode45` at crude tolerances for moderately stiff problems.

`ode15s` : this command solves stiff ODEs. It uses a variable step, variable order BDF (Backward Differentiation Formula) method that ranges from order 1 to 5. Use `ode15s` if `ode45` fails or is very inefficient.

`ode23s` : this command solves stiff ODEs with a Rosenbrock method of order (2,3). The `ode23s` solver evaluates the Jacobian during each step of the integration, so supplying it with the Jacobian matrix is critical to its reliability and efficiency

To obtain information on the performance of the different solvers use the option `opt = odeset('Stats','on')`. This will show the number of steps taken and the number of function calls.

The basic usage of `ode??`

Typing the command `help ode45` in *Octave* generates on the first few lines the calling options for `ode45()`, but all other `ode??()` commands are very similar, as are the MATLAB commands.

```
help ode45
-->
--   [T, Y] = ode45 (FUN, TRANGE, INIT)
 --  [T, Y] = ode45 (FUN, TRANGE, INIT, ODE_OPT)
 --  [T, Y, TE, YE, IE] = ode45 (...)
 --  SOLUTION = ode45 (...)
 --  ode45 (...)
```

The input arguments have to be given by the user. A call of the function ode45() will return results.

- input arguments

   FUN   is a function handle, inline function, or string containing the name of the function that defines the ODE: $\frac{d}{dt}y(t) = f(t, y(t))$ (or $\frac{d}{dt}\vec{y}(t) = f(t, \vec{y}(t))$). The function must accept two inputs, where the first is time t and the second is a column vector (or scalar) of unknowns y.

   TRANGE   specifies the time interval over which the ODE will be evaluated. Typically, it is a two-element vector specifying the initial and final times ($[t_{init}, t_{final}]$). If there are more than two elements, then the solution will be evaluated at these intermediate time instances.

   Observe that the algorithms ode??() will always first choose the intermediate times, using the adapted step sizes. If TRANGE is a two-element vector, these values are returned. If more intermediate times are asked for the algorithm will use a special interpolation algorithm to return the solution at the desired times. Asking for more (maybe many) intermediate times will **not** increase the accuracy. For increased accuracy use the options RelTol and AbsTol, see page 136.

   INIT   contains the initial value for the unknowns. If it is a row vector then the solution Y will be a matrix in which each column is the solution for the corresponding initial value in t_{init}.

   ODE_OPT   The optional fourth argument ODE_OPT specifies non-default options to the ODE solver. It is a structure generated by odeset(), see page 136.

- return arguments

   – If the function [T,Y] = ode??() is called with two return arguments, the first return argument is column vector T with the times at which the solution is returned. The output Y is a matrix in which each column refers to a different unknown of the problem and each row corresponds to a time in T.

   – If the function SOL = ode??() is called with one return argument, a structure with three fields: SOL.x are the times, SOL.y is the matrix of solution values and the string SOL.solver indicated which solver was used.

– If the function ode??() is called with no return arguments, a graphic is generated. Try ode45(@(t,y)y,[0,1],1).

– If using the Events option, then three additional outputs may be returned. TE holds the time when an Event function returned a zero. YE holds the value of the solution at time TE. IE contains an index indicating which Event function was triggered in the case of multiple Event functions.

Solving the ODE $\frac{d}{dt}y(t) = (1-t)y(t)$ for $0 \le t \le 5$ with initial condition $y(0) = 1$ is a one-liner.

```
[t,y] = ode45(@(t,y)(1-t)*y,[0,5],1);
plot(t,y,'+-')
```

The plot on the left in Figure 1.42 shows that the time steps used by ode45() are rather large and thus the solution seems to be inaccurate. The Dormand–Prince algorithm in the code ode45() used large time steps to achieve the desired accuracy and then returned the solution at those times only. It might be better to return the solution at more intermediate times, uniformly spaced. This can be specified in trange when calling ode45(), see the code below. Find the result on the right in Figure 1.42.

```
[t,y] = ode45(@(t,y)(1-t)*y,[0:0.1:5],1);
plot(t,y,'+-')
```

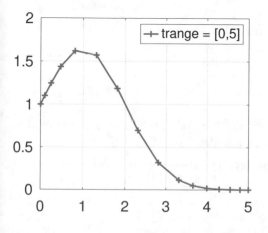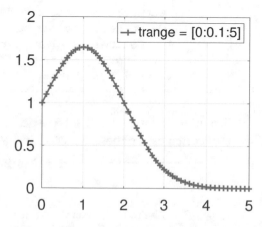

Figure 1.42: Solution of an ODE by ode45() at the computed times or at preselected times by choosing trange

Using options for the commands `ode??`

For these ODE solvers many options can and should be used. The command `odeset()` will generate a list of the available options and their default values. With `help odeset` obtain more information on these options. The available options differ slightly for *Octave* and `MATLAB`.

Below find a list of the options for the ODE solvers `ode??` in *Octave*, including the default values.

```
─────────────────────────── Octave odeset() ───────────────────────────
odeset()
-->
List of the most common ODE solver options.
Default values are in square brackets.
                AbsTol:  scalar or vector, >0, [1e-6]
                RelTol:  scalar, >0, [1e-3]
                   BDF:  binary, {["off"], "on"}
                Events:  function_handle, []
          InitialSlope:  vector, []
           InitialStep:  scalar, >0, []
              Jacobian:  matrix or function_handle, []
             JConstant:  binary, {["off"], "on"}
              JPattern:  sparse matrix, []
                  Mass:  matrix or function_handle, []
          MassSingular:  switch, {["maybe"], "no", "yes"}
              MaxOrder:  switch, {[5], 1, 2, 3, 4, }
               MaxStep:  scalar, >0, []
       MStateDependence:  switch, {["weak"], "none", "strong"}
             MvPattern:  sparse matrix, []
           NonNegative:  vector of integers, []
           NormControl:  binary, {["off"], "on"}
             OutputFcn:  function_handle, []
             OutputSel:  scalar or vector, []
                Refine:  scalar, integer, >0, []
                 Stats:  binary, {["off"], "on"}
            Vectorized:  binary, {["off"], "on"}
```

The options for the ODE solvers `ode??` in `MATLAB` are slightly different. Fortunately the most often used options `AbsTol` and `RelTol` are identical.

```
─────────────────────────── Matlab odeset() ───────────────────────────
odeset()
-->
              AbsTol:  [ positive scalar or vector {1e-6} ]
              RelTol:  [ positive scalar {1e-3} ]
         NormControl:  [ on | {off} ]
         NonNegative:  [ vector of integers ]
           OutputFcn:  [ function_handle ]
           OutputSel:  [ vector of integers ]
```

```
         Refine: [ positive integer ]
          Stats: [ on | {off} ]
    InitialStep: [ positive scalar ]
        MaxStep: [ positive scalar ]
            BDF: [ on | {off} ]
       MaxOrder: [ 1 | 2 | 3 | 4 | {5} ]
       Jacobian: [ matrix | function_handle ]
       JPattern: [ sparse matrix ]
     Vectorized: [ on | {off} ]
           Mass: [ matrix | function_handle ]
MStateDependence: [ none | {weak} | strong ]
      MvPattern: [ sparse matrix ]
    MassSingular: [ yes | no | {maybe} ]
    InitialSlope: [ vector ]
         Events: [ function_handle ]
```

The most frequently used options are `AbsTol` (default value 10^{-6}) and `RelTol` (default value 10^{-3}), used to specify the absolute and relative tolerances for the solution. At each time step the algorithm estimates the error(i) of the i-th component of the solution. Then the condition

$$|\text{error}(i)| <= \max\{\texttt{RelTol} * \text{abs}(y(i)), \texttt{AbsTol}(i)\}$$

has to be satisfied. Thus at least one of the absolute or relative error has to be satisfied. As a consequence it is rather useless to ask for a very small relative error, but keep the absolute error large. Both have to be made small.

The ODE $\frac{d}{dt}x(t) = 1 + x(t)^2$ with $x(0) = 0$ is solved by $x(t) = \tan(t)$. Thus we know the exact value of $x(\pi/4) = 1$. With the default values obtain

```
[t,x]  = ode23(@(t,x)1+x^2,[0,pi/4],0);
Error_Steps_default = [x(end)-1, length(t)-1]
-->
Error_Steps_default = [-3.6322e-05 25]
```

i.e. with 25 Heun steps the error is approximately $3.6 \cdot 10^{-5}$. With the options `AbsTol` and `RelTol` the error can be made smaller. The price to pay are more steps, i.e. a higher computational effort.

```
ode_opt =  odeset('AbsTol',1e-9,'RelTol',1e-9);
[t,x] = ode23(@(t,x)1+x^2,[0,pi/4],0,ode_opt);
Error_Steps_opt =   [x(end)-1 ,length(t)-1]
-->
Error_Steps_opt = [-2.3420e-10   495]
```

With the command `odeget()` one can read out specific options for the ODE solvers.

1.6.3   The command `lsode()` in *Octave*

Octave also provides the command `lsode()` to solve ordinary differential equations. The algorithms used by the command `lsode()` (Livermore Solver for Ordinary Differential Equations) were developed by Alan Hindmarsh [Hind93]. The code is very efficient and is a good starting option if working with *Octave*. There are minor, but annoying difference with `ode??()` though: the arguments in the function describing the differential equation have to be swapped.

$$
\begin{array}{rcl}
\text{ode??()} & \longleftrightarrow & \text{dx = f(t,x)} \\
\text{lsode()} & \longleftrightarrow & \text{dx = f(x,t)}
\end{array}
$$

For the algorithm in `lsode()` *Octave* allows to set many options:

- integration method (Adams, stiff, bdf)

- absolute tolerance

- relative tolerance

- initial step size

- minimal and maximal step size.

This is done by the command `lsode_options()`. If called without arguments the current values will be returned, e.g.

```
──────────────────────────── Octave ────────────────────────────
lsode_options()
-->
Options for LSODE include:

  keyword                                              value
  -------                                              -----
  absolute tolerance                                   1.49012e-08
  relative tolerance                                   1.49012e-08
  integration method                                   stiff
  initial step size                                    -1
  maximum order                                        -1
  maximum step size                                    -1
  minimum step size                                    0
  step limit                                           100000
```

Find more documentation in the on-line manual. As an example we might ask for a smaller absolute abolute and relative tolerance of 10^{-10} by

```
──────────────────────────── Octave ────────────────────────────
lsode_options("relative tolerance",1e-10)
lsode_options("absolute tolerance",1e-10)
```

1.6.4 Codes with fixed step size, Runge–Kutta, Heun, Euler

There are occasions when the fined tuned codes presented above do not perform well and one might have to fall back to the basics of solving differential equations numerically.

The classical Runge–Kutta method

One of the most often used methods is a Runge–Kutta method of order 4. It is often called the classical Runge–Kutta method. To apply one time step for the IVP

$$\dot{x} = f(t,x) \quad \text{with} \quad x(t_0) = x_0$$

with step size h from $t = t_i$ to $t = t_{i+1} = t_i + h$ use the computational scheme

$$
\begin{aligned}
k_1 &= f(t_i, x_i) \\
k_2 &= f(t_i + h/2, x_i + k_1 h/2) \\
k_3 &= f(t_i + h/2, x_i + k_2 h/2) \\
k_4 &= f(t_i + h, x_i + k_3 h) \\
x_{i+1} &= x_i + \frac{h}{6} (k_1 + 2k_2 + 2k_3 + k_4) \\
t_{i+1} &= t_i + h
\end{aligned}
$$

At four different positions the function $f(t,x)$ is evaluated, leading to four slopes k_i for the solution of the ODE. Then one time step is performed with a weighted average of the four slopes. This is visualized in Figure 1.43.

For the initial value problem $\frac{d}{dt} x(t) = x(t)^2 - 2t$ with $x_0 = x(0) = 0.75$ the calculations for one Runge–Kutta step of length $h = 1$ are given by

$$
\begin{aligned}
k_1 &= f(t_0, x_0) & &= f(0, 0.75) & &= 0.5625 \\
k_2 &= f(t_0 + \tfrac{h}{2}, x_0 + \tfrac{h}{2} k_1) & &= f(\tfrac{1}{2}, 1.03125) & &\approx 0.0634766 \\
k_3 &= f t_0 + \tfrac{h}{2}, (x_0 + \tfrac{h}{2} k_2) & &\approx f(\tfrac{1}{2}, 0.781738) & &\approx -0.388885 \\
k_4 &= f(t_0 + h, x_0 + h k_3) & &\approx f(1, 0.36111474) & &\approx -1.869596 \\
k &= \tfrac{1}{6}(k_1 + 2 k_2 + 2 k_3 + k_4) & &\approx -0.326319 \\
x(1) &\approx x(0) + h k & &\approx 0.75 + 1 (-0.326319) & &\approx 0.423681 .
\end{aligned}
$$

Similar, simpler ideas lead to the algorithms of Euler and Heun:

- Euler:

$$x_{i+1} = x_i + h f(t_i, x_i) \quad \text{and} \quad t_{i+1} = t_i + h$$

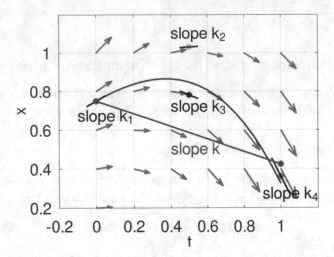

Figure 1.43: One step of the Runge–Kutta Method of order 4 for the ODE $\frac{d}{dt} x(t) = -x^2(t) - 2t$ with $x(0) = 0.75$ and step size $h = 1$

- Heun:

$$
\begin{aligned}
k_1 &= f(t_i, x_i) \\
k_2 &= f(t_i + h, x_i + k_1 h) \\
x_{i+1} &= x_i + \frac{h}{2}(k_1 + k_2) \quad \text{and} \quad t_{i+1} = t_i + h
\end{aligned}
$$

Using (tedious) Taylor expansions one can estimate the local and global discretization errors, depending on the step size h. These estimates assume that the function $f(t,x)$ is often differentiable, which is sometimes not the case. The algorithms in `ode??()` use this type of information to adapt the step sizes h to achieve the desired accuracy with the least possible computational effort.

Verfahren	step size h	local error	global error
Euler	$h = \frac{T}{n}$	$\approx C_E \cdot h^2$	$\approx C_E \cdot n \cdot h^2 = \tilde{C}_E\, h$
Heun	$h = \frac{T}{n}$	$\approx C_H \cdot h^3$	$\approx C_H \cdot n \cdot h^3 = \tilde{C}_H\, h^2$
Runge–Kutta	$h = \frac{T}{n}$	$\approx C_{RK} \cdot h^5$	$\approx C_{RK} \cdot n \cdot h^5 = \tilde{C}_{RK}\, h^4$

Table 1.12: Discretization errrors for the methods of Euler, Heun and Runge–Kutta

The above classical Runge–Kutta algorithm with fixed step size h can be implemented in *Octave*/MATLAB, see the code `ode_RungeKutta.m` in Figure 1.44. Similar codes `ode_Heun.m` and `ode_Euler.m` are available too.

```
┌──────────────────────── ode_RungeKutta.m ─────────────────────────┐
function [tout, yout] = ode_RungeKutta(Fun, t, y0, steps)
% [Tout, Yout] = ode_RungeKutta(fun, t, y0, steps)
%
%        Integrate a system of ordinary differential equations using
%        4th order Runge-Kutta formula.
%
% INPUT:
% Fun    - String containing name of user-supplied problem description.
%          Call: yprime = Fun(t,y)
%          t        - Vector of times (scalar).
%          y        - Solution column-vector.
%          yprime - Returned derivative column-vector;
%                   yprime(i) = dy(i)/dt.
% T      - vector of output times
%          T(1), initial value of t.
% y0     - Initial value column-vector.
% steps - steps to take between given output times
%
% OUTPUT:
% Tout   - Returned integration time points (column-vector).
% Yout   - Returned solution, one solution column-vector per tout-value.
%
% The result can be displayed by: plot(tout, yout).

% Initialization
y = y0(:);   yout = y'; tout = t(:);

% The main loop
for i = 2:length(t)
   h = (t(i)-t(i-1))/steps;
   tau = t(i-1);
   for j = 1:steps
      % Compute the slopes
      s1 = feval(Fun, tau,     y);          s1 = s1(:);
      s2 = feval(Fun, tau+h/2, y+h*s1/2); s2 = s2(:);
      s3 = feval(Fun, tau+h/2, y+h*s2/2); s3 = s3(:);
      s4 = feval(Fun, tau+h,   y+h*s3);   s4 = s4(:);
      tau = tau + h;
      y = y + h*(s1 + 2*s2+ 2*s3 + s4)/6;
   end%for
   yout = [yout; y.'];
end%for
└───────────────────────────────────────────────────────────────────┘
```

Figure 1.44: Code for Runge–Kutta with fixed step size

37 Example : The second order ODE

$$\frac{d^2}{dt^2}\, y(t) = -k\, \sin(y(t))$$

describes the angle $y(t)$ of a pendulum, possibly with large angles, since the approximation $\sin(y) \approx y$ is not used. This second order ODE is transformed to a system of order 1 by

$$\frac{d}{dt}\begin{pmatrix} y(t) \\ v(t) \end{pmatrix} = \begin{pmatrix} v(t) \\ -k\, \sin(y(t)) \end{pmatrix}.$$

This ODE leads to a function `pend()`, which is then used by `ode_RungeKutta()` to generate the solution for times $[0, 30]$ for different initial angles.

```
────────────────────────────── Pendulum.m ──────────────────────────────
Tend = 30;
%% on Matlab put the definition of the function in a file pend.m
function y = pend(t,x)
  k = 1;   y = [x(2);-k*sin(x(1))];
end%function

y0 = [0.1;0];         % small angle
% y0=[pi/2;0];        % large angle
% y0 = [pi-0.01;0]; % very large angle
t = linspace(0,Tend,100);
[t,y] = ode_RungeKutta('pend',t,y0,10); % Runge-Kutta
% [t,y] = ode_Euler('pend',t,y0,10);    % Euler
% [t,y] = ode_Heun('pend',t,y0,10);     % Heun
plot(t,180/pi*y(:,1))
xlabel('time'); ylabel('angle [Deg]')
```

$\Diamond$

38 Example : To illustrate the codes use the differential equation describing an electrical circuit with two diodes, see Figure 1.45. The behavior of the diode with current i and voltage u is given by a function

$$i = D\,(u) = \begin{cases} 0 & \text{for} \quad u \geq -u_s \\ R_D\,(u + u_s) & \text{for} \quad u < -u_s \end{cases}.$$

Observe that this function $D(u)$ consists of two straight lines with a jump for the slopes at $u = -u_s$. Based on Kirchhoff's law find the differential equations

$$\dot{u}_h(t) \quad = \quad \frac{1}{C_1}\,(-D\,(u_h(t) - u_{in}(t)) + D\,(u_{out}(t) - u_h(t)))$$

$$\dot{u}_{out}(t) \quad = \quad \dot{u}_{in}(t) - \frac{1}{C_2}\,D\,(u_{out}(t) - u_h(t)).$$

Set the initial conditions and the two functions in a script file and examine an input voltage of $u_{in}(t) = 10\, \sin(t)$.

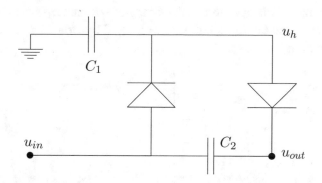

Figure 1.45: A voltage doubling circuit, using two diodes

Octave

```
Tend = 30; u0 = [0;0];

function curr = Diode(u)   % for Matlab put this in a file Diode.m
   Rd = 10; us = 0.7;
   curr = (Rd*(u+us)).*(u<-us);
end%function

function y = Circuit(t,u) % for Matlab put this in a file Circuit.m
   C1 = 1; C2 = 1;
   y = [-1/C1*(Diode(u(1)-10*sin(t))-Diode(u(2)-u(1)));
        10*cos(t)-1/C2*Diode(u(2)-u(1))];
end%function

tFix = linspace(0,Tend,100);
[tFix,uFix] = ode_RungeKutta(@Circuit,tFix,u0,1);        % Runge Kutta
opt = odeset('Stats','on')
[t45,u45]   = ode45(       @Circuit,[0,Tend],u0,opt); % ode45
figure(1); plot(tFix ,uFix,'+-')
           xlabel('time'); ylabel('voltages')
figure(2); plot(t45,u45,'+-')
           xlabel('time'); ylabel('voltages')
figure(3); hist(diff(t45))
           xlabel('step size'); ylabel('frequency')
-->
Number of successful steps: 140
Number of failed attempts:  94
Number of function calls:   1405
```

The output of the above code shows that ode45() used 1405 evaluations of the function Circuit(), while ode_RungeKutta() used only 400 evaluations. The results are very similar, see Figure 1.46. Whenever one of the voltages over the diodes is at -0.7 volts, the function Circuit() is not differentiable any more, and ode45() will choose smaller and smaller step sizes. This is visible in the histogram of the step sizes used by `ode45()` in Figure 1.47. Many steps of very small size were used by `ode45()`. For tighter tolerances the effect will be even larger.

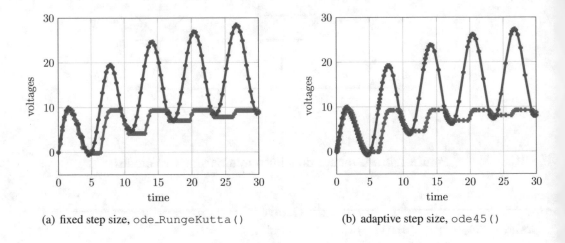

(a) fixed step size, `ode_RungeKutta()`

(b) adaptive step size, `ode45()`

Figure 1.46: Resulting voltages generated by two algorithms

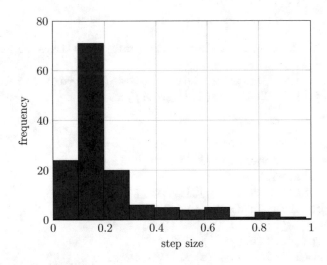

Figure 1.47: Histogram of the step sizes used by `ode45()`

◇

1.6.5   List of files

In the previous section the codes and data files in Table 1.13 were used.

filename	function
`Logistic.m`	script file to solve the logistic ODE
`VolterraLotka.m`	function file for the Volterra-Lotka model
`VolterraLotkaField.m`	script file to solve the Volterra-Lotka model
`VolterraLotkaC.cc`	C++ code file for the Volterra-Lotka model
`SIR.m`	ODE function for the SIR model
`SIR_Model.m`	script to run the SIR model
`Spring.m`	ODE function for the spring model
`Spring_Model.m`	script to rund the spring model
`ode_Euler.m`	algorithm of Euler, fixed step size
`ode_Heun.m`	algorithm of Heun, fixed step size
`ode_RungeKutta.m`	algorithm of Runge-Kutta, fixed step size
`Pendulum.m`	script file to solve the pendulum model
`DoubleTension.m`	sample code for the diode circuit

Table 1.13: Codes and data files for section 1.6

Chapter 2

Elementary Statistics With
Octave/MATLAB

2.1 Introduction

In this chapter a few MATLAB/*Octave* commands for statistics are listed and elementary sample codes are given. This should help you to get started using *Octave*/MATLAB for statistical problems. The short notes you are looking at right now should serve as a starting point and will **not** replace reading and understanding the built-in documentation of *Octave* and/or MATLAB. For users of MATLAB is is assumed that the statistics toolbox is available. For users of *Octave* is is assumed that the statistics package is available and loaded.

2.2 Commands to Load Data from Files

It is a common task that data is available in a file. Depending on the format of the data there are a few commands that help to load the data into MATLAB or *Octave*. They will be used often in these notes for the short sample codes. A list is shown in Table 2.1. Consult the built-in help to find more information on those commands.

2.3 Commands to Generate Graphics used in Statistics

In Table 2.2 find a few *Octave*/MATLAB commands to generate pictures used in statistics. Consult the built-in help to learn about the exact syntax. In the subsections below the most important graphics are illustrated by examples.

Supplementary Information The online version contains supplementary material available at [https://doi.org/10.1007/978-3-658-37211-8_2].

© The Author(s), under exclusive license to Springer Fachmedien Wiesbaden GmbH, part of Springer Nature 2022
A. Stahel, *Octave and MATLAB for Engineering Applications*, https://doi.org/10.1007/978-3-658-37211-8_2

`load()`	loading data in text format or binary formats
`dlmread()`	loading data in (comma) separated format
`dlmwrite()`	writing data in (comma) separated format
`textread()`	read data in text format
`strread()`	read data from a string
`fopen()`	open a file for reading or writing
`fclose()`	close a file for reading
`fread()`	read fron an open file
`fgetl()`	read one line from a file
`sscanf()`	formated reading from a string
`sprintf()`	formated writing to a string

Table 2.1: Commands to load data from a file

`hist()`	generate a histogram
`histc()`	compute histogram count
`histfit()`	generate histogram and fitted normal density
`bar()`	generate a bar chart
`barh()`	generate a horizontal bar chart
`bar3()`	generate a 3D bar chart
`pie()`	generate a pie chart
`stem()`	generate a 2D stem plot
`stem3()`	generate a 3D stem plot
`rose()`	generate an angular histogram
`stairs()`	generate a stairs plot
`boxplot()`	generate a boxplot

Table 2.2: Commands to generate statistical graphs

2.3.1 Histograms

With the command `hist()` you can generate histograms, as seen in Figure 2.1.

```
data = 3+2*randn(1000,1); % generate the random data
figure(1); hist(data)      % hisogram with default values
figure(2); hist(data,30)   % histogram with 30 classes
```

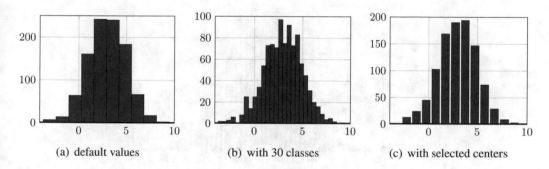

(a) default values   (b) with 30 classes   (c) with selected centers

Figure 2.1: Three histograms, based on the same data

It is possible to specify the centers of the classes and then compute the number of elements in the classes by giving the command `hist()` two return arguments. Then use `bar()` to generate the histogram. The code below chooses classes of width 0.5 between -2.5 and $+9.5$. Thus the first center is at -2.25, the last center at 9.25 and the centers are 0.5 apart. Find the result of the code below on the right in Figure 2.1.

```
[heights,centers] = hist(data,[-2.25:1:9.25]);
figure(3); bar(centers,heights)
```

The number of elements on the above classes can also be computed with the command `heights = histc(data,[-2.5:1.0:9.5])`. With this syntax the limits of the classes are specified. With a combination of the commands `unique()` and `hist()` count the number of entries in a vector.

```
a = randi(10,100,1) % generate 100 random integers between 1 and 10
[count, elem] = hist(a,unique(a)) % determine the entries (elem) and
                                  % their number (count)
bar(elem,count)                   % display the bar chart
```

With the command `histfit()` generate a histogram and the best matching normal distribution as a graph. Find the result of the code below in Figure 2.2.

```
data = round(50+20*randn(1,2000));
histfit(data)
xlabel('values'); ylabel('number of events')
```

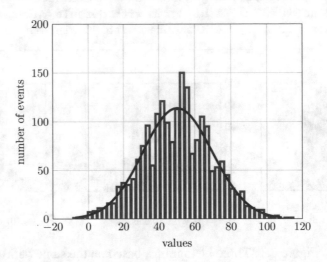

Figure 2.2: A histogram with matching normal distribution

2.3.2 Bar diagrams and pie charts

Using the commands `bar()` and `barh()` one can generate vertical and horizontal bar charts. The results of the code below is shown in Figure 2.3.

```
ages = 20:27;      students = [2 1 4 3 2 2 0 1];

figure(1); bar(ages,students)
           axis([19.5 27.5 0 5])
           xlabel('age of students'); ylabel('number of students')

figure(2); barh(ages,students)
           axis([0 5 19.5 27.5])
           ylabel('age of students'); xlabel('number of students')
```

Using the commands `pie()` and `pie3()` one can generate pie charts. With the correct options set labels and some of the slices can be drawn slightly removed from the pie. The result of the code below is shown in Figure 2.4.

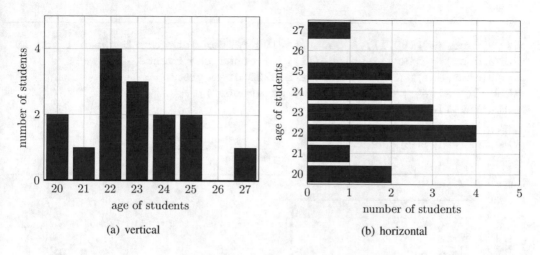

(a) vertical
(b) horizontal

Figure 2.3: Bar diagram

```
strength = [55 52  36 28 13 16];
Labels = {'SVP','SP','FDP','CVP','GR','Div'}
figure(1); pie(strength)
figure(2); pie(strength,[0 1 0 0 0 0],Labels)
figure(3); pie3(strength,Labels)
```

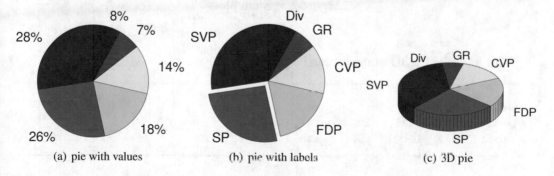

(a) pie with values
(b) pie with labels
(c) 3D pie

Figure 2.4: Different pie charts

2.3.3 More plots

With a stem plot a vertical line with a small marker at the top can be used to visualize data. The code below first generates a set of random integers, and then uses a combination of unique() and hist() to determine the frequency (number of occurrences) of those numbers.

```
ii = randi(10,100,1); % generate 100 random integers between 1 and 10
[anz,cent] = hist(ii,unique(ii))  % count the events
stem(cent,anz)          % generate a 2D stem graph
xlabel('value'); ylabel('number of events');
axis([0, 11, -1, max(anz)+1])
-->
anz =    12     5    10    12    12     9    11    11     7    11
cent =    1     2     3     4     5     6     7     8     9    10
```

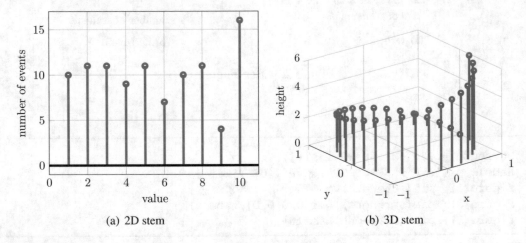

(a) 2D stem   (b) 3D stem

Figure 2.5: Stem plots

With `stem3()` a 3D stem plot can be generated.

```
theta = 0:0.2:6;
stem3 (cos (theta), sin (theta), theta);
xlabel('x'); ylabel('y'); zlabel('height')
```

MATLAB/Octave provide commands to generate angular histograms and stairstep graphs.

```
dataRaw = randi([5 10],1,400);
figure(1); rose(dataRaw,8)     % generate rose plot
           title('angular histogram with 8 sectors')
           [data,cent] = hist(dataRaw,unique(dataRaw)) % count events

figure(2); stairs(cent,data)  % generate stairstep plot
           xlabel('value'); ylabel('number of events'); xlim([5 10]);
```

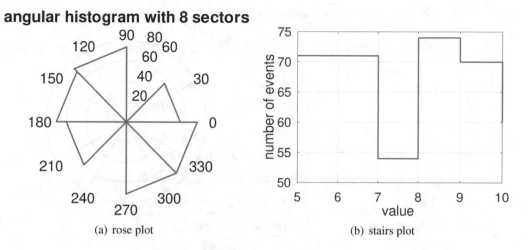

(a) rose plot

(b) stairs plot

Figure 2.6: Rose plot and stairs plot

With the command `boxplot()` you can generate a plot showing the median, the first and third quartile as a box and the extreme values. The commands to compute the relevant values are shown in the next section. Observe that there are different ways to compute the positions of the quartiles and some implementations of `boxplot()` detect and mark outliers. By using an optional argument you can select which points are considered as outliers. Consult the documentation in *Octave*. Boxplots can also be displayed horizontally or vertically, as shown in Figure 2.7. Find the codes the generate the boxplots on page 155.

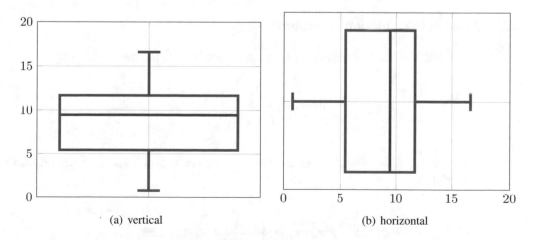

(a) vertical

(b) horizontal

Figure 2.7: Vertical and horizonal boxplots

2.4 Data Reduction Commands

In Table 2.3 find a few *Octave*/MATLAB commands to extract information from data sets.

`mean()`	mean value of a data set
`std()`	standard deviation of a data set
`var()`	variance of a data set
`median()`	median value of a data set
`mode()`	determine the most frequently occurring value
`quantile()`	determine arbitrary quantiles
`LinearRegression()`	perform a linear regression
`regress()`	perform a linear regression
`lscov()`	generalized least square estimation, with weights
`polyfit()`	perform a linear regression for a polynomial
`ols()`	perform an ordinary linear regression
`gls()`	perform an generalized linear regression
`cov()`	covariance matrix
`corr()`	linear correlation matrix
`corrcoef()`	correlation coefficient

Table 2.3: Commands for data reduction

2.4.1 Basic data reduction commands

- For a vector $x \in \mathbb{R}^n$ the command `mean(x)` determines the mean value by

$$\text{mean}(x) = \bar{x} = \frac{1}{n} \sum_{j=1}^{n} x_j \ .$$

- For a vector $x \in \mathbb{R}^n$ the command `std(x)` determines the standard deviation by the formula

$$\text{std}(x) = \sqrt{\text{var}(x)} = \left(\frac{1}{n-1} \sum_{j=1}^{n} (x_j - \bar{x})^2 \right)^{1/2} \ .$$

By default `std()` will normalize by $(n-1)$, but using an option you may divide by n, e.g. by using $\text{std}(x, 1)$.

- For a vector $x \in \mathbb{R}^n$ the command `var(x)` determines the variance by the formula

$$\text{var}(x) = (\text{std}(x))^2 = \frac{1}{n-1} \sum_{j=1}^{n} (x_j - \bar{x})^2 \,.$$

By default `var()` will normalize by $(n-1)$, but using an option you may divide by n, e.g. by using `var(x,1)`.

- For a vector $x \in \mathbb{R}^n$ the command `median(x)` determines the median value. For a sorted vector is is given by

$$\text{median}(x) = \begin{cases} x_{(n+1)/2} & \text{if } n \text{ is odd} \\ \frac{1}{2}(x_{n/2} + x_{n/2+1}) & \text{if } n \text{ is even} \end{cases}\,.$$

- For a vector $x \in \mathbb{R}^n$ the command `mode(x)` determines the most often occurring value in x. The commands `mode(randi(10,20,1))` generate 20 random integer values between 1 and 10, and then the most often generated value.

- With the command `quantile()` you can compute arbitrary quantiles. Observe that there are different methods to determine the quantiles, leading to different results! Consult the built-in documentation in *Octave* by calling `help quantile`.

- Often quartile (division by four) or percentiles (division by 100) have to be determined. The command `quantile()` with a proper set of parameters does the job. You may also use `prctile()` to determine the values for the quartile (default) or other values for the divisions.

Demo_Boxplot.m

```
N = 10;                 % number of data points
data1 = 20*rand(N,1);
Mean = mean(data1)
Median = median(data1)
StdDev = std(data1)    % uses a division by (N-1)
Variance = StdDev^2
Variance2 = mean((data1-mean(data1)).^2)   % uses a division by N
Variance3 = sum((data1-mean(data1)).^2)/(N-1)

figure(1); Quartile1 = boxplot(data1)
           set(gca,'XTickLabel',{' '})      % remove labels on x axis
           c_axis = axis(); axis([0.5 1.5 c_axis(3:4)])

figure(2); boxplot(data1,0,'+',0)   % Octave
           %boxplot(data1,'orientation','horizontal')   % Matlab
           set(gca,'YTickLabel',{' '}) % remove labels on y axis
           c_axis = axis(); axis([c_axis(1:2),0.5 1.5])

Quartile2  = quantile(data1,[0 0.25 0.5 0.75 1])
```

```
Quantile10 = quantile(data1,0:0.1:1)

data2 = randi(10,[100,1]);
ModalValue = mode(data2)  % determine the value occuring most often
```

It is possible to put multiple boxplots in one graph, and label the axis according to the data. In Figure 2.8 the abreviated names of weekdays are used to label the horizontal axis.

```
% generate the random data, with some structure
N = 20;   data = zeros(N,7);
for i = 1:7
   data(:,i) = 3+4*sin(i/4)+randn(N,1);
end%for

boxplot(data);
set(gca(),'xtick',[1:7],...
          'xticklabel',{'Mo','Tu','We','Th','Fr','Sa','Su'});
```

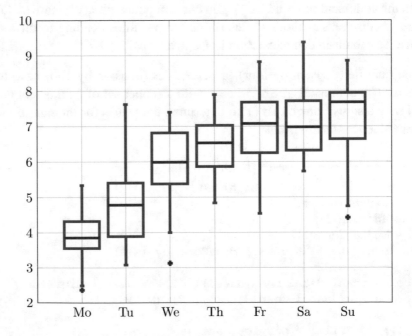

Figure 2.8: Multiple boxplots in one figure

2.4.2   Data reduction commands for pairs of vectors

For covariance and correlation coefficients the first step is to subtract the mean value of the components from a vector, i.e. the new mean value is zero.

- The covariance of two vectors $\vec{x}, \vec{y} \in \mathbb{R}^n$

$$\operatorname{cov}(x, y) = \frac{1}{n-1} \sum_{j=1}^{n} (x_j - \operatorname{mean}(x)) \cdot (y_j - \operatorname{mean}(y))$$

$$= \frac{1}{n-1} \sum_{j=1}^{n} (x_j\, y_j - \operatorname{mean}(x)\, \operatorname{mean}(y)) \; .$$

By default `cov()` will normalize by $(n-1)$, but using an option you may divide by n, e.g. by using `cov(x,y,1)`. If $x = y$ obtain

$$\operatorname{cov}(x, x) = \frac{1}{n-1} \sum_{j=1}^{n} (x_j - \operatorname{mean}(x))^2 = \operatorname{var}(x) \; .$$

- The correlation coefficient of two vectors $\vec{x}, \vec{y} \in \mathbb{R}^n$

$$\operatorname{corr}(x, y) = \frac{\operatorname{cov}(x, y)}{\operatorname{std}(x) \cdot \operatorname{std}(y)}$$

$$= \frac{\langle (\vec{x} - \operatorname{mean}(\vec{x})), (\vec{y} - \operatorname{mean}(\vec{y})) \rangle}{\|\vec{x} - \operatorname{mean}(\vec{x})\| \, \|\vec{y} - \operatorname{mean}(\vec{y})\|}$$

$$= \frac{\sum_{j=1}^{n} (x - \operatorname{mean}(x))_j \cdot (y - \operatorname{mean}(y))_j}{(\sum_{j=1}^{n} (x_j - \operatorname{mean}(x))^2)^{1/2} (\sum_{j=1}^{n} (y_j - \operatorname{mean}(y))^2)^{1/2}} \; .$$

Observe that if the average value of the components of both vectors are zero, then there is a geometric interpretation of the correlation coefficient as the angle between the two vectors.

$$\operatorname{corr}(\vec{x}, \vec{y}) = \frac{\langle \vec{x}, \vec{y} \rangle}{\|\vec{x}\| \, \|\vec{y}\|} = \cos(\alpha) \; .$$

This correlation coefficient can also be computed with by `corrcoef()`.

```
x = linspace(0,pi/2,20)'; y = sin(x);  % generate some artificial data
MeanValues   = [mean(x), mean(y)]
Variances    = [var(x), var(y)]
StandardDev  = [std(x), std(y)]
Covariance   = cov(x,y)
Correlation  = corr(x,y)
-->
MeanValues   =   0.78540    0.62944
Variances    =   0.23922    0.10926
StandardDev  =   0.48910    0.33055
Covariance   =   0.15794
Correlation  =   0.97688
```

2.4.3 Data reduction commands for matrices

Many of the above commands can be applied to matrices. Use each column as one data vector. Assume that $\mathbf{M} \in \mathbb{R}^{N \times m}$ is a matrix of m column vectors with N values in each column.

- `mean(M)` compute the average of each column. The result is a row vector with m components.
- `std(M)` compute the standard deviation of each column. The result is a row vector with m components.
- `var(M)` compute the variance of each column. The result is a row vector with m components.
- `median(M)` compute the median value of each column. The result is a row vector with m components.

To describe the effect of `cov()` and `corr()` the first step is to assure that the average of each column equals zero.

```
Mm = M - ones(N,1)*mean(M);
```

Observe that this operation does not change the variance of the column vectors.

- `cov(M)` determines the $m \times m$ covariance matrix

$$\text{cov}(M) = \frac{1}{N-1} \mathbf{Mm}' \cdot \mathbf{Mm} .$$

- The $m \times m$ correlation matrix contains all correlation coefficients of the m column vectors in the matrix $\mathbf{M}$. To compute this, first make sure that the norm of each column vector equals 1, i.e. the variance of the column vectors is normalized to 1.

```
Mm1 = Mm / diag(sqrt(sum(Mm.^2)));
```

Determine the $m \times m$ (auto)correlation matrix `corr(M)` by

$$\text{corr}(M) = \mathbf{Mm1}' \cdot \mathbf{Mm1} .$$

Observe that the diagonal entries are 1, since the each column vector correlates perfectly with itself.

2.5 Performing Linear Regression

In this section a few command to use apply linear regression are shown. More background and real examples are presented in Section 3.2, starting on page 219.

The method of least squares, linear or nonlinear regression is one of the most often used tools in science and engineering. MATLAB/*Octave* provide multiple commands to use these algorithms.

2.5.1 **Using the command** `LinearRegression()`

The command `LinearRegression()` was written by the author of these notes.

- For *Octave* the command is contained in the optimization package `optim`. It is also available with the source codes for these notes.

- The command can be used with MATLAB too, but you need a MATLAB version `LinearRegression.m`, available with the source codes for these notes. Put the file in a directory visible by MATLAB.

With this command you can apply the method of least square to fit a curve (or surface) to a given set of data points. The curve does **not** have to be a linear function, but a linear combination of (almost arbitrary) functions. In the code below a straight line is fitted to some points on a curve $y = \sin(x)$. Thus we try to find the optimal values for a and m such that

$$\chi^2 = \sum_j (a + m\,x_j - y_j)^2 \quad \text{is minimal.}$$

The code to perform the this linear regression is given by

```
% generate the artificial data
x = linspace(0,2,10)'; y = sin(x);

% perform the linear regression, aiming for a straight line
F = [ones(size(x)),x];
[p,e_var,r,p_var] = LinearRegression(F,y);
Parameters_and_StandardDeviation = [p sqrt(p_var)]
estimated_std = sqrt(mean(e_var))
-->
Parameters_and_StandardDeviation =    0.202243   0.091758
                                      0.477863   0.077345
estimated_std =   0.15612
```

The above result implies that the best fitting straight line is given by

$$y = a + m\,x = 0.202243 + 0.477863\,x\,.$$

Assuming that the data is normally distributed one can show that the values of a and m normally distributed too. For this example the estimated standard deviation of a is given by 0.09 and the standard deviation of m is 0.08. The standard deviation of the residuals $r_j = a + m\,x_j - y_j$ is estimated by 0.16. This is visually confirmed by Figure 2.9(a), generated by

```
y_reg = F*p;
plot(x,y,'+', x , y_reg)
```

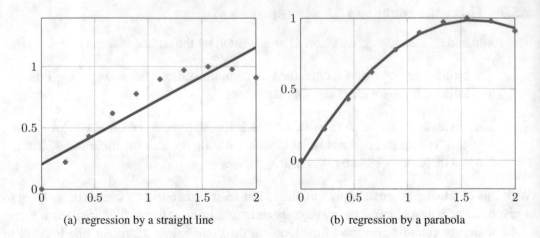

(a) regression by a straight line   (b) regression by a parabola

Figure 2.9: Results of two linear regressions

With linear regression one may fit different curves to the given data. The code below generates the best matching parabola and the resulting Figure 2.9(b).

```
% perform the linear regression, aiming for a parabola
F = [ones(size(x)),x, x.^2];
[p,e_var,r,p_var] = LinearRegression(F,y);

Parameters_and_StandardDeviation = [p sqrt(p_var)]
estimated_std = sqrt(mean(e_var))

y_reg = F*p;
plot(x,y,'+', x , y_reg)
-->
Parameters_and_StandardDeviation =   -0.026717    0.015619
                                      1.250604    0.036370
                                     -0.386371    0.017506
estimated_std =   0.019865
```

Since the parabola is a better match for the points on the curve $y = \sin(x)$ find smaller estimates for the standard deviations of the parameters and residuals.

It is possible perform linear regression with functions of multiple variables. The function

$$z = p_1 \cdot 1 + p_2 \cdot x + p_3 \cdot y$$

describes a plane in space $\mathbb{R}^3$. A surface of this type is fit to a set of given points (x_j, y_j, z_j) by the code below, resulting in Figure 2.10. The columns of the matrix $\mathbf{F}$ have to contain the values of the basis functions 1, x and y at the given data points.

```
N = 100;  x =  2*rand(N,1);  y =  3*rand(N,1);
         z =   2 + 2*x- 1.5*y + 0.5*randn(N,1);

F = [ones(size(x)),  x ,  y];
p = LinearRegression(F,z)

[x_grid, y_grid] = meshgrid([0:0.1:2],[0:0.2:3]);
z_grid = p(1) + p(2)*x_grid + p(3)*y_grid;

figure(1);  plot3(x,y,z,'*')
           hold on
           mesh(x_grid,y_grid,z_grid)
           xlabel('x'); ylabel('y'); zlabel('z');
           hold off
-->
p =   1.7689   2.0606   -1.4396
```

Since only very few (N=100) points were used the exact parameter values $\vec{p} = (+2, +2, -1.5)$ are note very accurately reproduced. Increasing N will lead to more accurate results for this simulation, or decrease the size of the random noise in +0.5*randn(N,1).

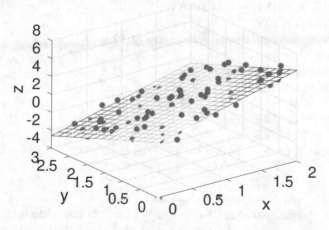

Figure 2.10: Result of a 3D linear regression

The command LinearRegression() does not determine the confidence intervals (CI) for the parameters, but it returns the estimated standard deviations, resp. the variances. With these the confidence intervals can be computed, using the Student-t distribution. To determine the CI modify the above code slightly. Select a level of significance, e.g. $\alpha = 0.05$, and then use the inverse of the Student-t distribution and the variance to determine the confidence interval.

```
[p,~,~,p_var] = LinearRegression(F,z);
alpha = 0.05;
p_CI = p + tinv(1-alpha/2,N-3)*[-sqrt(p_var) +sqrt(p_var)]
-->
p_CI =   +1.6944  +2.2357
         +1.8490  +2.2222
         -1.5869  -1.3495
```

The result implies that the 95% confidence intervals for the parameters p_i are given by

$$
\begin{aligned}
+1.6944 &< p_1 < +2.2357 \\
+1.8490 &< p_2 < +2.2222 \qquad \text{with a confidence level of 95\%.} \\
-1.5869 &< p_3 < -1.3495
\end{aligned}
$$

2.5.2 Using the command `regress()`

MATLAB and *Octave* provide the command `regress()` to perform linear regressions. The following code determines the best matching straight line to the given data points.

```
x = linspace(0,2,10)'; y = sin(x);
F = [ones(size(x)),x];
[p, p_int, r, r_int, stats] = regress(y,F);
parameters = p
parameter_intervals = p_int
estimated_std = std(r)
-->
parameters =    0.20224
                0.47786
parameter_intervals =  -0.0093515  +0.4138380
                       +0.2995040  +0.6562220
estimated_std =  0.14719
```

The values of the optimal parameters (obviously) have to coincide with the result generated by `LinearRegression()`. Instead of the standard deviations for the parameters `regress()` returns the confidence intervals for the parameters. The above numbers imply for the straight line $y = a + m\,x$

$$
\begin{aligned}
-0.0093 &< a < 0.4138 \\
0.300 &< m < 0.656
\end{aligned}
\qquad \text{with a confidence level of 95\%.}
$$

The value of the confidence level can be adjusted by calling `regress()` with a third argument, see `help regress`.

2.5.3 Using the commands `lscov()`, `polyfit()` or `ols()`

The command `lscov()` is rather similar to the above, but can return the covariance matrix for the determined parameters.

If your aim is to fit a polynomial to the data, you may use `polyfit()`. The above example for a parabola (polynomial of degree 2) is solved by

```
[p,s] = polyfit(x,y,2);
p
-->
p =  -0.386371   1.250604  -0.026717
```

Observe that the coefficient of the polynomial are returned in deceasing order. Since the commands `regress()` and `LinearRegression()` are more flexible and provide more information your author's advice is to use those, even if `polyfit()` would work.

With *Octave* one may also use the command `ols()`, short for Ordinary Least Square. But as above, there is no advantage over using `LinearRegression()`.

```
p = ols(y,F)
-->
p =  -0.026717  1.250604 -0.386371
```

2.6 Generating Random Numbers

In Table 2.4 find commands to generate random numbers, given by different distributions.

`rand()`	uniform distribution
`randi()`	random integers
`randn()`	normal distribution
`rande()`	exponentially distributed
`randp()`	Poisson distribution
`randg()`	gamma distribution
`normrnd()`	normal distribution
`binornd()`	binomial distribution
`exprnd()`	exponential distribution
`trnd()`	Student-t distribution
`discrete_rnd()`	discrete distribution

Table 2.4: Commands for generating random numbers

As an example generate $N = 1000$ random numbers given by a binomial distribution with $n = 9$ trials and $p = 0.8$. Thus each of the 1000 random numbers will be an integer between 0 and 9. Find the result of the code below in Figure 2.11 and compare with Figure 2.12(d), showing the result for the exact (non random) distribution.

```
N = 1000;  data = binornd(9, 0.8, N, 1);    % generate random numbers
[height,centers] = hist(data,unique(data))  % data for the histogram
bar(centers,height/sum(height))
xlabel('value'); ylabel('experimental probability')
title('Binomial distribution with n=9, p=0.8')
```

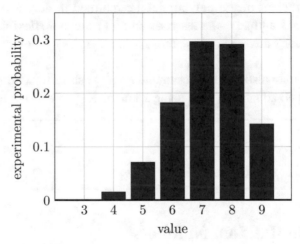

Figure 2.11: Histogram of random numbers, generated by a binomial distribution with $n = 9$ and $p = 0.8$

2.7   Commands to Work with Probability Distributions

MATLAB/*Octave* provides functions to compute the values of probability density functions (PDF), and the cumulative distribution functions (CDF). In addition the inverse of the CDF are provided, i.e. solve $\text{CDF}(x) = y$ for x. As examples examine the following short code segments, using the normal distribution.

- To determine the values of the PDF for a normal distribution with mean 3 and standard deviation 2 for x values between -1 and 7 use

```
x = linspace(-1,7);  p = normpdf(x,3,2);
plot(x,p)
```

- To determine the corresponding values of the CDF use `cp = normcdf(x,3,2)`.

- To determine for what values of x the value of the CDF equals 0.025 and 0.975 use

```
norminv([0.025,0.975],3,2)
-->
-0.91993    6.91993
```

The result implies that 95% of all values are between $-0.91 \approx -1$ and $+6.91 \approx 7$. This is consistent with the approximative rule of thumb $\mu \pm 2\sigma$.

For most distributions `MATLAB` and *Octave* provide multiple commands to work with the density functions and cumulative distribution functions, see Table 2.7. The first part of the name of the command consists of an abbreviated name of the distribution and the second part spells out the operation to bc applied. As an example consider the normal distribution and then use the command `normpdf()`, `normcdf()`, `norminv()`, `normrnd()` and `normstat()` to work with the normal distribution.

Name of command	Function
*pdf()	probability density function
*cdf()	cumulative density function
*inv()	inverse of the cumulative density function
*rnd()	generate random numbers
*stat()	compute mean and variance
*test()	testing of hypothesis

Table 2.5: Functions for distributions

In addition one may use the commands `cdf()` and `pdf()` to compute values of the probability density function. As example consider the result of `cdf('normal',0,0,1)`, leading to 0.500 .

2.7.1 Discrete distributions

In Table 2.7.1 find a few discrete distributions, and the graphs in 2.12 a few graphs.

For any discrete distribution the mean value μ and the variance σ^2 are determined by

$$\mu = \sum_j \mathrm{pdf}(x_j) \cdot x_j \quad \text{and} \quad \sigma^2 = \sum_j \mathrm{pdf}(x_j) \cdot (x_j - \mu)^2 .$$

Name of distribution	Function	μ	σ
Discrete	`discrete_pdf(x,v,p)`		
Bernoulli	`discrete_pdf(x,[0 1],[1-p,p])`	p	$\sqrt{p(1-p)}$
Binomial	`binopdf(x,n,p)`	np	$\sqrt{np(1-p)}$
Poisson	`poisspdf(x,lambda)`	λ	$\sqrt{\lambda}$
Hypergeometric	`hygepdf(x,T,M,n)`	$n\frac{m}{T}$	

Table 2.6: Discrete distributions, mean value μ and standard deviation σ

Bernoulli distribution and general discrete distributions

With the functions `discrete_pdf()` and `discrete_cdf()`[1] you can generate discrete probability distributions. To generate a Bernoulli distribution with probability 1/3

$$P(X=0) = \frac{2}{3} \quad \text{and} \quad P(X=1) = \frac{1}{3}$$

use the code below, leading to Figure 2.12(a). There is no need to normalize the total probability to 1, i.e. in the code below `discrete_pdf(x,[0 1],[2 1])` would work just as well.

```
x   = -1:2;   p  = discrete_pdf(x,[0 1],[2/3 1/3]);
              cp = discrete_cdf(x,[0 1],[2/3 1/3]);
figure(1); stem(x,p,'b');      hold on
           stairs(x,cp,'k'); hold off
           title('Bernoulli distribution with p=1/3')
           axis([-1 2 -0.1 1.1]);
           legend('pdf','cdf','location','northwest');
```

The Bernoulli distribution can also be considered a special case of the binomial distribution, with $n = 1$.

Throwing a regular dice also leads to a discrete distribution, each of the possible results 1, 2, 3, 4, 5 and 6 will show with probability 1/6. Find the result in Figure 2.12(b).

```
x   = 0:7; p  = discrete_pdf (x,1:6,ones(6,1)/6);
           cp = discrete_cdf (x,1:6,ones(6,1)/6);
figure(2); stem(x,p,'b');      hold on
           stairs(x,cp,'k'); hold off
           title('discrete distribution, throwing dice')
           axis([0 7 -0.1 1]);
           legend('pdf','cdf','location','northwest');
```

---

[1]The current version of MATLAB does not provide these two commands. Ask this author for a MATLAB compatible versions.

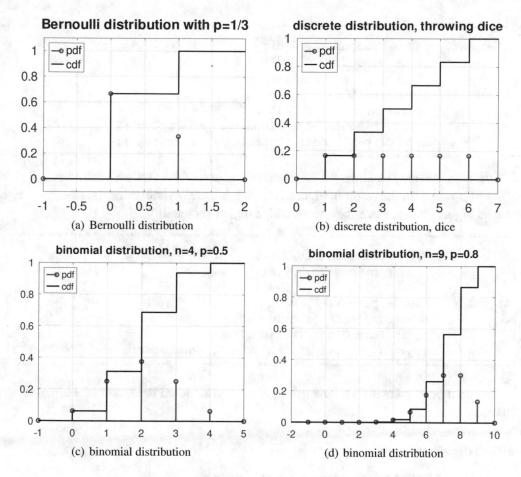

Figure 2.12: Graphs of some discrete distributions

Binomial distribution

The binomial distribution is generated by n independent Bernoulli trials with identical probability p, i.e. for each of the n independent events you obtain the result 1 with probability p. Then add up the results. This leads to

$$X = \sum_{j=1}^{n} X_j$$

$$P(X = i) = p(i) = \binom{n}{i} p^i \cdot (1-p)^{n-i} = \frac{n!}{i! \cdot (n-i)!} p^i \cdot (1-p)^{n-i} .$$

The code below generates the PDF and CDF for a binomial distribution with 4 events and the individual probability $p = 0.5$. Find the results in Figure 2.12(c). The resulting distribution is symmetric about the mean value 2.

```
x   = -1:5;   p  = binopdf(x,4,0.5);
              cp = binocdf(x,4,0.5);

figure(3); stem(x,p,'b');     hold on
           stairs(x,cp,'k'); hold off
           title('binomial distribution, n=4, p=0.5')
           legend('pdf','cdf','location','northwest');
```

Similarly you may examine the distribution function of 9 draws with an individual probability of $p = 0.8$. In the result in Figure 2.12(d) it is clearly visible that the result is skewed towards higher values, since they occur with a higher probability..

```
x   = -1:10;   p  = binopdf(x,9,0.8);
               cp = binocdf(x,9,0.8);

figure(4); stem(x,p,'b');     hold on
           stairs(x,cp,'k'); hold off
           title('binomial distribution, n=9, p=0.8')
           legend('pdf','cdf','location','northwest');
```

The command `binostat` determines mean and standard deviation of a binomial distribution.

Poisson distribution

A Poisson distribution with parameter $\lambda > 0$ is given by

$$P(X = i) = p(i) = \frac{\lambda^i}{i!} e^{-\lambda} .$$

Find the graph of the Poisson distribution with $\lambda = 2.5$ in Figure 2.13, generated by

```
x   = -1:10; lambda = 2.5;
p  = poisspdf(x,lambda);
cp = poisscdf(x,lambda);

figure(5); stem(x,p,'b');     hold on
           stairs(x,cp,'k'); hold off
           title('Poisson distribution, \lambda=2.5')
           legend('pdf','cdf','location','northwest');
```

2.7.2   Continuous distributions

For all continuous distributions one may compute the average value μ, the standard deviation σ and the cumulative distribution function by integrals. The mode and median of the

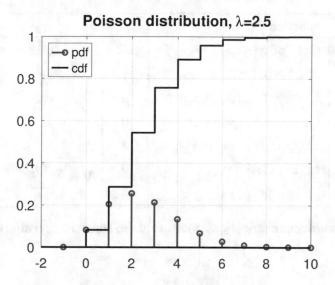

Figure 2.13: A Poisson distribution with $\lambda = 2.5$

distributions are characterized as special values of the probability density function (PDF) or the cumulative distribution function (CDF).

$$\mu = \int_{-\infty}^{+\infty} \text{pdf}(x) \cdot x \, dx$$

$$\sigma^2 = \int_{-\infty}^{+\infty} \text{pdf}(x) \cdot (x - \mu)^2 \, dx$$

$$\text{cdf}(x) = \int_{-\infty}^{x} \text{pdf}(s) \, ds$$

$$\text{cdf(median)} = 0.5$$

$$\text{pdf(mode)} = \max_{x \in \mathbb{R}} \{\text{pdf}(x)\}$$

Uniform distribution

The uniform distribution on the interval $[A, B]$ is characterized by

$$\text{pdf}(x) = \begin{cases} \frac{1}{B-A} & \text{for } A \leq x \leq B \\ 0 & \text{otherwise} \end{cases} \quad \text{and} \quad \text{cdf}(x) = \begin{cases} 0 & \text{for } x \leq A \\ \frac{x-A}{B-A} & \text{for } A \leq x \leq B \\ 1 & \text{for } B \leq x \end{cases}.$$

Find the result for a uniform distribution on the interval $[0, 1]$ in Figure 2.14(a).

Name	Function	μ	σ	median
Uniform	`unifpdf(x,0,1)`	1/2	$1/\sqrt{12}$	1/2
Uniform	`unifpdf(x,A,B)`	$\frac{A+B}{2}$	$\frac{B-A}{\sqrt{12}}$	$\frac{A+B}{2}$
Normal	`normpdf(x,mu,sigma)`	μ	σ	μ
Std. Normal	`stdnormal_pdf(x)`	0	1	0
Exponential	`exppdf(x,lambda)`	λ	λ	$\lambda \ln 2$
Student-t	`tpdf(x,n)`	0	$\sqrt{\frac{n}{n-2}}$ if $n>2$	0
χ^2 distribution	`chi2pdf(x,n)`	n	$\sqrt{2n}$	$\approx n\left(1-\frac{2}{9n}\right)^3$

Table 2.7: Continuous distributions, mean value μ, standard deviation σ and median

```
x   = linspace(-0.5,1.5);
p   = unifpdf(x,0,1);       cp = unifcdf(x,0,1);

figure(1); plot(x,p,'b',x,cp,'k')
           title('uniform distribution')
           axis([-0.5 1.5 -0.1 1.2])
           legend('pdf','cdf','location','northwest')
```

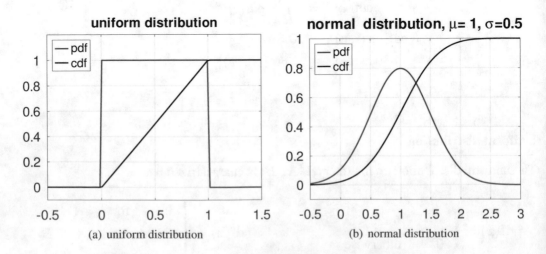

(a) uniform distribution　　　　　　　(b) normal distribution

Figure 2.14: Graphs of a uniform and a normal distribution

Normal distribution

The normal distribution with mean μ and standard devition σ is given by

$$
\begin{aligned}
\mathrm{pdf}(x) &= \frac{1}{\sigma\sqrt{2\pi}}\exp(-\frac{(x-\mu)^2}{2\sigma^2}) \\
\mathrm{cdf}(x) &= \int_{-\infty}^{x}\mathrm{pdf}(s)\,ds = \frac{1}{2}\left(1+\mathrm{erf}(\frac{x-\mu}{\sqrt{2}\sigma})\right).
\end{aligned}
$$

Find the result for a normal distribution with mean $\mu = 1$ and standard deviation $\sigma = \frac{1}{2}$ in Figure 2.14(b), generated by

```
x  = linspace(-0.5,3.0);
p  = normpdf(x,1,0.5);
cp = normcdf(x,1,0.5);

figure(1); plot(x,p,'b',x,cp,'k')
           title('normal distribution, \mu= 1, \sigma=0.5')
           axis([-0.5 3.0 -0.1 1.])
           legend('pdf','cdf','location','northwest')
```

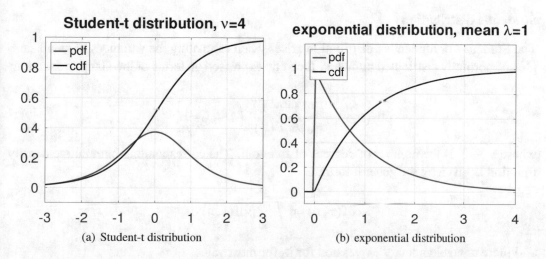

(a) Student-t distribution (b) exponential distribution

Figure 2.15: Graphs of the Student–t and an exponential distribution

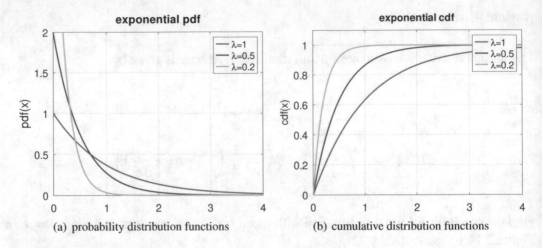

(a) probability distribution functions   (b) cumulative distribution functions

Figure 2.16: A few exponential distributions

Exponential distribution

The exponential distribution[2] with mean λ in MATLAB and *Octave* is given by

$$\text{pdf}(x) = \frac{1}{\lambda} \exp(-x/\lambda) \quad \text{and} \quad \text{cdf}(x) = 1 - \exp(-x/\lambda)$$

and are computed by `exppdf(x, lambda)` and `expcdf(x, lambda)`. Find the graphs for $\lambda = 1, 0.5$ and 0.2 in Figure 2.16. Some typical application of the exponential distribution are the length of phone calls, length of wait in a line, lifetime of components, . . .

Student-t distribution

The Student-t or Student's t distribution arises when estimating the variances of small samples of normally distributed numbers. It can be expressed on terms of the Gamma function[3] Γ by

$$\text{pdf}(x, \nu) = \frac{\Gamma(\frac{\nu+1}{2})}{\sqrt{\nu\pi}\,\Gamma(\frac{\nu}{2})} \left(1 + \frac{x^2}{\nu}\right)^{-\frac{\nu+1}{2}}$$

where $\nu \in \mathbb{N}$ is the number of degrees of freedom. The corresponding cumulative density function is given by the general formula

$$\text{cdf}(x, \nu) = \int_{-\infty}^{x} \text{pdf}(s, \nu)\, ds$$

and there is no elementary expression for it, for most values of ν.

---

[2]Some references use the factor $1/\lambda$ instead of λ, i.e. $\text{pdf}(x) = \lambda \exp(-x\lambda)$ and $\text{cdf}(x) = 1 - \exp(-x\lambda)$.

[3]The Gamma function is an extension of the well known factorial function, $\Gamma(n+1) = n! = \prod_{i=1}^{n} i$.

- The probability density function resembles a normal distribution with mean 0 and standard deviation 1, but it is wider and and a lower maximal value. The standard deviation is not equal to 1.

- As the number of degrees of freedom ν increases it converges to a standard normal distribution, see Figure 2.17.

- For some small values of ν there are explicit formulas, shown in Table 2.8.

Figure 2.17 was generated by

```
x  = linspace(-3,3);
p1 = tpdf(x,1); p2 = tpdf(x,2); p10 = tpdf(x,10); pn = normpdf(x,0,1);

figure(1); plot(x,p1,x,p2,x,p10,x,pn)
           title('Student-t distribution'); axis([-3 3 -0.1 0.5])
           legend('\nu=1','\nu=2','\nu=10','normal',...
           'location','northwest')
```

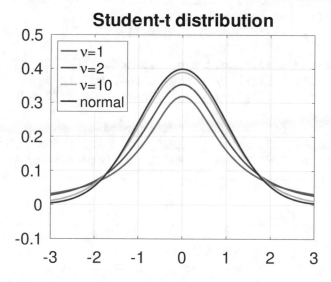

Figure 2.17: A few Student-t distributions and a normal distribution

χ^2 distribution

The χ^2–distribution with parameter n (degrees of freedom) is defined for $x > 0$ and given by

$$\mathrm{pdf}(x) = \frac{1}{2^{n/2}\Gamma(\frac{n}{2})}\, x^{\frac{n}{2}-1}\exp(-\frac{x}{2})$$

$$\mathrm{cdf}(x) = \int_0^x \mathrm{pdf}(s)\, ds = \frac{1}{\Gamma(\frac{n}{2})}\,\gamma(\frac{n}{2},\frac{x}{2})$$

ν	$\mathrm{pdf}(x, \nu)$	$\mathrm{cdf}(x, \nu)$
1	$\frac{1}{\pi\,(1+x^2)}$	$\frac{1}{2} + \frac{1}{\pi}\,\arctan(x)$
2	$\frac{1}{(2+x^2)^{3/2}}$	$\frac{1}{2} + \frac{x}{2\,\sqrt{2+x^2}}$
3	$\frac{6\,\sqrt{3}}{\pi\,(3+x^2)^2}$	
∞	$\frac{1}{\sqrt{2\pi}}\,\exp(-\frac{x^2}{2})$	$\frac{1}{2}\,(1 + \mathrm{erf}(\frac{x}{\sqrt{2}}))$

Table 2.8: Student-t distribution for some small values of ν

where $\gamma(s, t)$ is the lower incomplete gamma function. The mode (maximal value) is attained at $\max\{n - 2,\, 0\}$. Find the result for a a few χ^2 distributions in Figure 2.18, generated by

```
x = linspace(0,4);
pdf1 = chi2pdf(x,1); cdf1 = chi2cdf(x,1);
pdf2 = chi2pdf(x,2); cdf2 = chi2cdf(x,2);
pdf3 = chi2pdf(x,3); cdf3 = chi2cdf(x,3);
pdf5 = chi2pdf(x,5); cdf5 = chi2cdf(x,5);

figure(1); plot(x,pdf1,x,pdf2,x,pdf3,x,pdf5)
           ylabel('pdf(x)'); title('\chi^2 pdf');
           legend('n=1','n=2','n=3','n=5'); axis([0,4,0,1])

figure(2); plot(x,cdf1,x,cdf2,x,cdf3,x,cdf5)
           ylabel('cdf(x)'); title('\chi^2 cdf'); axis([0,4,0,1.1]);
           legend('n=1','n=2','n=3','n=5','location','northwest');
```

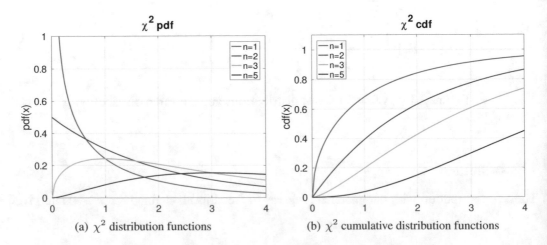

(a) χ^2 distribution functions   (b) χ^2 cumulative distribution functions

Figure 2.18: A few χ^2 distributions, PDF and CDF

2.8 Commands for Confidence Intervals and Hypothesis Testing

In this section a few command to determine confidence intervals and testing of hypothesis are shown. My personal preference is clearly to work with confidence intervals. Is is (too) easy to abuse the commands for hypothesis testing and computing p–values.

2.8.1 Confidence intervals

A confidence interval contains the true parameter μ to be examined with a certain level of confidence[4], as illustrated in Figure 2.19. One has to chose a level of significance α. Then the confidence interval has to contain the true parameter μ with a level of confidence (probability) of $p = 1 - \alpha$. Often used values for α are $0.05 = 5\%$ and $0.01 = 1\%$.

$$0 < \alpha < 1 \quad : \quad \text{level of significance}$$
$$p = 1 - \alpha \quad : \quad \text{level of confidence}$$

99% chance this interval contains true value μ

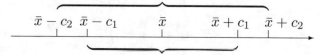

95% chance this interval contains true value μ

Figure 2.19: Confidence intervals at levels of significance $\alpha = 0.05$ and $\alpha = 0.01$

For the following examples we repeatedly use a data set to illustrate the commands. The numbers may represent the number of defects detected on a sample of 17 silicon wafers, selected at random from a large production.

```
 ─────────────────────── WaferDefects.txt ───────────────────────
  7 16 19 12 15 9 6 16 14 7 2 15 23 15 12 18 9
```

Estimating the mean value μ, with (supposedly) known standard deviation σ

Assume to have data with an unknown mean value μ, but a known standard deviation σ. This is a rather unusual situation, in most cases the standard deviation is not known and one

---

[4]A subtile detail, see [DownClar97, p. 192] or https://en.wikipedia.org/wiki/Confidence_interval .

- The true mean value μ is an (unknown) constant, not a random variable.

- A confidence interval with a level of confidence of 95% does not imply that the true mean μ is in this interval with a probability of 95% .

- The determined confidence interval $[\bar{x} - c, \bar{x} + c]$ is a random expression, 95% of the determined intervals will contain the true mean value μ.

has to use the similar computations in the next section 2.8.1. For sake of simplicity start with a known value of σ.

A sampling of n data points leads to x_i and you want to determine the mean value μ. According to the central limit theorem the random variable

$$\bar{X} = \frac{1}{n} \sum_{i=1}^{n} X_i$$

is approximated by a normal distribution with mean μ and standard deviation $\sigma/\sqrt{n}$. Now we seek a value u such that the green area in Figure 2.20(a) equals $(1-\alpha)$, where α is a small positive value. This implies that the true (unknown) value of μ is with a high probability in the green section in Figure 2.20(a). This is a two–sided confidence interval. If we want to know whether the true value of μ is below (or above) a certain threshold we use a similar argument to determine the one–sided confidence interval in Figure 2.20(b).

- To determine a two–sided confidence interval for the significance level α examine Figure 2.20(a) and proceed as follows:

  1. Determine u such that

$$
\begin{aligned}
(1 - \alpha) &= P(-u < U < u) = \frac{1}{\sqrt{2\pi}} \int_{-u}^{u} e^{-x^2/2} \, dx \\
\frac{\alpha}{2} &= P(U < -u) = \frac{1}{\sqrt{2\pi}} \int_{-\infty}^{-u} e^{-x^2/2} \, dx = \mathrm{cdf}(-u) \\
&= P(+u < U) = \frac{1}{\sqrt{2\pi}} \int_{+u}^{+\infty} e^{-x^2/2} \, dx = 1 - \mathrm{cdf}(+u) \; .
\end{aligned}
$$

     With MATLAB/*Octave* this value may be determined by either one of the commands `u = - norminv(alpha/2)` or `u = norminv(1-alpha/2)`.

  2. Then determine the estimator $\bar{x} = \frac{1}{n} \sum_{i=1}^{n} x_i$ and the two–sided confidence interval is given by $[\bar{x} - \frac{u\sigma}{\sqrt{n}}, \bar{x} + \frac{u\sigma}{\sqrt{n}}]$, i.e.

$$P(\bar{x} - \frac{u\sigma}{\sqrt{n}} < x < \bar{x} + \frac{u\sigma}{\sqrt{n}}) = 1 - \alpha \; .$$

- For the above example use

```
data = load('WaferDefects.txt');
N = length(data)
% we use the estimated variance as the supposedly given value
sigma = std(data); % this is NOT a realistic situation
alpha = 0.05 % choose the significance level
u = -norminv(alpha/2)
x_bar = mean(data);
ConfidenceInterval = [x_bar - u*sigma/sqrt(N) ,
                      x_bar + u*sigma/sqrt(N)]
-->
ConfidenceInterval = [10.082   15.212]
```

You may also use the command `ztest()` from Section 2.8.2 to compute the confidence interval.

The above code can be reused with a smaller significance level `alpha = 0.01`, leading to a wider confidence interval of $[9.2760, 16.0182]$.

- To determine the one–sided confidence interval for a significance level α examine Figure 2.20(b) and proceed as follows:

  1. Determine u such that

  $$
  (1 - \alpha) = P(U < u) = \frac{1}{\sqrt{2\pi}} \int_{-\infty}^{u} e^{-x^2/2} \, dx = \mathrm{cdf}(u)
  $$

  $$
  \alpha = P(u < U) = \frac{1}{\sqrt{2\pi}} \int_{u}^{+\infty} e^{-x^2/2} \, dx = 1 - \mathrm{cdf}(u).
  $$

  With MATLAB/*Octave* this value may be determined by `norminv(1-alpha)`.

  2. Determine the estimator $\bar{x} = \frac{1}{n} \sum_{i=1}^{n} x_i$. Now one–sided confidence interval is given by $[-\infty, \bar{x} + \frac{u\sigma}{\sqrt{n}}]$, i.e.

  $$
  P(x < \bar{x} + \frac{u\sigma}{\sqrt{n}}) = 1 - \alpha.
  $$

- For the above example we may use the following code.

```
data = load('WaferDefects.txt');
N = length(data)
% we use the estimated variance as the supposedly given value
sigma = std(data); % this is NOT a realistic situation
alpha = 0.05 % choose the significance level
u = norminv(1-alpha)
x_bar = mean(data);
UpperLimit = x_bar + u*sigma/sqrt(N)
-->
UpperLimit = 14.800
```

Estimating the mean value μ, with unknown standard deviation σ

Assuming you have data X_i all given by the same normal distribution $N(\mu, \sigma)$ with mean μ and standard deviation σ. Now the unbiased estimators

$$
\bar{X} = \frac{1}{n} \sum_{i=1}^{n} X_i \quad \text{and} \quad S^2 = \frac{1}{n-1} \sum_{i=1}^{n} (X_i - \bar{X})^2
$$

are **not independent**. The distribution of the random variable

$$
Z = \frac{\mu - \bar{X}}{S/\sqrt{n}}
$$

is a Student-t distribution with $n - 1$ degrees of freedom, see Section 2.7.2.

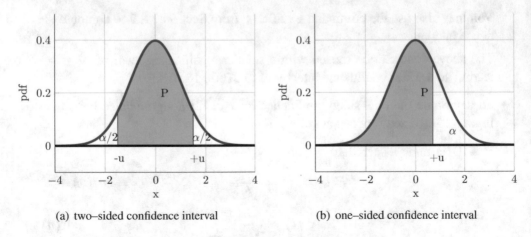

(a) two–sided confidence interval   (b) one–sided confidence interval

Figure 2.20: Two- and one–sided confidence intervals

- To determine a two–sided confidence interval with significance level α examine Figure 2.20(a) and use

$$
\begin{aligned}
(1 - \alpha) \quad &= \quad P(-u < Z < +u) = P\left(-u < \frac{\mu - \bar{X}}{S/\sqrt{n}} < +u\right) \\
&= \quad P\left(-u \frac{S}{\sqrt{n}} < \mu - \bar{X} < +u \frac{S}{\sqrt{n}}\right) = P\left(\bar{X} - u \frac{S}{\sqrt{n}} < \mu < \bar{X} + u \frac{S}{\sqrt{n}}\right)
\end{aligned}
$$

The value of u is determined by

$$
\frac{\alpha}{2} = \int_{-\infty}^{-u} \text{pdf}(x)\, dx = \text{cdf}(-u)
$$

and computed by `u = - tinv(alpha/2,n-1)`. With the estimators

$$
\bar{x} = \frac{1}{n} \sum_{i=1}^{n} x_i \quad \text{and} \quad \sigma^2 = \frac{1}{n-1} \sum_{i=1}^{n} (x_i - \bar{x})^2
$$

the two–sided confidence interval is given by

$$
\left[\bar{x} - u \frac{\sigma}{\sqrt{n}}, \ \bar{x} + u \frac{\sigma}{\sqrt{n}} \right].
$$

- For the above example use

```
data = load('WaferDefects.txt');
N = length(data)
alpha = 0.05 % choose the significance level
u = -tinv(alpha/2,N-1)
```

```
x_bar = mean(data);
sigma = std(data);
ConfidenceInterval = [x_bar - u*sigma/sqrt(N) ,
                      x_bar + u*sigma/sqrt(N)]
-->
ConfidenceInterval = [9.8727    15.4215]
```

Observe that this confidence interval is slightly wider than the one with (supposedly) known standard deviation σ. This is reasonable, since we have less information at our disposition.

- To determine the one–sided confidence interval for the significance level α examine Figure 2.20(b) and proceed as follows:

   1. Determine u such that

      $$(1 - \alpha) = P(U < u) = \int_{-\infty}^{u} \text{pdf}(x) \, dx = \text{cdf}(u) \, .$$

      With MATLAB/*Octave* this value may be determined by `tinv(1-alpha,n-1)`.
   2. With the estimators

      $$\bar{x} = \frac{1}{n} \sum_{i=1}^{n} x_i \quad \text{and} \quad \sigma^2 = \frac{1}{n-1} \sum_{i=1}^{n} (x_i - \bar{x})^2$$

      the one–sided confidence interval is given by $[-\infty, \bar{x} + \frac{u\sigma}{\sqrt{n}}]$, i.e.

      $$P(x < \bar{x} + \frac{u\sigma}{\sqrt{n}}) = 1 - \alpha \, .$$

- For the above example use

```
data = load('WaferDefects.txt');
N = length(data)
alpha = 0.05 % choose the significance level
u = tinv(1-alpha,N-1)
x_bar = mean(data);
sigma = std(data);
UpperLimit =  x_bar + u*sigma/sqrt(N)
-->
UpperLimit =  14.932
```

Observe that for large samples ($n \gg 1$) the Student-t distribution is close to the standard normal distribution and the results for the confidence intervals for known or unknown standard deviation σ differ very little.

Estimating the variance for nomaly distributed random variables

Assume that X_i are random variables with a normal distribution $N(\mu, \sigma)$, i.e. with mean μ and standard deviation σ. An unbiased estimator for the variance σ^2 is given by

$$S^2 = \frac{1}{n-1} \sum_{i=1}^{n} (X_i - \bar{X})^2 \, .$$

This is a random variable whose distribution is related to the χ^2 distribution. The modified variable

$$Y = \frac{(n-1)S^2}{\sigma^2} = \frac{1}{\sigma^2} \sum_{i=1}^{n} (X_i - \bar{X})^2$$

follows a χ^2 distribution with $n - 1$ degrees of freedom. To determine confidence intervals we have to observe that this distribution is not symmetric, see Section 2.7.2. Obviously we can not obtain negative values. The values of $\chi^2_{\alpha/2, n-1}$ and $\chi^2_{1-\alpha/2, n-1}$ are characterized by

$$(1-\alpha) \;=\; P(\chi^2_{\alpha/2, n-1} < Y < \chi^2_{1-\alpha/2, n-1}) = \int_{\chi^2_{\alpha/2, n-1}}^{\chi^2_{1-\alpha/2, n-1}} \mathrm{pdf}(x)\, dx$$

$$\frac{\alpha}{2} \;=\; \int_{0}^{\chi^2_{\alpha/2, n-1}} \mathrm{pdf}(x)\, dx \;=\; \mathrm{cdf}(\chi^2_{\alpha/2, n-1})$$

$$\frac{\alpha}{2} \;=\; \int_{\chi^2_{1-\alpha/2, n-1}}^{+\infty} \mathrm{pdf}(x)\, dx \;=\; 1 - \mathrm{cdf}(\chi^2_{1\alpha/2, n-1})$$

and thus can be computed by either one of the commands `chi2inv(alpha/2,n-1)` or `chi2inv(1-alpha/2,n-1)`. Since

$$(1-\alpha) \;=\; P\left(\chi^2_{\alpha/2, n-1} < \frac{(n-1)S^2}{\sigma^2} < \chi^2_{1-\alpha/2, n-1} \right)$$

$$=\; P\left(\frac{\chi^2_{\alpha/2, n-1}}{(n-1)S^2} < \frac{1}{\sigma^2} < \frac{\chi^2_{1-\alpha/2, n-1}}{(n-1)S^2} \right)$$

$$=\; P\left(\frac{(n-1)S^2}{\chi^2_{1-\alpha/2, n-1}} < \sigma^2 < \frac{(n-1)S^2}{\chi^2_{\alpha/2, n-1}} \right)$$

find a confidence interval for the variance σ^2

$$\left[\frac{(n-1)S^2}{\chi^2_{1-\alpha/2, n-1}} \, , \; \frac{(n-1)S^2}{\chi^2_{\alpha/2, n-1}} \right]$$

or for the standard deviation σ

$$\left[\frac{\sqrt{n-1}\, S}{\sqrt{\chi^2_{1-\alpha/2, n-1}}} \, , \; \frac{\sqrt{n-1}\, S}{\sqrt{\chi^2_{\alpha/2, n-1}}} \right] \, .$$

For the above example use

```
data = load('WaferDefects.txt');
N = length(data);
alpha = 0.05; % choose the significance level
chi2_low  = chi2inv(alpha/2,N-1);
chi2_high = chi2inv(1-alpha/2,N-1);
sigma = std(data)
ConfidenceIntervalVariance = sigma^2*(N-1)*[1/chi2_high , 1/chi2_low]
ConfidenceIntervalStd = sqrt(ConfidenceIntervalVariance)
-->
sigma = 5.3961
ConfidenceIntervalVariance = [16.151    67.444]
ConfidenceIntervalStd      = [4.0188    8.2124]
```

Observe that the confidence interval for the standard deviation is not symmetric about the estimated value of 5.3961.

Estimating the parameter p for a binomial distribution

For random variable X_i with a binomial distribution with parameter $0 < p < 1$ use

$$\bar{P} = \frac{\bar{k}}{n} = \frac{1}{n} \sum_{i=1}^{n} X_i$$

as an unbiased estimator for the parameter p. To construct a confidence interval for p we seek a lower limit p_l and an upper limit p_u such that

$$P(p_l < p < p_u) = 1 - \alpha .$$

For this to happen we need to solve the equations

$$P(S_n > \bar{k}) = \sum_{i=\bar{k}+1}^{n} \binom{n}{i} p_l^i (1 - p_l)^{n-i} = 1 - \mathrm{cdf}(\bar{k}, p_l) = \frac{\alpha}{2}$$

$$P(S_n \leq \bar{k}) = \sum_{i=1}^{\bar{k}-1} \binom{n}{i} p_u^i (1 - p_u)^{n-i} = \mathrm{cdf}(\bar{k}, p_u) = \frac{\alpha}{2} .$$

Thus we have to solve the equations $1 - \mathrm{cdf}(\bar{k}, p_l) = \alpha/2$ and $\mathrm{cdf}(\bar{k}, p_u) = \alpha/2$ for the unknowns p_l and p_u. This can be done using the command fzero(). Use help fzero to obtain information about this command, used to solve a single equation. For the command fzero() we have to provide the interval in which p is to be found, e.g. $0 \leq p \leq 1$.

Assuming that out of 1000 samples only 320 objects satisfy the desired property. The estimator for p is $\bar{p} = \frac{320}{1000} = 0.32$. Then use the code below to determine the two–sided confidence interval $[p_l, p_u]$ at significance level $\alpha = 0.05$.

```
N = 1000; Yes = 320; alpha = 0.05;
p_low = fzero(@(p)1-binocdf(Yes-1,N,p)-alpha/2,[0,1]);
p_up  = fzero(@(p)binocdf(Yes,N,p)-alpha/2,[0,1]);
Interval_Binom = [p_low p_up]
-->
Interval_Binom = [0.29115  0.34991]
```

Since $N = 1000$ is large and p is neither close to 1 or 0 we can approximate the binomial distribution by a normal distribution with mean 0.32 and standard deviation $\sigma = \sqrt{\frac{p(1-p)}{N}}$. The resulting confidence interval has to be similar to the above, as confirmed by the code below.

```
N = 1000; Yes = 320; alpha = 0.05;
% use an approximative normal distribution
p = Yes/N;
u = -norminv(alpha/2);
sigma = sqrt(p*(1-p)/N)
Interval_Normal = [ p-u*sigma  p+u*sigma ]
-->
Interval_Normal =  [0.29109   0.34891]
```

In the above example a two–sided interval is constructed. There are applications when a one–sided interval is required. Examine a test of 100 samples, with only 2 samples failing. Now determine an upper limit for the fail rate, using a confidence level of $\alpha = 0.01$. The parameter p (the fail rate) is too large if the probability to detect 2 or less fails is smaller than α. Thus the upper limit p_u satisfies the equation

$$P(S_n \le 2) = \sum_{i=0}^{2} \binom{100}{i} p_u^i (1-p_u)^{100-i} = \text{cdf}(2, p_u) = \alpha .$$

The code below shows that the fail rate is smaller than $\approx 8\%$, with a probability of $1 - \alpha = 99\%$.

```
N = 100; Fail = 2; alpha = 0.01;
p_up  = fzero(@(p)binocdf(Fail,N,p)-alpha,[0,1]);
-->
p_up = 0.081412
```

Rerunning the above code with $\alpha = 5\%$ leads to a fail rate smaller than $\approx 6\%$, with a probability of 95%. The maximal fail rate is smaller now, since we accept a lower probability. An approximation by a normal distribution is not justified in this example, since p is rather close to 0. The (wrong) result for the fail rate would be $\approx 5.3\%$ for $\alpha = 1\%$.

2.8.2 Hypothesis testing, p–value

MATLAB and *Octave* provide a set of command to use the method of testing a hypothesis, see Table 2.9. The commands can apply one–sided and two–sided tests, and may determine a confidence interval.

ztest()	testing for the mean, with known σ
ttest()	testing for the mean, with unknown σ
binotest()	testing for p, using a binomial distribution

Table 2.9: Commands for testing a hypothesis

A coin flipping example

When flipping a coin you (usually) assume that the coin is fair, i.e. "head" and "tail" are equally likely to show, or the probability p for "head" is $p = 0.5$. This is a **null hypothesis**.

$$\text{Null hypothesis} \quad H_0 \quad : \quad p = \frac{1}{2}$$

The corresponding **alternative hypothesis** is

$$\text{Alternative hypothesis} \quad H_1 \quad : \quad p \neq \frac{1}{2} \, .$$

By flipping a coin 20 times you want to decide whether the coin is fair. Choose a level of significance α and determine the resulting **domain of acceptance** A. In this example the domain of acceptance is an interval containing 10. If the actual number of heads is in A you accept the hypothesis $p = \frac{1}{2}$, otherwise you reject it. The probability of rejecting H_0, even if is is true, should be α, i.e. α is the probability of committing a type 1 error. You might also commit a type 2 error, i.e. accept H_0 even if it is not true. The probability of committing a type 2 error is β and $1 - \beta$ is called the **power of the test**. The relations are shown in Table 2.10.

	H_0 is true	H_1 is true
H_0 accepted	$1 - \alpha$	β
	correct	type 2 error
H_0 rejected	α	$1 - \beta$
	type 1 error	correct

Table 2.10: Errors when testing a hypothesis with level of significance α

Obviously the choice of the level of significance has an influence on the result.

- If $0 < \alpha < 1$ is very small:

  – The domain of acceptance A will be large, and we are more likely to accept the
    hypothesis, even if it is wrong. Thus we might make a type 2 error.

  – The probability to reject a true hypothesis is small, given by α. Thus we are not
    very likely to make a type 1 error.

- If $0 < \alpha < 1$ is large:

  – The domain of acceptance A will be small, and we are more likely to reject the
    hypothesis, even if it is true. Thus we might make a type 1 error.

  – The probability to accept a false hypothesis is small. Thus we are not very likely
    to make a type 2 error.

- The smaller the value α of the level of significance, the more likely the hypothesis
  H_0 is accepted. Thus there is a smallest value of α, leading to rejection of H_0.

The smallest value of α for which the null hypothesis H_0 is rejected, based on the
given sampling, is called the **p–value**.

Testing for the mean value μ, with (supposedly) known standard deviation σ, `ztest()`

Assume to have normally distributed data with an unknown mean value μ, but known standard deviation σ, just as in Section 2.8.1. The null hypothesis to be tested is that the mean
value equals a given value μ_0. For a given level of significance α the domain of acceptance A is characterized by

Null hypothesis H_0	mean $\mu = \mu_0$
Alternative hypothesis H_1	mean $\mu \neq \mu_0$

$$1 - \alpha = P(\bar{X} \in A) = P(\mu_0 - a \leq \bar{X} \leq \mu_0 + a),$$

where $\bar{X} = \frac{1}{n} \sum_{i=1}^{n} X_i$ is an unbiased estimator of the true mean value μ. Using the
central limit theorem we know that the random variable $U = \frac{\bar{X} - \mu}{\sigma/\sqrt{n}}$ follows a standard
normal distribution. Using this compute a by `a = -norminv(alpha/2)`, i.e.

$$\frac{\alpha}{2} = \int_{-\infty}^{-a} \mathrm{pdf}(x)\, dx$$

and then the domain of acceptance is given by

$$A = [\mu_0 - a\frac{\sigma}{\sqrt{n}},\ \mu_0 + a\frac{\sigma}{\sqrt{n}}].$$

The p–value P is characterized by

$$
\begin{aligned}
1 - \alpha_{min} &= P(|\bar{X} - \mu_0| \geq |\bar{x} - \mu_0|) \\
P = \alpha_{min} &= P(\bar{X} < \mu_0 - |\bar{x} - \mu_0|) + P(\bar{X} > \mu_0 + |\bar{x} - \mu_0|) \\
&= 2\,P(\bar{X} < \mu_0 - |\bar{x} - \mu_0|)
\end{aligned}
$$

and thus can be computed by $P = 2 \cdot \text{tcdf}(-|\bar{x} - \mu_0| \frac{\sqrt{n}}{\sigma}, n - 1)$. This is easily coded in MATLAB/*Octave*.

As example we want to test the hypothesis that the average number of defects in the wafers for the data set introduced on page 175 is given by 14.

```
data = load('WaferDefects.txt');
mu = mean(data) % the mean of the sampling
n = length(data);
mu0 = 14;      % test against this mean value
sigma = std(data);    % assumed value of the standard deviation
alpha = 0.05; % choose level of significance

a = -tinv(alpha/2,n-1);
if abs(mu-mu0)<a*sigma/sqrt(n) disp('H_0 not rejected, might be true')
else                          disp('H_0 rejected, probably false')
end%if
P_value = 2*tcdf(-abs(mu-mu0)*sqrt(n)/sigma,n-1)
-->
mu =   12.647
H_0 not rejected, might be true
P_value =   0.31662
```

Observe that the average μ of the sample is well below the tested value of $\mu_0 = 14$. Since the size of the sample is rather small, we do not have enough evidence to reject the hypothesis. This **does not** imply that the hypothesis is true. The computed p–value is larger then the chosen level of significance $\alpha = 0.05$, which also indicated that the hypothesis can not be rejected.

With the command `ttest()` you can test this hypothesis. It will also compute the confidence interval by the methods from Section 2.8.1.

```
[H, PVAL, CI] = ttest(data,mu0)
-->
H    = 0
PVAL = 0.31662
CI   = 9.8727    15.4215
```

In the documentation of `ttest` find the explanation for the results.

- Since $H = 0$ the null hypothesis not rejected, i.e. it might be true.

- Since the p–value of 0.32 is larger than $\alpha = 0.05$ the null hypothesis is not rejected.

- The two–sided confidence interval $[9.8727\,,\,15.4215]$ is determined by the method in Section 2.8.1.

The default value for the level of significance is $\alpha = 0.05 = 5\%$. By providing more arguments you can used different values, e.g.

```
[H, PVAL, CI] = ttest(data,mu0,'alpha',0.01).
```

One–sided testing for the mean value μ, with unknown standard deviation σ

One can also apply one sided tests. Assume that we claim the the actual mean value μ is below a given value μ_0. For a given level of significance α the domain of acceptance A is characterized by

Null hypothesis H_0	mean $\mu \leq \mu_0$
Alternative hypothesis H_1	mean $\mu > \mu_0$

$$1 - \alpha = P(\bar{X} \in A) = P(\bar{X} \leq \mu_0 + a)\,.$$

Using this compute a by `a = tinv(1-alpha,n-1)`, i.e.

$$1 - \alpha = \int_{-\infty}^{a} \text{pdf}(x)\,dx \quad \text{or} \quad \alpha = \int_{a}^{\infty} \text{pdf}(x)\,dx$$

and then the domain of acceptance is given by

$$A = [-\infty\,,\,\mu_0 + a\,\frac{\sigma}{\sqrt{n}}]\,.$$

The $P = \alpha_{min}$ value is characterized by

$$\alpha_{min} = P(\bar{X} \geq \bar{x}) = 1 - P(\bar{X} \leq \mu_0 + (\bar{x} - \mu_0))$$

and thus can be computed by $P = 1 - \text{tcdf}((\bar{x} - \mu_0)\frac{\sqrt{n}}{\sigma}, n - 1)$. This is easily coded in MATLAB/*Octave*.

As example we want to test the hypothesis that the average number of defects in the wafers for the data set introduced on page 175 is smaller than 14.

```
data = load('WaferDefects.txt');
mu = mean(data)      % the mean of the sampling
n = length(data);
mu0 = 14;            % test against this mean value
sigma = std(data);   % value of the standard deviation
alpha = 0.05;        % choose level of significance

a = tinv(1-alpha,n-1);
```

```
if mu < mu0+a*sigma/sqrt(n)  disp('H_0 not rejected, might be true')
else                         disp('H_0 rejected, probably false')
end%if
P_value = 1 - tcdf((mu-mu0)*sqrt(n)/sigma,n-1)
-->
mu =  12.647
H_0 not rejected, might be true
P_value =  0.84169
```

Observe that the average μ of the sample is well below the tested value of $\mu_0 = 14$. Thus the hypothesis is very likely to be correct, which is confirmed by the above result.

With the command `ttest()` you can test this hypothesis. It will also compute the one–sided confidence interval.

```
[H, PVAL, CI] = ttest (data,mu0,'tail','right')
-->
H    = 0
PVAL = 0.84169
CI   = 10.362  -Inf
```

- Since $H = 0$ the null hypothesis is not rejected, i.e. it might be true.

- Since the p–value of 0.84 is larger than $\alpha = 0.05$ the null hypothesis is not rejected.

- The two–sided confidence interval $[10.362, +\infty]$ is determined by the method in Section 2.8.1.

The default value for the level of significance is $\alpha = 0.05 = 5\%$. By providing more arguments you can used different values, e.g.
`[H, PVAL, CI] = ttest(data,mu0,'tail','right','alpha',0.01).`

A two–sided test for the parameter p of a binomial distribution

By flipping a coin 1000 times you observe 475 "heads". Now there are different hypothesis that you might want to test for the parameter p, the ratio of "heads" and total number of flips.

Situation	Hypothesis	Test
"head" and "tail" are equally likely	$p = \frac{1}{2}$	two–sided
"head" is less likely than "tail"	$p \leq \frac{1}{2}$	one–sided
"head" is more likely than "tail"	$p \geq \frac{1}{2}$	one–sided

The methods and commands to be examined below will lead to statistical answers to the above questions.

Assume to have data given by a binomial distribution with parameter p, i.e. we have N data points and each point has value 1 with probability p and 0 otherwise. The null hypothesis to be tested is that the parameter p equals a given value p_0. Let $k = \sum_{i=1}^{N} x_i$ and $\bar{x} = \frac{k}{N} = \frac{1}{N} \sum_{i=1}^{N} x_i$ be the result of a sample, then $\bar{x}$ is an estimator of p. For a given

Null hypothesis H_0	$p = p_0$
Alternative hypothesis H_1	$p \neq p_0$

level of significance α the domain of acceptance A is characterized by

$$1 - \alpha \;\; \leq \;\; P(\bar{x} \in A) = P(A_{low} \leq \bar{x} \leq A_{high})$$
$$\alpha \;\; \geq \;\; P(\bar{x} \notin A) = P(\bar{x} < A_{low}) + P(A_{high} < \bar{x}) = \mathrm{cdf}(A_{low}) + 1 - \mathrm{cdf}(A_{high})$$

where `cdf()` is the cumulative density function for the binomial distribution with parameter p_0. This condition translates to

$$\mathrm{cdf}(A_{low}) \leq \frac{\alpha}{2} \quad \text{and} \quad 1 - \mathrm{cdf}(A_{high}) \leq \frac{\alpha}{2} \, .$$

Since the binomial distribution is a discrete distribution we can not insist on the limiting equality, but have to work with inequalities, leading to

$$\texttt{binocdf}(N \cdot A_{low}, N, p_0) \leq \frac{\alpha}{2} \quad \text{and} \quad 1 - \texttt{binocdf}(N \cdot A_{high}, N, p_0) \leq \frac{\alpha}{2} \, .$$

Using MATLAB/*Octave* commands this can be solved by

$$A_{low} = \texttt{binoinv}(\frac{\alpha}{2}, N, p_0)/N \quad \text{and} \quad A_{high} = \texttt{binoinv}(1 - \frac{\alpha}{2}, N, p_0)/N \, .$$

The null hypothesis $H_0 : p = p_0$ is accepted if

$$A_{low} \leq \frac{k}{N} \leq A_{high} \, .$$

Since the p–value is also given by the total probability of all events less likely than the observed k, we have an algorithm to determine P.

1. Compute the probability that the observed number k of "heads" shows by $p_k = \texttt{binopdf}(k, N, p_0)$.

2. Of all $p_j = \texttt{binopdf}(j, N, p_0)$ add those that are smaller or equal to p_k.

3. This can be packed into a few lines of code.

```
p_k = binopdf(k,n,p0);
p_all = binopdf([0:n],n,p0);
p = sum(p_all(find(p_all <= p_k)));
```

As an example consider flipping a coin 1000 times and observe 475 "heads". To test whether this coin is fair make the hypothesis $p = 0.5$ and test with a significance level of $\alpha = 0.05$.

```
N = 1000        % number of coins flipped
p0 = 0.5;       % hypothesis to be tested
alpha = 0.05;   % level of sigificance
Heads = 475     % number of observed heads


A_low  = (binoinv(alpha/2,N,p0))/N;
A_high = binoinv(1-alpha/2,N,p0)/N;
DomainOfAcceptance = [A_low,A_high]
if (Heads/N >= A_low) && (A_high >= Heads/N)
            disp('H_0 not rejected, might be true')
else
            disp('H_0 rejected, probably false')
end%if
p_k = binopdf(Heads,N,p0);   p_all = binopdf([0:N],N,p0);
P_value = sum(p_all(find(p_all <= p_k)))
-->
DomainOfAcceptance = [0.469   0.531]
H_0 not rejected, might be true
P_value =   0.12121
```

The above can be compared with the confidence interval determined in Section 2.8.1, see page 181.

```
p_low = fzero(@(p)1-binocdf(Heads-1,N,p)-alpha/2,[0,1]);
p_up  = fzero(@(p)binocdf(Heads,N,p)-alpha/2,[0,1]);
Interval_Binom = [p_low p_up]
-->
Interval_Binom = [0.44366   0.50649]
```

Since the value of $p = \frac{1}{2}$ is inside the interval of confidence we conclude that the coin might be fair. Observe that the confidence interval is built around the estimated expected value $p = 0.475$, while the domain of acceptance is built around the tested value $p_0 = \frac{1}{2}$.

The above result can be generated by the *Octave* command binotest()[5].

```
[h,p_val,ci] = binotest(475,1000,0.5)
-->
h = 0
p_val =   0.12121
ci =    [0.44366   0.50649]
```

---

[5]The function binotest.m is available in the statistics package with version 1.3.0 or newer. There is also a version of the code for MATLAB, available with the source code of these notes. Rename the file binotestMatlab.m to binotest.m.

With MATLAB the command binofit() determines the estimator for the parameter p and the confidence interval. Observe that the hypothesis is not tested and the p–value not computed. Thus the command binofit() uses results from Section 2.8.1 on page 181.

```
[p,ci] = binofit(475,1000,0.05)
-->
p   =  0.4750
ci  =  0.4437     0.5065
```

One–sided test for the parameter p for a binomial distribution

The above method can also be used for one–sided tests. For a given level of significance α

Null hypothesis H_0	$p \leq p_0$
Alternative hypothesis H_1	$p > p_0$

the domain of acceptance A is characterized by

$$1 - \alpha \ \leq \ P(\bar{x} \in A) = P(\bar{x} \leq A_{high})$$
$$\alpha \ \geq \ P(\bar{x} \notin A) = P(A_{high} < \bar{x}) = 1 - \text{cdf}(A_{high}) \ .$$

This condition translates to $1 - \text{cdf}(A_{high}) \leq \alpha$, leading to $1 - \texttt{binocdf}(N \cdot A_{high}, N, p_0) \leq \alpha$. Using MATLAB/*Octave* commands to can be solved by

$$A_{high} = \texttt{binoinv}(1 - \alpha, N, p_0)/N \ .$$

The null hypothesis $H_0 : p \leq p_0$ is accepted if $\frac{k}{N} \leq A_{high}$. Since the p–value is defined as the smallest value of the level of significance α for which the null hypothesis is rejected use

$$P = \alpha_{min} = 1 - \texttt{binocdf}(k - 1, N, p_0) \ .$$

For the above coin flipping example we claim that the coin is less likely to show "heads" than "tail".

```
N = 1000        % number of coins flipped
p0 = 0.5;       % hypothesis to be tested
alpha = 0.05;   % level of sigificance
Heads = 475     % number of observed heads

A_high = binoinv(1-alpha,N,p0)/N;
DomainOfAcceptance = [0,A_high]
if (A_high >= Heads/N)  disp('H_0 not rejected, might be true')
else                    disp('H_0 rejected, probably false')
end%if
```

```
P_value = 1-binocdf(Heads-1,N,p0)
-->
DomainOfAcceptance = [0  0.52600]
H_0 not rejected, might be true
P_value =  0.94663
```

The result states that the coin might be more likely to show "heads". The observed value of $p \approx 0.475$ is well within the domain of acceptance $A = [0\,, 0.526]$.

The above result can be generated by the *Octave* command binotest().

```
[h,p_val,ci] = binotest(475,1000,0.5,'tail','left')
-->
h = 0
p_val =  0.94663
ci =    [0 0.50150]
```

Obviously we can also test whether the coin is less likely to show "heads". Since the

Null hypothesis H_0	$p \geq p_0$
Alternative hypothesis H_1	$p < p_0$

arguments are very similar to the above we just show the resulting code.

```
A_low = binoinv(alpha,N,p0)/N;
DomainOfAcceptance = [A_low,1]
if (A_low <= Heads/N)    disp('H_0 not rejected, might be true')
else                     disp('H_0 rejected, probably false')
end%if
P_value = binocdf(Heads,N,p0)
-->
DomainOfAcceptance = [0.474 1]
H_0 not rejected, might be true
P_value =  0.060607
```

The result states that the coin might be less likely to show "heads". But the observed value of $p \approx 0.475$ is barely within the domain of acceptance $A = [0.474\,, 1]$. The p–value of $P \approx 0.06$ is just above $\alpha = 5\%$. If we increase α slightly, i.e. be more tolerant towards errors of the first type, the hypothesis would be rejected.

The above result can be generated by the *Octave* command binotest.

```
[h,p_val,ci] = binotest(475,1000,0.5,'tail','right')
-->
h = 0
p_val =  0.060607
ci =    [0.44860  1]
```

Observe that in the previous examples for 475 "heads" on 1000 coin flips none of the three null hypotheses $p = \frac{1}{2}$, $p \leq \frac{1}{2}$ or $p \geq \frac{1}{2}$ is rejected. This is clearly illustrating that we do **not** prove that one of the hypotheses is correct. All we know is that they are not very likely to be false.

Testing for the parameter p for a binomial distribution for large N

If N and $N p_0 (1 - p_0)$ are large (e.g. $N > 30$ and $N p_0 (1 - p_0) > 10$) the binomial distribution of $Y = \frac{1}{N} \sum_{i=1}^{N} X_i$ with parameter p_0 can be approximated by a normal distribution with mean p_0 and standard deviation $\sigma = \sqrt{\frac{p_0 (1-p_0)}{N}}$. Thus we can replace the binomial distribution in the above section by this normal distribution and recompute the domains of acceptance and the p–values. The formulas to be used are identical to the ones in Section 2.8.2. For the confidence intervals use the tools from Section 2.8.1.

- Two–sided test with null hypothesis H_0 : $p = p_0$.

```
N = 1000       % number of coins flipped
p0 = 0.5;      % hypothesis to be tested
alpha = 0.05;  % level of sigificance
Heads = 475    % number of observed heads
sigma = sqrt(p0*(1-p0)/N);  % standard deviation for p

u = -norminv(alpha/2);
DomainOfAcceptance = [p0-u*sigma,p0+u*sigma]
if (abs(Heads/N-p0)<u*sigma)
        disp('H_0 not rejected, might be true')
else    disp('H_0 rejected, probably false')
end%if
P_value = 2*normcdf(-abs(Heads/N-p0)/sigma)
-->
DomainOfAcceptance = [0.469   0.531]
H_0 not rejected, might be true
P_value =   0.11385
```

- One–sided test with null hypothesis H_0 : $p \leq p_0$.

```
u = -norminv(alpha);
DomainOfAcceptance = [0 p0+u*sigma]
if (Heads/N<p0+u*sigma)
        disp('H_0 not rejected, might be true')
else
        disp('H_0 rejected, probably false')
end%if
P_value = 1-normcdf((Heads/N-p0)/sigma)
-->
```

```
DomainOfAcceptance = [0   0.52601]
H_0 not rejected, might be true
P_value =  0.94308
```

- One–sided test with null hypothesis $H_0 : p \geq p_0$.

```
u = -norminv(alpha);
DomainOfAcceptance = [p0-u*sigma 1]
if (Heads/N>p0-u*sigma)
        disp('H_0 not rejected, might be true')
else
        disp('H_0 rejected, probably false')
end%if
P_value = normcdf((Heads/N-p0)/sigma)
-->
DomainOfAcceptance = [0.47399 1]
H_0 not rejected, might be true
P_value =  0.056923
```

- All of the above results are very close to the numbers obtained by the binomial distribution in Sections 2.8.2 and 2.8.2. This is no surprise, since $N = 1000$ is large enough and $N p_0 (1 - p_0) = 250 \gg 10$ and thus the normal distribution is a good approximation of the binomial distribution.

2.9   List of Files for Statistics

In the previous section the codes and data files in Table 2.11 were used.

filename	function
Demo_Boxplot.m	script to generate boxplots and quantiles
LinearRegression.m	function to perfom linear regression, *Octave*
LinearRegression.m	for MATLAB in subdirectory Matlab
WaferDefects.txt	data file for some of the demos
binotest.m	function to test for the binomial parameter p, *Octave*
binotest.m	for MATLAB in subdirectory Matlab

Table 2.11: Codes and data files for chapter 2

Chapter 3

Engineering Applications

In this chapter a few engineering applications of MATLAB/*Octave* are presented. In each the question or problem is formulated and then solved with the help of *Octave*/MATLAB. For some of the necessary mathematical tools brief explanations are provided. But the notes are assuming that the reader is familiar with the Mathematics and Physics of a typical engineering curriculum.

This small set of applications with solutions shall help you to use *Octave* or MATLAB to solve **your** engineering problems. The selection is strongly influenced by your authors personal preference and some of the topics are based on works by students. For each section find the *Octave*/MATLAB skills to be used to solve the problems.

- 3.1: Numerical Integration and Magnetic Fields
  Based on the law of Biot–Savart the magnetic field of a wire carrying a current is computed. The Helmholtz configuration of two circular coils is examined carefully.

  - Numerical integration
  - Generate a vector field

- 3.2: Linear and Nonlinear Regression
  The basic notations for linear and nonlinear regression are explained and possibles sources of errors discussed. This is followed by a few real world applications.

  - Linear regression, using `LinearRegression()` and `regress()`
  - Nonlinear regression, using `leasqr()` and `fsolve()`

- 3.3. Regression with Constraints
  Using a linear regression with a constraint fitting of a straight line to data points is performed, using the true geometric distance. Then a plane is fitted to data points and algorithms to fit an ellipse to data points are presented.

- 3.4: Computing Angles on an Embedded Device
  Using integer arithmetic only arbitrary functions are approximated. This is then used to computed angles on an embedded device.

Supplementary Information The online version contains supplementary material available at [https://doi.org/10.1007/978-3-658-37211-8_3].

© The Author(s), under exclusive license to Springer Fachmedien Wiesbaden GmbH, part of Springer Nature 2022
A. Stahel, *Octave and MATLAB for Engineering Applications*, https://doi.org/10.1007/978-3-658-37211-8_3

- – Integer arithmetic with data type int16 and similar
- – Simulation of operations on a micro controller
- – Visualization and analysis of approximation errors

- 3.5: Analysis of Stock Performance, Value of a Stock Option
A probabilistic analysis of stock performance as presented, leading to the Black–Scholes–Merton approach to put a price tag on a stock option.

  - – Formatted reading from a file
  - – Probabilistic analysis of data
  - – Monte Carlo simulation

- 3.6: Motion Analysis of a Circular Disk
The motion and deformation of a watch caliber falling on the ground is analyzed and visualized.

  - – Reading data from a file
  - – Generating an animation on screen
  - – Use external programs to generate a movie, to be played by any movie player

- 3.7: Analysis of a Vibrating Cord
The performance of a vibrating string based force sensor is examined. The motion of a damped vibrating cord is analyzed and the quality factor determined.

  - – Fitting of an exponentially decaying vibration to measured data
  - – Calling external programs within *Octave*/MATLAB
  - – Construction of *Octave* commands by combining strings and then evaluating the command.

- 3.8: An Example for Fourier Series
The motion of a beam struck by a hammer is measured by acceleration sensors on beam and hammer. The resulting frequency spectrum is computed, as function of time.

  - – Reading and displaying data
  - – Use FFT to determine the frequency spectrum on different time slices

- 3.9: Grabbing Data from the Screen and Spline Interpolation
At first mouse clicks on the display are converted to data, then regularly spaced data is generated by interpolation.

  - – Reading data from screen, either in an MATLAB/*Octave* window (ginput) or even anywhere on the screen (xinput).
  - – Interpolation of data.

- 3.10: Intersection of Circles and Spheres, GPS
  An algorithm to determine intersection points of circles and spheres is presented. Then the over-determined system of the intersection of many circles and spheres is examined and reduced to a least square problem. This is then the basis for a short presentation of some of the Mathematics for the GPS (Global Positioning System).

- 3.11: Scanning a 3–D Object with a Laser
  A solid is scanned by a laser from two different angles. Based on this data the shape of the solid is reconstructed.

    – Evaluation irregularly spaced data on a uniform grid.

    – Merging two pictures into one picture.

- 3.12: Transfer Functions, Bode and Nyquist plots
  Elementary operations with transfer functions.

    – Bode and Nyquist plots.

    – Control theory, combination of transfer functions, feedback.

3.1   Numerical Integration and Magnetic Fields

In this section a few methods to evaluate definite integrals numerically are presented.

- In subsection 3.1.1 the trapezoidal rule and Simpson's rule are used to evaluate integrals for functions given by data points.

- In subsection 3.1.2 commands to evaluate integrals of functions given by an expression (formula) are presented.

- In subsection 3.1.3 some ideas and *Octave*/MATLAB codes on how to integrate over domains in the plane $\mathbb{R}^2$ are introduced.

- In subsections 3.1.4 and beyond the magnetic field of circular conductors is examined, using numerical integration. Particular attention is given to the Helmholtz configuration.

3.1.1   Basic integration methods if data points are given

Trapezoidal integration using `trapz()`

The simples integration in *Octave* is based on the trapezoidal rule and implemented in the command `trapz()`. Examine the online help. The code below computes

$$\int_0^\pi \sin(x)\ dx$$

with 100 subintervalls of equal length.

```
x = 0:pi/100:pi;
y = sin(x);
trapz(x,y)
```

The returned value of 1.9998 is rather close to the exact value of 2.

Numerical analysis indicates that the approximation error of the trapezoidal rule is proportional to h^2, where h the the length of the subintervals. This is confirmed by the graphic created by the code below.

- for $n = 10,\ 20,\ 40,\ \dots,\ 10 \cdot 2^9$ the integral is computed with n subintervals.

- The error is then plotted with double logarithmic scales. Since

$$\text{error} \ \approx \ c \cdot h^2$$
$$\ln(\text{error}) \ \approx \ \ln(c) + 2\ \ln h = \ln(c) + 2\ \ln\frac{\pi}{n} = \ln(c) + 2\ln\pi - 2\ \ln n$$

the result should be a straight line with slope -2. This is confirmed in Figure 3.2.

```
Nrun = 10;   n = zeros(1,Nrun);   err = zeros(1,Nrun);
for k = 1:Nrun
  n(k)  = 10*2^(k-1);
  x = linspace(0,pi,n(k)+1);
  err(k) = abs(2-trapz(x,sin(x)));
end%for

loglog(n,err);
```

Using the linear regression commands

```
F = [ones(size(n))', log(n')];
p = LinearRegression(F,log(err'))
-->
p = 0.4988
   -2.0002
```

conclude that

$$\ln(\text{err}(n)) \approx 0.5 - 2 \ln(n) \quad \text{and} \quad \text{err}(n) \approx \frac{1.6}{n^2}$$

and this confirms the error estimate for the trapezoidal integration method.

   With the command `cumtrapz()` (cumulative trapezoidal) we can not only compute the integral over the complete interval, but also the values of the integral at the intermediate points, i.e. the code

```
x  = 0:pi/100:pi;
y  = sin(x);
ys = cumtrapz(x,y);
plot(x,ys)
xlabel('position x'); ylabel('integral of sin(x)'); grid on
```

will compute

$$\int_0^x \sin(s)\,ds$$

for 101 values of x evenly distributed from 0 to π. The result is shown in Figure 3.1.

Simpson integration

In your (numerical) analysis course you should have learned about the Simpson method for numerical integrals. Below find an implementation in *Octave*. We have the following requirements for the code:

- Simpsons integration formula for an even number of subintervals has to be applied. The code can only handle subintervals of equal length.

- The function can be given either as a function name or by a list of values.

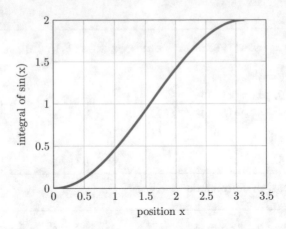

Figure 3.1: Cumulative trapezoidal integration of $\sin(x)$

---

simpson.m

```
function res = simpson(f,a,b,n)
%% simpson(integrand,a,b,n) compute the integral of the function f
%% on the interval [a,b] with using Simpsons rule
%% use n subintervals of equal length , n has to be even
%% otherwise n+1 is used
%% f is either a function handle, e.g  @sin or a vector of values
if isa(f,'function_handle')
  n = round(n/2+0.1)*2; %% assure even number of subintervals
  h = (b-a)/n;
  x = linspace(a,b,n+1);
  f_x = x;
  for k = 0:n
    f_x(k+1) = feval(f,x(k+1));
  end%for
else
  n = length(f);
  if (floor(n/2)-n/2==0)
    error('simpson: odd number of data points required');
  else
    n = n-1;
    h = (b-a)/n;
    f_x = f(:)';
  end%if
end%if
w = 2*[ones(1,n/2); 2*ones(1,n/2)]; w = w(:);
w = [w;1]; w(1)=1;  % construct the simpson weights
res = (b-a)/(3*n)*f_x*w;
```

This Simpson integration can now be tested similarly to the above tests. The convergence rate of the trapezoidal and Simpson integration is visualized in Figure 3.2.

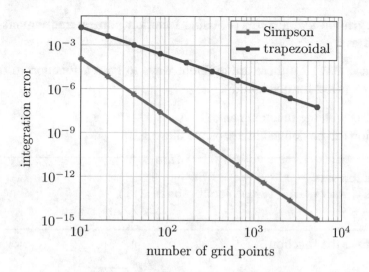

Figure 3.2: Error of trapezoidal and Simpsons integration method

```
Nrun = 10;   n = zeros(1,Nrun); err = zeros(1,Nrun);
for k = 1:Nrun
   n(k)    = 10*2^(k-1);
   err(k) = abs(2-simpson(@sin,0,pi,n(k)));
end%for
loglog(n,err,'-+',n,errTrap,'-*');
xlabel('number of grid points'); ylabel('integration error'); grid on
legend('Simpson','trapezoidal')

F - [ones(size(n))' log(n')];
p = LinearRegression(F,log(err'))
-->
p =    0.1365
      -4.0130
```

The result of the regression confirms that the error is proportional to h^4. This is confirmed in Figure 3.2. Observe that the accuracy of the integration methods can not be better than machine accuracy. This effect is starting to show in the lower right corner of Figure 3.2.

3.1.2   Basic integration methods for given functions

In this section some of the functions provided by *Octave* and/or MATLAB are illustrated. For more details consult the appropriate manuals.

Adaptive integration using `quad()`**, using function names, anonymous functions or function handles**

With *Octave* and MATLAB there are different ways to pass a function to the integration routine `quad()` (short for quadrature).

- As a string with the function name
  A function can be defined in a separate file, e.g.

```
                              ff.m
function res = ff(x)
   res  = x.*sin (1./x).*sqrt(abs(x-1));
end
```

This defines the function

$$\mathrm{ff}(x) = x \cdot \sin(\frac{1}{x}) \cdot \sqrt{|x-1|} \ .$$

Observe that the code for the function is written with vectorization in mind, i.e. it can be evaluated for many arguments with one function call.

```
ff([0.2:0.3:2])
-->
-0.1715   0.3215   0.3395   0.2744   0.5800   0.7892   0.9589
```

This is essential for the integration routines and required for speed reasons.

  - With *Octave* the definition of this function can be before the function call in an *Octave* script or in a separate file ff.m.
  - With MATLAB the definition of this function has to be in a separate file ff.m or at the end of a MATLAB script.

An integration over the interval $[0,3]$ can then be performed by

```
result = quad('ff',0,3)
-->
ans = 1.9819
```

- As a function handle
  The above integral can also be computed by using a function handle.

```
ff = @(x) x.*sin(1./x).*sqrt(abs(x-1));
quad(ff,0,3)
-->
ans = 1.9819
```

- Function handles are very convenient to determine integrals depending on parameters. To compute the integrals

$$\int_0^2 \sin(\lambda t)\, dt$$

for λ valucs between 1 and 2 use

```
for lambda = [1:0.1:2]
    quad(@(t)sin(lambda*t),0,2)
end%for
```

Using the function `integral()`

For the integration by `integral()` the function has to be passed as a handle. The limits of integration can be $-\infty$ or $+\infty$. To determine

$$\int_0^{+\infty} \cos(t)\, \exp(-t)\, dt$$

use

```
integral(@(t)cos(t).*exp(-t),0,Inf)
-->
ans = 0.5000
```

The function `integral()` can be used with optional parameters, specified as pairs of the string with the name of the option and the corresponding value. The most often used options are:

- `AbsTol` to specify the absolute tolerance. The default value is $\mathtt{AbsTol} = 10^{-10}$.

- `RelTol` to spccify the relative tolerance. The default value is $\mathtt{AbsTol} = 10^{-6}$.

- The adaptive integration is refined to determine the value Q of the integral, until the condition
$$\mathtt{errol} \leq \max\{\mathtt{AbsTol}, \mathtt{RelTol} \cdot |Q|\}$$
is satisfied, i.e. either the relative or absolute error are small enough.

- `Waypoints` to specify a set of points at which the function to be integrated might not be continuous. This can be used instead of multiple calls of `integral()` on sub-intervals.

As an example for the function `integral()` examine the integral

$$\int_0^2 |\cos(x)|\, dx = 2 - \sin(2)\,.$$

Observe that this function is not differentiable at $x = \frac{\pi}{2}$. Thus the high order of approximation (e.g. for Simpson) are not valid and convergence problems are to be expected.

```
int_exact = 2 - sin(2);
int_1 = integral(@(x)abs(cos(x)),0,2)
int_2 = integral(@(x)abs(cos(x)),0,2,'AbsTol',1e-12,'RelTol',1e-12)
int_3 = integral(@(x)abs(cos(x)),0,2,'Waypoints',pi/2)
Log10_Errors = log10(abs([int_1,int_2,int_3] - int_exact))
-->
int_1      =     1.0907
int_2      =     1.0907
int_3      =     1.0907
Log10_Errors  =   -7.9982   -14.3313   -15.6536
```

The results illustrate the usage of the tolerance parameters. Specifying the special point $x = \frac{\pi}{2}$ generates a more accurate result with less computational effort.

Comparison of integration commands in *Octave*/MATLAB

Above three different commands for numerical integration are shown. For any given integration one has to choose the best method. In Table 3.1 find a brief description.

`trapz()`, `cumtrapz()`	uses trapezoidal rule
	to be used for discretized values only
	uneven spacing is possible
`simpson()`	uses Simpson's method
	for discretized values of an anonymous function or
	by an even number of subintervals of equal length
`quad()`	uses **Quadpack**, adaptive algorithm
	function name or handle has to be given
`integral()`	uses adaptive algorithm, based on Gauss
	function handle has to be given

Table 3.1: Integration commands in *Octave*

For each situation the best algorithm has to be chosen. A comparison is given in Table 3.2.

The commands `quad()` or `integral()` should be used whenever possible.

More functions available in *Octave*

In Section 23.1 of the *Octave* manual a few more integration functions are documented.

quad : Numerical integration based on Gaussian quadrature.

	trapz()	simpson()	quad()	integral()
accuracy	poor	intermediate	excellent	excellent
built–in error control	no	no	yes	yes
use if values only given	yes	yes	no	no
use for unequal subintervals	yes	no		
use for given function	no	yes	yes	yes

Table 3.2: Comparison of the integration commands in *Octave*

quadv : Numerical integration using an adaptive vectorized Simpson's rule.

quadl : Numerical integration using an adaptive Lobatto rule.

quadgk : Numerical integration using an adaptive Gauss-Konrod rule.

quadcc : Numerical integration using adaptive Clenshaw-Curtis rules.

integral : A compatibility wrapper function that will choose between quadv and quadgk depending on the integrand and options chosen.

3.1.3   Integration over domains in $\mathbb{R}^2$

Integrals over rectangular domains in Figure 3.3(a) $\Omega = [a, b] \times [c, d] \subset \mathbb{R}^2$ are computed by nested 1-D integrations

$$Q = \iint_\Omega f(x, y) \, dA = \int_a^b \left(\int_c^d f(x, y) \, dy \right) dx .$$

MATLAB/*Octave* provides commands to perform this double integration. The function $f(x, y)$ depends on two arguments, the corresponding MATLAB/*Octave* code has to accept matrices for x and y as arguments and return a matrix of the same size.

- integral2: The basic call is Q = integral2(f,a,b,c,d). Possible options have to be specified by the string with the name of the option and the corresponding value.

- quad2d: The basic call is Q = quad2d(f,a,b,c,d). Possible options have to be specified by the string with the name of the option and the corresponding value.

39 Example :   To integrate the function $f(x, y) = x^2 + y$ over the rectangular domain $1 \le x \le 2$ and $0 \le y \le 3$ use

```
quad2d(@(x,y)x.^2+y,1,2,0,3)
integral2(@(x,y)x.^2+y,1,2,0,3)
-->
ans = 11.500
ans = 11.500
```

◇

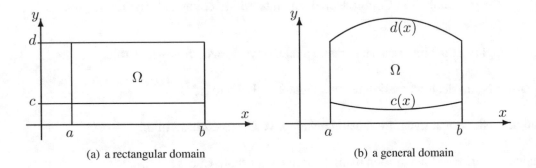

(a) a rectangular domain   (b) a general domain

Figure 3.3: Domains in the plane $\mathbb{R}^2$

Integrals over non rectangular domains in Figure 3.3(b) can be given by $a \le x \le b$ and x dependent limits for the y values $c(x) \le y \le d(x)$. The integrals are computed by nested 1-D integrations

$$Q = \iint_{\Omega} f(x,y)\, dA = \int_a^b \left(\int_{c(x)}^{d(x)} f(x,y)\, dy \right) dx\ .$$

MATLAB/*Octave* provides commands to perform this double integration. The only change is that the upper and lower limits are provided as functions of x.

Integrals over domains where the left and right limit are given as functions of y (i.e. $a(y) \le x \le b(y)$) are covered by swapping the order of integration,

$$Q = \iint_{\Omega} f(x,y)\, dA = \int_c^d \left(\int_{a(y)}^{b(y)} f(x,y)\, dx \right) dy\ .$$

40 Example :   Consider the triangular domain with corners $(0,0)$, $(1,0)$ and $(0,1)$. Thus the limits are $0 \le x \le 1$ and $0 \le y \le 1 - x$. To integrate the function

$$f(x,y) = \frac{(1+x+y)^2}{\sqrt{x^2+y^2}}$$

use

$$\iint\limits_{\Omega} f(x,y)\, dA = \int_0^1 \left(\int_0^{1-x} \frac{(1+x+y)^2}{\sqrt{x^2+y^2}}\, dy \right) dx .$$

This is readily implemented using `integral2()` or `quad2d()`.

```
fun = @(x,y) 1./( sqrt(x + y) .* (1 + x + y).^2 );
ymax = @(x) 1 - x;
Q1 = quad2d(fun,0,1,0,ymax)
Q2 = integral2(fun,0,1,0,ymax)
-->
Q1 = 0.2854
Q2 = 0.2854
```

◇

41 Example : The previous example can also be solved using polar coordinates r and ϕ. Express the area element dA in polar coordinates by

$$dA = dx \cdot dy = r \cdot dr\, d\phi .$$

For the function use

$$f_p(r, \phi) = f(x,y) = f(r\cos(\phi), r\sin(\phi)) .$$

The upper limit $y = 1 - x$ of the domain has to expressed in terms of r and ϕ by

$$y = 1 - x \quad \Longrightarrow \quad r\sin(\phi) = 1 - r\cos(\pi) \quad \Longrightarrow r_{\max}(\phi) = \frac{1}{\cos(\phi) + \sin(\phi)} .$$

Then the double integral

$$\iint\limits_{\Omega} f_p(r, \phi)\, dA = \int_0^{\pi/2} \left(\int_0^{r_{max}(\phi)} f_p(r, \phi)\, r\, dr \right) d\phi .$$

is computed, using `integral2()` or `quad2d()`.

```
polarfun = @(theta,r) fun(r.*cos(theta),r.*sin(theta)).*r;
rmax = @(theta) 1./(sin(theta) + cos(theta));
Q1 = quad2d(polarfun,0,pi/2,0,rmax)
Q2 = integral2(polarfun,0,pi/2,0,rmax)
-->
Q1 = 0.2854
Q2 = 0.2854
```

43 Remark :

- MATLAB/*Octave* provide dblquad() for integration over domains in $\mathbb{R}^2$.

- For triple integrals MATLAB/*Octave* provide the command integral3() with a syntax similar to integral2().

$\Diamond$

3.1.4   From Biot–Savart to magnetic fields

The *Octave* command for numerical integration will be used to determine the magnetic field of a circular conductor. The situation is shown in Figure 3.4. If a short segment $\vec{ds}$ of a conductor carries a current I then the contribution $d\vec{H}$ to the magnetic field is given by the law of Biot–Savart

$$d\vec{H} = \frac{I}{4\,\pi\,r^3}\;\vec{ds} \times \vec{r}$$

where $\vec{r}$ is the vector connecting the point on the conductor to the points at which the field is to be computed.

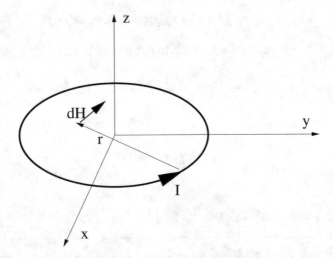

Figure 3.4: Circular conductor for which the magnetic field has to be computed

A parameterization of the circle is given by

$$\begin{pmatrix} R\,\cos\,\phi \\ R\,\sin\,\phi \\ 0 \end{pmatrix} \qquad \text{where}\quad 0 \leq \phi \leq 2\,\pi$$

leading to a line segment $d\vec{s}$ of

$$d\vec{s} = \begin{pmatrix} -R\sin\phi \\ R\cos\phi \\ 0 \end{pmatrix} d\phi \,.$$

The field $\vec{H}$ generated will have a radial symmetry and we may examine the field in the xz–plane only, i.e. at points $(x, 0, z)$. We find

$$\vec{r} = \begin{pmatrix} x \\ 0 \\ z \end{pmatrix} - \begin{pmatrix} R\cos\phi \\ R\sin\phi \\ 0 \end{pmatrix}$$

and

$$\begin{aligned} r^2 &= (R\cos\phi - x)^2 + R^2\sin^2\phi + z^2 \\ &= R^2 - 2\,x\,R\cos\phi + x^2 + z^2 \,. \end{aligned}$$

To apply Biot–Savart we need the expression

$$d\vec{s} \times \vec{r} = \begin{pmatrix} -R\sin\phi \\ R\cos\phi \\ 0 \end{pmatrix} \times \begin{pmatrix} x - R\cos\phi \\ -R\sin\phi \\ z \end{pmatrix} d\phi = \begin{pmatrix} z\,R\cos\phi \\ z\,R\sin\phi \\ R^2 - x\,R\cos\phi \end{pmatrix} d\phi$$

and thus obtain an integral for the 3 components of the field $\vec{H}$ at the point $(x, 0, z)$

$$\vec{H}(x, 0, z) = \frac{I}{4\pi} \int_0^{2\pi} \frac{1}{(R^2 - 2\,x\,R\cos\phi + x^2 + z^2)^{3/2}} \begin{pmatrix} z\,R\cos\phi \\ z\,R\sin\phi \\ R^2 - x\,R\cos\phi \end{pmatrix} d\phi \,. \tag{3.1}$$

For each of the three components of the field we find one integral to be computed.

In the following sections we examine the special situation $R = 1$ and $I = 1$.

3.1.5 Field along the central axis and the Helmholtz configuration

Along the z–axis we use $x = 0$ and the above integral simplifies to

$$\vec{H} = \frac{I}{4\pi} \int_0^{2\pi} \frac{1}{(R^2 + z^2)^{3/2}} \begin{pmatrix} z\,R\cos\phi \\ z\,R\sin\phi \\ R^2 \end{pmatrix} d\phi \,.$$

Verify that the x- and y component of $\vec{H}$ vanish, as they should because of the radial symmetry. For the z component $H_z(z)$ obtain

$$H_z(z) = \frac{I}{4\pi} \int_0^{2\pi} \frac{R^2}{(R^2 + z^2)^{3/2}} \, d\phi = \frac{I}{2} \frac{R^2}{(R^2 + z^2)^{3/2}} \, .$$

For very small and large values of z the above may be simplified to

$$H_{z \ll R} \approx \frac{I}{2} \frac{1}{R} \quad \text{and} \quad H_{z \gg R} \approx \frac{I}{2} \frac{R^2}{z^3} \, .$$

The above approximations allow to compute the field at the center of the coil and show that the field along the center axis converges to 0 like $1/z^3$.

For many applications is is important that the magnetic field should be constant over the domain to be examined. The above computations show that the field generated by a single coil is far from constant. For a Helmholtz configuration place two of the above coils, at the heights $z = \pm h$. The value of h has to be such that the field around the center is as homogeneous as possible. To examine this situation we shift one coil up by h and another coil down by $-h$ and then examine the resulting field. On the left in Figure 3.5 we find the results if the two coils are close together, on the right if they are far apart. Neither situation is optimal for a homogeneous magnetic field.

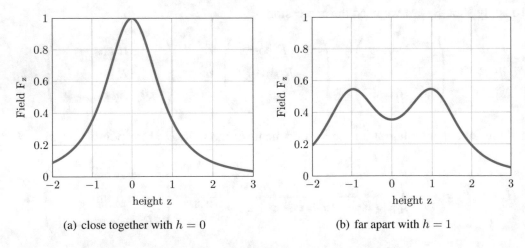

(a) close together with $h = 0$ (b) far apart with $h = 1$

Figure 3.5: Magnetic field along the central axis, using $R = 1$

Examine the field G generated by both coils and thus G is given as the sum of the two fields by the individual coils at height $z = \pm h$.

$$G(z) = H_z(z + h) + H_z(z - h) = \frac{I}{2} \left(\frac{R^2}{(R^2 + (z - h)^2)^{3/2}} + \frac{R^2}{(R^2 + (z + h)^2)^{3/2}} \right) \, .$$

The field at $z = 0$ is as homogeneous as possible if as many terms as possible in the Taylor expansion vanish.

$$G(z) \approx G(0) + \frac{dG(0)}{dz} z + \frac{1}{2} \frac{d^2 G(0)}{dz^2} z^2 + \frac{1}{6} \frac{d^3 G(0)}{dz^3} z^3 + \dots$$

Since H_z is an even function know that the first derivative is an odd function and the second derivative is an even function. Thus

$$\frac{d\,G(0)}{dz} = \frac{d}{dz}\,H(h) + \frac{d}{dz}\,H(-h) = 0$$

and

$$\frac{d^2\,G(0)}{dz^2} = \frac{d^2}{dz^2}\,H_z(h) + \frac{d^2}{dz^2}\,H_z(-h) = 2\,\frac{d^2}{dz^2}\,H_z(h)\,.$$

The optimal solution is characterized as zero of the second derivative of $H_z(z)$.

$$\frac{d^2}{dz^2}\,H_x(z) = \frac{I\,R^2}{2}\,\frac{d^2}{dz^2}\left(\frac{1}{(R^2+z^2)^{3/2}}\right) = 0$$

This leads to

$$\frac{d}{dz}\,\frac{1}{(R^2+z^2)^{3/2}} = \frac{-3\cdot 2\,z}{2\,(R^2+z^2)^{5/2}} = \frac{-3\,z}{(R^2+z^2)^{5/2}}$$

$$\frac{d^2}{dz^2}\,\frac{1}{(R^2+z^2)^{3/2}} = \frac{-3\,(R^2+z^2)^{5/2} + 3\,z\,\frac{5}{2}(R^2+z^2)^{3/2}\,2\,z}{(R^2+z^2)^5}$$

$$= \frac{-3\,(R^2+z^2) + 3\,z\,5\,z}{(R^2+z^2)^{7/2}} = 3\,\frac{4\,z^2 - R^2}{(R^2+z^2)^{7/2}}\,.$$

Thus the second derivative of $G(0)$ vanishes if $h = \pm\frac{R}{2}$. This implies the the distance between the centers of the coil should be equal to the Radius R. This is confirmed by the results in Figure 3.6. Figures 3.5 and 3.6 are generated by *Octave*/MATLAB with the help of an anonymous function `HzAxis()` to compute the field along the axis with the commands

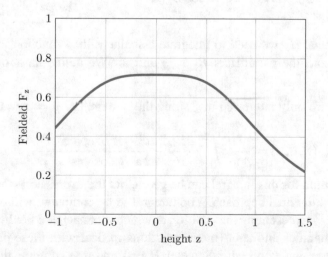

Figure 3.6: Magnetic field along the central axis, Helmholtz configuration with $h = R/2$

Demo_HzAxis.m

```
R = 1;
HzAxis = @(z)R^2/2*(R^2+z.^2).^(-3/2);
z = linspace(-2,3,101);

dz = 0.0;
figure(1); plot(z,HzAxis(z-dz)+HzAxis(z+dz))
           axis([-2,3,0,1])
           grid on; xlabel('height z'); ylabel('Field Fz');
dz = 1;
figure(2); plot(z,HzAxis(z-dz)+HzAxis(z+dz))
           axis([-2,3,0,1])
           grid on; xlabel('height z'); ylabel('Field Fz');
dz = 1/2;
figure(3); plot(z,HzAxis(z-dz)+HzAxis(z+dz))
           axis([-1,1.5,0,1])
           grid on; xlabel('height z'); ylabel('Field Fz');
```

3.1.6   Field in the plane of the conductor

In the plane $z = 0$ of the conductor the field will show a radial symmetry again, the field at the point $(x, y, 0)$ is given by a rotation of the field at the point $(\sqrt{x^2 + y^2}, 0, 0)$. Thus we compute the field along the x–axis only. Based on equation (3.1) we have to compute three integrals.

$$\vec{H}(x,0,z) = \frac{I}{4\pi} \int_0^{2\pi} \frac{1}{(R^2 - 2xR\cos\phi + x^2 + z^2)^{3/2}} \begin{pmatrix} zR\cos\phi \\ zR\sin\phi \\ R^2 - xR\cos\phi \end{pmatrix} d\phi \ .$$

For the z component H_z we need to integrate a scalar valued function, depending on the variable angle ϕ and the parameters R, x, y and z. We define an anonymous function dHz().

Currently we are only interested in $_z$ along the x–axis, i.e. $y = z = 0$ and we want to compute

$$H_z(x,0,0) = \frac{I}{4\pi} \int_0^{2\pi} \frac{R^2 - xR\cos\phi}{(R^2 - 2xR\cos\phi + x^2)^{3/2}} \, d\phi \ .$$

An analytical formula for this integral can be given, but the expression is very complicated. Thus we prefer a numerical approach. The integral to be computed will depend on the parameters x, y and R representing the position at which the magnetic field will be computed. We use function handles and anonymous functions to deal with these parameters for the integration. We examine a coil with diameter $R = 1$ and first compute the field for values $-0.5 \le x \le 0.8$ and then for $1.2 \le x \le 3$. The results are shown in Figure 3.7. Observe that the magnetic field is large if we are close to the wire at $x \approx R = 1$ and the z component changes sign outside of the circular coil. As x increases H_z converges to 0. This is confirmed by physical facts.

```
─────────────────────── Demo_Hz_xAxis.m ───────────────────────
dHz = @(al,R,x,z)R*(R-x.*cos(al))./(R^2-2*R*x.*cos(al)+x.^2+z.^2).^1.5;
x   = -0.5:0.05:0.8;    Fz = zeros(size(x));
for k = 1:length(x)
   fz    = @(al)dHz(al,1,x(k),0);    % define the anonymous function
   Fz(k) = quad(fz,0,2*pi)/(4*pi);   % integrate
end%for

figure(1);   plot(x,Fz,'b')
             grid on;   axis([-0.5 0.8 0 1.2]);
             xlabel('position x');   ylabel('Field F_z');

x2 = 1.2:0.05:3;       Fz2 = zeros(size(x2));
for k = 1:length(x2)
   fz = @(al)dHz(al,1,x2(k),0);      % define the anonymous function
   Fz2(k) = quad(fz,0,2*pi)/(4*pi);  % integrate
end%for

figure(2);   plot(x,Fz,'b',x2,Fz2,'b')
             grid on;   axis([-0.5 3 -0.6 1.2]);
             xlabel('position x');   ylabel('Field F_z');
```

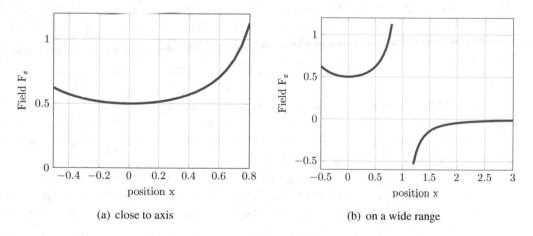

(a) close to axis   (b) on a wide range

Figure 3.7: Magnetic field along the x axis

3.1.7   Field in the xz–plane

In the xz plane we have to examine both components of the field and thus we need to compute $H_x(x,0,y)$ too. Using equation (3.1) find

$$H_x(x,0,z) = \frac{I}{4\pi} \int_0^{2\pi} \frac{z\,R\,\cos\phi}{(R^2 - 2\,x\,R\,\cos\phi + x^2 + z^2)^{3/2}}\,d\phi\,.$$

We write an anonymous function dHx() for the expression to be integrated. Then we proceed as follows:

1. Choose the domain of $-0.4 \leq x \leq 0.7$ and $-1 \leq z \leq 2$. Generate vectors with the values of x and z for which the field $\vec{H}$ will be computed.

2. Generate a mesh of points with the command meshgrid().

3. Create the empty matrices Hx and Hz for the components of the field.

4. Use a for loop to fill in the values of both components of $\vec{H}$, using the integrals based on (3.1).

VectorFields.m

```
dHz=@(al,R,x,z)R*(R-x.*cos(al))./(R^2-2*R*x.*cos(al)+x.^2 +z.^2).^1.5;
dHx=@(phi,R,x,z)R*z.*cos(phi)./ (R^2-2*R*x.*cos(phi)+x.^2 +z.^2).^1.5;

z = -1:0.2:2; x = -0.4:0.1:0.7;
[xx,zz] = meshgrid(x,z);   x = xx(:); z = zz(:); % convert to vectors
Hx = zeros(size(x)); Hz = Hx;

for k = 1:length(x)
   fx    = @(al)dHx(al,1,x(k),z(k));   % define the anonymous function
   Hx(k) = quad(fx,0,2*pi)/(4*pi);     % integrate
   fz    = @(al)dHz(al,1,x(k),z(k));   % define the anonymous function
   Hz(k) = quad(fz,0,2*pi)/(4*pi);     % integrate
end%for
```

The next step is to visualize the vector field $\vec{H}$ in the xz plane in the center of the circular coil. To arrive at Figure 3.8 we use the command quiver() to display the vector field. Depending on which version of *Octave*/MATLAB you are using, you might have to adapt the scaling factors in the quiver() command.

In the left part of Figure 3.8 we find magnetic fields of drastically different sizes, in particular close to the conductor. For a good visualization it is often useful to normalize all vectors to have equal length. Find the result of the code below in the right part of Figure 3.8.

```
subplot(1,2,1)
quiver(x,z,Hx,Hz,1.5)
grid on; axis([-0.4 0.8 -1 2])

scal = 1./sqrt(Hx(:).^2+Hz(:).^2);  % length of each vector
Hx = scal.*Hx; Hz = scal.*Hz;       % normalize length

subplot(1,2,2)
quiver(x,z,Hx,Hz,0.6)
grid on; axis([-0.4 0.8 -1 2])
```

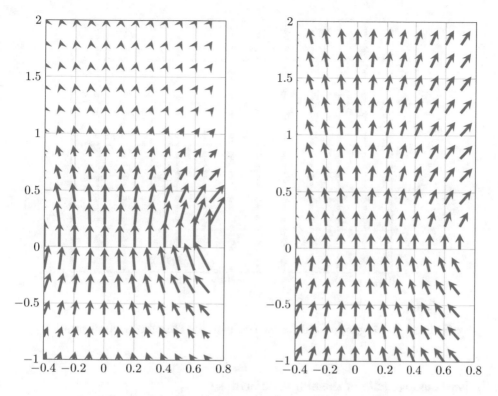

Figure 3.8: Magnetic vector field in a plane, actual length and normalized

3.1.8 The Helmholtz configuration

To obtain a homogeneous field in the center of two coils we have to place them at a distance equal to the radii of the circular coils. We can compute and visualize the field of this configuration using the same ideas and codes as in the section above. Find the result in Figure 3.9.

─────── **VectorFieldsHelmholtz.m** ───────

```
x = -0.3:0.05:0.3; z = x;
h = 0.5; % optimal distance for Helmholtz configuration
[xx,zz] = meshgrid(x,z);    x = xx(:); z = zz(:);
Hx = zeros(size(x)); Hz = Hx;
for k = 1:length(x)
   fx    = @(al)(dHx(al,1,x(k),z(k)-h)+dHx(al,1,x(k),z(k)+h));
   Hx(k) = quad(fx,0,2*pi)/(4*pi);
   fz    = @(al)(dHz(al,1,x(k),z(k)-h)+dHz(al,1,x(k),z(k)+h));
   Hz(k) = quad(fz,0,2*pi)/(4*pi);
end%for

quiver(x,z,Hx,Hz,0.6) % new scaling
grid on; xlabel('x'); ylabel('z'); axis([-0.3 0.3 -0.3 0.3])
```

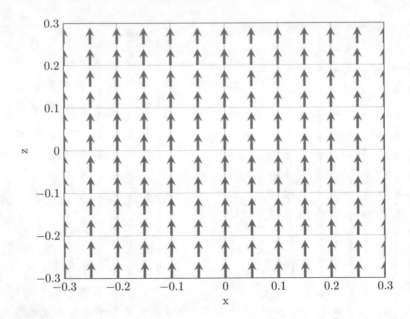

Figure 3.9: Magnetic field for two coils in the Helmholtz configuration

Analysis of homogeneity of the magnetic field

The main purpose of the Helmholtz configuration is to provide a homogeneous field at the center of the coils. Thus we want the all components of the field to be constant. It is not obvious how to quantify the deviation from a constant magnetic field. We suggest three options:

1. Variation in z component only
   The z component H_z is clearly the largest component. Thus we might examine the variations of H_z only. One possible method is to generate level curves for the relative deviation.

$$\text{relative deviation in } H_z \quad = \left| \frac{H_z(x,y,z) - H_z(0,0,0)}{H_z(0,0,0)} \right|$$

2. Variation in all components
   We might also take the other components into account and examine the deviation vector

$$\vec{H}(x,y,z) - \vec{H}(0,0,0) = \begin{pmatrix} H_x(x,y,z) \\ H_y(x,y,z) \\ H_z(x,y,z) \end{pmatrix} - \begin{pmatrix} 0 \\ 0 \\ H_z(0,0,0) \end{pmatrix}$$

and then generate level curves for the relative deviation

$$\text{relative deviation in } \vec{H} \quad = \frac{\|\vec{H}(x,y,z) - \vec{H}_z(0,0,0)\|}{H_z(0,0,0)} .$$

3. Variation in the strength of the magnetic field
   If only the strength $\|\vec{H}\|$ matters and the direction of the magnetic field is irrelevant we might examine level curves for

$$\text{relative deviation in strength } \|\vec{H}\| \quad = \quad \left| \frac{\|\vec{H}(x,y,z)\| - H_z(0,0,0)}{H_z(0,0,0)} \right| .$$

The above ideas can be implemented in *Octave*, leading to the result in Figure 3.10 for the relative deviation in H_z. The deviation is examined in the xz–plane, i.e. for $y = 0$. The result in Figure 3.10 for the relative deviation shows that the relative deviation is rather small at the center $(x, z) \approx (0, 0)$ but increases shaply at the corners of the examined domain of $-0.3 \le x \le 0.3$ and $-0.3 \le z \le 0.3$.

HelmholtzSurface.m

```
n   = sqrt(length(Hz))
HzM = reshape(Hz,n,n);
Hz0 = HzM(floor(n/2)+1,floor(n/2)+1)
reldev = abs(HzM-Hz0)/Hz0;
mesh(xx,zz,reldev)
xlabel('x'); ylabel('z')
```

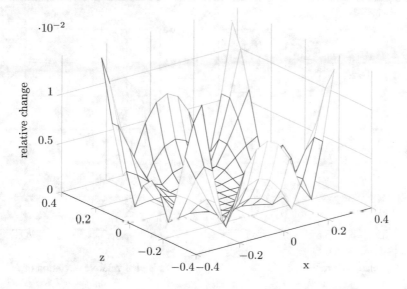

Figure 3.10: Relative changes of H_z in the plane $y = 0$

Level curves for relative deviations

To estimate the domain for which the relative deviation is small use the tool of level curves. In Figure 3.11(a) find the level curves for relative deviation in H_z at levels 0.001, 0.002,

0.005 and 0.01 . Observe that the deviation remains small along a few lines, even if we move away from the center. This is confirmed by Figure 3.10. Use a rather fine mesh to obtain smoother curves. One might have to pay attention to the computation time. If a grid on $n \times n$ points is examined, then $2\,n^2$ integrals have to be computed. For the sample code below this translates to $2 \cdot 51^2 = 5202$ integrals.

HelmholtzContours.m

```
n = 51; x = linspace(-0.3,0.3,n); z = x;
h = 0.5; % optimal distance for Helmholtz configuration
[xx,zz] = meshgrid(x,z);        x = xx(:); z = zz(:);
Hx = zeros(size(x)); Hz = Hx;
for k = 1:length(x)
   fx     = @(al)(dHx(al,1,x(k),z(k)-h)+dHx(al,1,x(k),z(k)+h));
   Hx(k)  = quad(fx,0,2*pi)/(4*pi);
   fz     = @(al)(dHz(al,1,x(k),z(k)-h)+dHz(al,1,x(k),z(k)+h));
   Hz(k)  = quad(fz,0,2*pi)/(4*pi);
end%for
HzM = reshape(Hz,n,n);
Hz0 = HzM(floor(n/2)+1,floor(n/2)+1)
reldev = abs(HzM-Hz0)/Hz0;
contour(xx,zz,reldev,[0.001,0.002,0.005,0.01])
axis('equal'); grid on ;  xlabel('x'); ylabel('z');
```

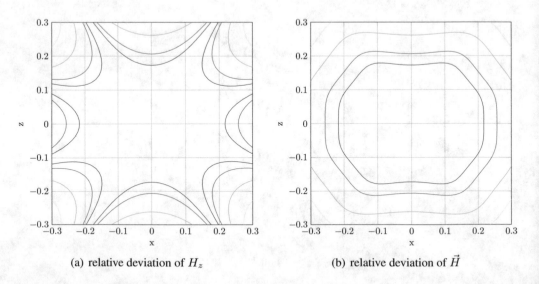

(a) relative deviation of H_z (b) relative deviation of $\vec{H}$

Figure 3.11: Level curves for the relative changes of H_z and $\vec{H}$ at levels 0.001, 0.002, 0.005 and 0.01

It might be a better approach to examine the relative deviation in $\vec{H}$ and not the z component only. Find the code below and the resulting Figure 3.11(b). Observe that the relative deviation of $\vec{H}$ increases uniformly as we move away from the center. For Helmholtz coils with radius $R = 1$ find that in an approximate cylinder with radius 0.2 and height 0.4 the relative deviation in the magnetic field is smaller than 0.001, thus the field is rather homogeneous. This is the main advantage of the Helmholtz configuration.

```
HxM = reshape(Hx,n,n);
HDev = sqrt(HxM.^2 + (HzM-Hz0).^2)/Hz0;
contour(xx,zz,HDev,[0.001,0.002,0.005,0.01])
axis('equal');grid on
xlabel('x'); ylabel('z');
```

3.1.9 List of codes and data files

In the previous section the codes and data files in Table 3.3 were used.

filename	function
`simpson.m`	function file, implementing Simpson's algorithm
`Demo_Hz_zAxis.m`	script for z–component of field along z–axis
`Demo_Hz_xAxis.m`	script for z–component of field along x–axis
`VectorFields.m`	script for vector field in xz–plane
`VectorFieldsHelmholtz.m`	script for vector field of Helmholtz configuration
`HelmholtzSurface.m`	script for deviation for Helmholtz configuration
`HelmholtzContours.m`	script for level curves of Helmholtz configuration

Table 3.3: Codes and data files for section 3.1

3.2 Linear and Nonlinear Regression

One of the most common engineering problems is to find optimal parameters for your model to be as close as possible to some measured data. This very often leads to a regression problem and there is a vast literature (e.g. [Stah99],[MontPeckVini12],[Hock05],[Bevi69]) and many pieces of code are available. Obviously *Octave* and MATLAB also provides a set of tools for this task. This section shall serve as an introduction on when and how to use those codes. The structure of the section is as follows:

- First we show the example of a straight line regression, the most common case. Only basic *Octave* commands are used. See Section 3.2.1.

- Then a generally applicable matrix notation is introduced to examine all types of linear regression problems. See Section 3.2.2.

- Then we examine the variance (accuracy) of the parameters to be determined. This aspect is often not given the deserved attention by engineers. See Section 3.2.3.

- Using a real example we illustrate how the basis ideas might lead to serious problems (and wrong results), as is the case for many real world problems. We point out how to avoid or eliminate those problems. Some of the mathematical background (QR factorization) is given. See Sections 3.2.6 and 3.2.7.

- Then we present some information on weighted regression problems and the resulting algorithm. See Section 3.2.8.

- All of the above will lead to the code in `LinearRegression.m`, see Section 3.2.10. This code is part of the optimization package of *Octave*. The code is also available with the source codes for these notes.

- Then all part of the above puzzle will be used to examine four real world, non obvious applications of linear regression.

  1. The performance of a magnetic motor is examined as function of two parameters. In this example we use a linear regression with two independent variables, i.e. we seek a function of two variables. See Section 3.2.11

  2. Using two acceleration sensors one can design an orientation sensor. To calibrate this sensor we use linear regression. See Section 3.2.12.

  3. With an AFM microscope the surface of ball bearing can be examined. With linear regression we determine the exact shape of the bills. See Section 3.2.13.

  4. Using linear regression with a piecewise linear function we examine a system consisting of two springs. In this example the regression is combined with a nonlinear optimization problem. See Section 3.2.14.

- In Section 3.2.15 the commands `leasqr()` and `fsolve()` are used to solve nonlinear regression problems, illustrated by simple examples. This is the applied to a real problem in Sections 3.2.16 and 3.2.17. In Section 3.2.19 another application is examined. The importance of obtaining good starting values is illustrated.

For most of the sample code you need the optimization package loaded in *Octave*. Use the command `pkg list` to display the installed packages, the one marked by a star $*$ are currently loaded. To load the optimization package use the command `pkg load optim`.

3.2.1 Linear regression for a straight line

For n given points (x_i, y_i) in a plane we try to determine a straight line $y(x) = p_1 \cdot 1 + p_2 \cdot x$ to match those points as good as possible. One good option is to examine the residuals

$r_i = p_1 \cdot 1 + p_2 \cdot x_i - y_i$. Using matrix notation we find

$$\vec{r} = \mathbf{F} \cdot \vec{p} - \vec{y} = \begin{bmatrix} 1 & x_1 \\ 1 & x_2 \\ 1 & x_3 \\ \vdots & \vdots \\ 1 & x_n \end{bmatrix} \cdot \begin{pmatrix} p_1 \\ p_2 \end{pmatrix} - \begin{pmatrix} y_1 \\ y_2 \\ y_3 \\ \vdots \\ y_n \end{pmatrix}.$$

Linear regression corresponds to minimization of the norm of $\vec{r}$, i.e. minimize

$$\|\vec{r}\|^2 = \|\mathbf{F} \cdot \vec{p} - \vec{y}\|^2 = \langle \mathbf{F} \cdot \vec{p} - \vec{y}, \, \mathbf{F} \cdot \vec{p} - \vec{y} \rangle.$$

Consider $\|\vec{r}\|^2$ as a function of p_1 and p_2. At the minimal point the two partial derivatives have to vanish. This leads to a system of linear equations for the vector $\vec{p}$.

$$\mathbf{X} \cdot \vec{p} = \left(\mathbf{F}^T \cdot \mathbf{F} \right) \cdot \vec{p} = \mathbf{F}^T \cdot \vec{y}.$$

This can easily be implemented in *Octave*, leading to the result in Figure 3.12(a) and a residual of $\|\vec{r}\|^2 \approx 1.23$.

```
x = [0; 1; 2; 3.5; 4]; y = [-0.5; 1; 2.4; 2.0; 3.1];

F = [ones(size(x)) x]
p = (F'*F)\(F'*y)
residual = norm(F*p-y)

xn = [-1 5]; yn = p(1)+p(2)*xn;
plot(x,y,'*r',xn,yn);
xlabel('independent variable x'); ylabel('dependent variable y')
```

3.2.2 General linear regression, matrix notation

The above idea carries over to a linear combination of functions $f_j(x)$ for $1 \le j \le m$. For a vector $\vec{x} = (x_1, x_2, \ldots, x_k)^T$ we examine a function of the form

$$f(\vec{x}) = \sum_{j=1}^{m} p_j \cdot f_j(\vec{x}).$$

The optimal values of the parameter vector $\vec{p} = (p_1, p_2, \ldots, p_m)^T$ have to be determined. Thus we try to minimize the expression

$$\chi^2 = \|\vec{r}\|^2 = \sum_{i=1}^{n} (f(x_i) - y_i)^2 = \sum_{i=1}^{n} \left(\left(\sum_{j=1}^{m} p_j \cdot f_j(x_i) \right) - y_i \right)^2$$

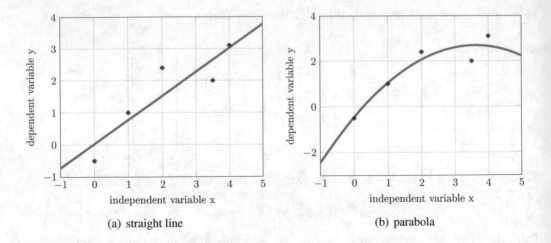

(a) straight line   (b) parabola

Figure 3.12: Regression of a straight line and a parabola to the same data points

Using a vector and matrix notation this can be written in the form

$$\vec{p} = \begin{pmatrix} p_1 \\ p_2 \\ \vdots \\ p_m \end{pmatrix} \quad , \quad \vec{y} = \begin{pmatrix} y_1 \\ y_2 \\ \vdots \\ y_n \end{pmatrix} \quad , \quad \mathbf{F} = \begin{bmatrix} f_1(x_1) & f_2(x_1) & \dots & f_m(x_1) \\ f_1(x_2) & f_2(x_2) & \dots & f_m(x_2) \\ \vdots & \vdots & \ddots & \vdots \\ f_1(x_n) & f_2(x_n) & \dots & f_m(x_n) \end{bmatrix}$$

we have to minimize the expression

$$\|\vec{r}\|^2 = \|\mathbf{F} \cdot \vec{p} - \vec{y}\|^2 = \langle \mathbf{F} \cdot \vec{p} - \vec{y} \,, \mathbf{F} \cdot \vec{p} - \vec{y} \rangle \,,$$

leading again to the necessary condition

$$\mathbf{X} \cdot \vec{p} = \left(\mathbf{F}^T \cdot \mathbf{F} \right) \cdot \vec{p} = \mathbf{F}^T \cdot \vec{y} \,.$$

This system of n linear equations for the unknown vector of parameters $\vec{p} \in \mathbb{R}^m$ is called a **normal equation**. Once we have the optimal parameter vector $\vec{p}$, compute the values of the regression curve with a matrix multiplication.

$$(\mathbf{F} \cdot \vec{p})_i = \sum_{j=1}^{m} p_j \cdot f_j (x_i)$$

As an example fit a parabola

$$y = p_1 \cdot 1 + p_2 \cdot x + p_3 \cdot x^2$$

through the points given in the above example. For this example the matrix $\mathbf{F}$ is given by

$$\mathbf{F} = \begin{bmatrix} 1 & x_1 & x_1^2 \\ 1 & x_2 & x_2^2 \\ 1 & x_3 & x_3^2 \\ 1 & x_4 & x_4^2 \\ 1 & x_5 & x_5^2 \end{bmatrix} = \begin{bmatrix} 1 & 0 & 0 \\ 1 & 1 & 1 \\ 1 & 2 & 2^2 \\ 1 & 3.5 & 3.5^2 \\ 1 & 4 & 4^2 \end{bmatrix}.$$

This can be coded, leading to the result in Figure 3.12(b) and a residual of $\|\vec{r}\|^2 \approx 0.89$. This residual is, as expected, smaller than the residual for a straight line fit.

```
x = [0; 1; 2; 3.5; 4];   y = [-0.5; 1; 2.4; 2.0; 3.1];

F = [ones(length(x),1) x x.^2]
p = (F'*F)\(F'*y)
residual = norm(F*p-y)


xn = [-1:0.1:5]';
yn = p(1) + p(2)*xn + p(3)*xn.^2;
plot(x,y,'*r',xn,yn)
xlabel('independent variable x'); ylabel('dependent variable y')
```

3.2.3 Estimation of the variance of parameters, confidence intervals

Using the above results (for the parabola fit) we can determine the residual vector

$$\vec{r} = \mathbf{F} \cdot \vec{p} - \vec{y}$$

and then the mean and variance $V = \sigma^2$ of the y–errors can be estimated. The estimation is valid if all y–errors are assumed to be of equal size, i.e. we assume a-priori that the errors are given by a normal distribution.

```
residual = F*p-y;
mean(residual)
sum(residual.^2)/(length(residual)-3)   % 3 parameters in parabola
```

The mean should equal zero and the standard deviation $\sigma \approx \sqrt{0.39} \approx 0.63$ is an estimate for the errors in the y–values. Smaller values of σ indicate that the values of y are closer to the regression curve.

In most applications the values of the parameters p_j contain the essential information. It is often important to know how reliable the obtained results are, i.e. we want to know the variance of the determined parameter values p_j. To this end consider the normal equation

$$\left(\mathbf{F}^T \cdot \mathbf{F}\right) \cdot \vec{p} = \mathbf{F}^T \cdot \vec{y}$$

and thus the explicit expression for $\vec{p}$

$$\vec{p} = \left(\mathbf{F}^T \cdot \mathbf{F}\right)^{-1} \cdot \mathbf{F}^T \cdot \vec{y} = \mathbf{M} \cdot \vec{y} \tag{3.2}$$

or

$$p_j = \sum_{i=1}^{n} m_{j,i}\, y_i \quad \text{for} \quad 1 \leq j \leq m$$

where

$$\mathbf{M} = [m_{j,i}]_{1 \leq j \leq m, 1 \leq i \leq n} = \left(\mathbf{F}^T \cdot \mathbf{F}\right)^{-1} \cdot \mathbf{F}^T$$

is a $m \times n$–matrix, where $m < n$ (more columns than rows).

This explicit representation of p_j allows[1] to compute the variance $\mathrm{var}(p_j)$ of the parameters p_j, using the estimated variance σ^2 of the y–values. The result is given by

$$\mathrm{var}(p_j) = \sum_{i=1}^{n} m_{j,i}^2\, \sigma^2 \quad \text{where} \quad \sigma^2 = \frac{1}{n-m} \sum_{i=1}^{n} r_i^2$$

Once we know the standard deviation and assume a normal distribution one can readily[2] determine the 95% **confidence interval** for the parameters, i.e. with a probability of 95% the actual value of the parameter is between $p_i - 1.96\,\sqrt{\mathrm{var}_i}$ and $p_i + 1.96\,\sqrt{\mathrm{var}_i}$.

All the above computations can be packed in a function file `LinearRegression.m`[3] to compute the optimal values of the parameters and the estimated variances.

```
function [p,e_var,r,p_var] = LinearRegression1(F,y)
%% temporary code, DO NOT USE FOR PRODUCTION
p = (F'*F)\(F'*y);          % estimate the values of the parameters
residual = F*p-y;           % compute the residual vector
r = norm(residual);         % and its norm
e_var = sum(residual.^2)/(rF-cF);  % variance of the y-errors

M = inv(F'*F)*F';
M = M.*M;                   % square each entry in the matrix M
p_var = sum(M,2)*e_var;     % variance of the parameters
```

---

[1] If z_k are **independent** random variables given by a normal distribution with variances $\mathrm{var}(z_k)$, then a linear combination of the z_i also leads to a normal distribution. The variances are given by the following rules:

$$\begin{aligned}
\mathrm{var}(z_1 + z_2) &= \mathrm{var}(z_1) + \mathrm{var}(z_2) \\
\mathrm{var}(\alpha_1 z_1) &= \alpha_1^2\, \mathrm{var}(z_1) \\
\mathrm{var}\left(\sum_i \alpha_i z_i\right) &= \sum_i \alpha_i^2\, \mathrm{var}(z_i)
\end{aligned}$$

[2] Use

$$\int_{-1.96\,\sigma}^{+1.96\,\sigma} \frac{1}{\sqrt{2\pi}} \exp(-x^2/(2\sigma^2))\, dx \approx 0.95$$

This value can be computed by `fsolve(@(x)normcdf(x)-normcdf(-x)-0.95,2)`.

[3] A better implementation is shown in Figure 3.20 on page 244, thus we use a temporary function name.

The function `LinearRegression()` now allows to solve the straight line problem leading to Figure 3.12(a) with only a few lines of code.

```
x = [0; 1; 2; 3.5; 4];   y = [-0.5; 1; 2.4; 2.0; 3.1];
F = [ones(length(x),1) x];
[p,y_v,r,p_v] = LinearRegression(F,y)
sigma = sqrt(p_v)
alpha = 0.05;
p95 = p + norminv(1-alpha/2)*[-sigma +sigma]
```

The result implies that the equation for the optimal straight line is

$$y = 0.025 + 0.75 \cdot x$$

where the constant contribution (0.025) has a standard deviation of 0.55 and the standard deviation of the slope (0.75) is given by 0.21. Thus with a probability of 95% we find for the parameters in $y(x) = \alpha + \beta x$

$$-1.05 = 0.025 - 1.96 \cdot 0.55 < \alpha < 0.025 + 1.96 \cdot 0.55 = +1.10$$
$$+0.33 = 0.75 - 1.96 \cdot 0.21 < \beta < 0.75 + 1.96 \cdot 0.21 = +1.17$$

Thus the tolerance for the parameters α and β is this example is huge. This information should prevent you from showing too many digits when analyzing measured data.

The above assumes that the distribution of the parameters are normal distributions. Actually the distribution to use is a Student's t-distribution with $n - 2$ degrees of freedom and the code should be modified to

```
p95 = p + tinv(1-alpha/2,length(x)-2)*[-sigma  +sigma]
```

leading to

$$-1.720517 < \alpha < 1.770517$$
$$0.073118 < \beta < 1.426882$$

If we would have many data points (not only 5 as in the above example) the normal distribution and the Student's t- distribution differ very little. Then the estimated confidence intervals will differ very little.

To fit a parabola through the same points replace one line of code by

```
F = [ones(length(x),1) x x.^2];
```

3.2.4 Estimation of variance of the dependent variable

For the regression of a straight line $y(x) = \alpha + \beta x$ we can also estimate the expected variance of the y–values. We assume that all measured values y_i of y share a common standard deviation of σ. We use the notations $S_x = \sum_i x_i$, $S_{xx} = \sum_i x_i^2$ and $S_{xy} = \sum_i x_i y_i$ and the explicit formulas

$$\Delta = n \cdot S_{xx} - S_x^2 \;\;,\;\; \alpha = \frac{1}{\Delta}\left(S_{xx}S_y - S_x S_{xy}\right) \;\;,\;\; \beta = \frac{1}{\Delta}\left(n S_{xy} S_y - S_x S_y\right).$$

Thus use[4]

$$
\begin{aligned}
y &= \alpha + \beta x = \frac{1}{\Delta}\left(S_{xx}S_y - S_xS_{xy}\right) + \frac{1}{\Delta}\left(n\cdot S_{xy} - S_xS_y\right) x \\
&= \frac{1}{\Delta}\sum_i \left(S_{xx} - S_x x_i + n x x_i - x S_x\right) y_i
\end{aligned}
$$

and the computational rules for variances lead[5] to

$$
\begin{aligned}
V(y) &= \frac{1}{\Delta^2}\sum_i \left(S_{xx} - S_x x_i + n x x_i - x S_x\right)^2 \sigma^2 \\
&= \text{elementary, tedious algebra} \\
&= \frac{\sigma^2}{\Delta}\left(n\,(x - \frac{S_x}{n})^2 + \frac{1}{n}\left(n S_{xx} - S_x^2\right)\right) = \sigma^2\left(\frac{1}{n} + \frac{n}{\Delta}(x - \bar{x})^2\right).
\end{aligned}
$$

This is the variance of the computed y values on the straight line. Since a new measurement adds another contribution to the variance we obtain a width of the confidence band by

$$\left(1 + \frac{1}{n} + \frac{n}{\Delta}(x - \bar{x})^2\right)^{1/2}\sigma.$$

The above formula is correct when fitting a straight line through given data points. For general linear regressions we have to use the general formulas. Let $\vec{y}_m$ denote the set of measured y values and $\vec{y}_p$ the vector of predicted y values, using the result of a linear regression. Based on equation (3.2) and $\vec{y}_p = \mathbf{F}\cdot\vec{p}$ we find

$$\vec{y}_p = \mathbf{F}\cdot(\mathbf{F}^T\cdot\mathbf{F})^{-1}\cdot\mathbf{F}^T\,\vec{y}_m = \mathbf{B}\,\vec{y}_m. \tag{3.3}$$

Estimate the variance of the straight line

Based on this explicit formula (3.3) for the values of y_i we can estimate the variances of the components of the computed values of y_i.

$$V(y_i) = \sum_{k=1}^{n} b_{i,k}^2\,\sigma^2.$$

This is implemented in the function `LinearRegression()` and the values are returned in the variable `fit_var`. With this expression one can determine the confidence band for the straight line.

---

[4]Observe that the simple formula $y = \alpha + \beta x$ does **not** lead to $V(y) = V(\alpha) + V(\beta)x^2$ since the two parameters α and β are not statistically independent.

[5]Ask this author for a printout with all the glorious details shown.

Estimate the variance of a future measurement

If we want to predict the variance of a new data point at x_i there are two contributions to the variance: the independent variance of the measurement and the variance from the straight line. Since the events are assumed to be independent, we have to add the variances, leading to

$$\sigma^2 + V(y_i) = \sigma^2 + \sum_{k=1}^{n} b_{i,k}^2 \sigma^2 .$$

With `LinearRegression()` this can be computed by `e_var+fit_var` and used in the code below, leading to Figure 3.13(a). There find the raw data, the best fitting straight line and the 95% confidence band for straight line and possible new values. The confidence band can be displayed by multiplying the square root of the estimated variance by 1.96 and adding/subtracting this from the fitted value. Observe that the confidence band for the straight line is considerably narrower than the one for the expected new values of y.

```
n = 100; x = sort(rand(n,1)*5-1);   y = 1+0.05*x + 0.1*randn(size(x));
F = [ones(n,1),x(:)];  % straight line regression
[p,e_var,r,p_var,y_var] = LinearRegression(F,y);
yFit = F*p;
fac = fsolve(@(x)normcdf(x)-normcdf(-x)-0.95,2); % 95% level, fac=1.96

figure(1);
plot(x,y,'+b',x,yFit,'-g',...
x,yFit+fac*sqrt(y_var),'--r',x,yFit+fac*sqrt(e_var+y_var),'--k',...
x,yFit-fac*sqrt(y_var),'--r',x,yFit-fac*sqrt(e_var+y_var),'--k')
title('straight line by linear regression'); grid on
legend('data','fit','+/-95% line','+/-95% data',...
       'location','northwest');
```

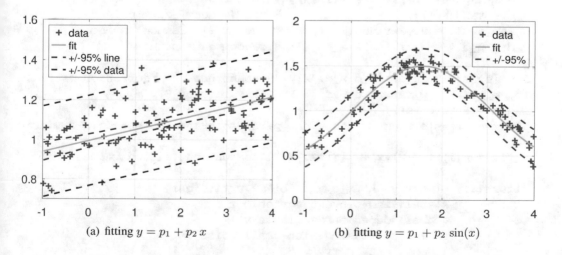

(a) fitting $y = p_1 + p_2 x$

(b) fitting $y = p_1 + p_2 \sin(x)$

Figure 3.13: Regressions with the fitted data and the 95% confidence bands

The above result is not restricted to straight line regressions, e.g. we may fit a curve $y(x) = p_1 + p_2 \sin(x)$ through a set of data points and display similar results in Figure 3.13(b).

```
n = 100; x = sort(rand(n,1)*5-1); y = 1+0.5*sin(x)+0.1*randn(size(x));
F = [ones(n,1),sin(x(:))];
[p,e_var,r,p_var,y_var] = LinearRegression(F,y);
yFit = F*p;
figure(2); plot(x,y,'+b',x,yFit,'-g',...
               x,yFit+1.96*sqrt(e_var+y_var),'--r',...
               x,yFit-1.96*sqrt(e_var+y_var),'--r')
          legend('data','fit','+/-95%'); grid on
```

3.2.5 An elementary example

As an illustrative example examine a true curve $y = x - x^2$ for $0 \le x \le 1$. Then some normally distributed noise is added, leading to data points (x_i, y_i) and shown in Figure 3.14. Using the command LinearRegression() a parabola $y(x) = p_1 1 + p_2 x + p_3 x^2$ is fitted to the generated values y_i. The result are the estimated values for the parameters p_i and the estimated standard deviation.

Finding the optimal parameters, their standard deviation and the confidence intervals

On the first few lines in the code below the data is generated, using a normally distributed noise contribution.. Then LinearRegression() is applied to determine the solutions.

```
N = 20 ;                          % number of data points
x = linspace(0,1); y = x.*(1-x);  % the "true" data
x_d = rand(N,1) ;                 % the (random) data points
noise_small = 0.02*randn(N,1);    % the small (random) noise
y_d = x_d.*(1-x_d) + noise_small; % random data points

F = [ones(size(x_d)) x_d x_d.^2]; % construct the regression matrix
[p, e_var, r, p_var, fit_var] = LinearRegression(F,y_d);
sigma = sqrt(p_var);
parameters = [p sigma] % show the parameters and the standard deviation

y_fit = p(1) + p(2)*x + p(3)*x.^2; % compute the fitted values

figure(1); plot(x,y,'k',x_d,y_d,'b+',x,y_fit,'g')
          xlabel('independent variable x')
          ylabel('dependent variable y')
          title('regression with small noise')
          legend('true curve','data','best fit')
-->
parameters =   -0.00066395   0.00805592
```

```
                    +0.98982013    0.04322260
                    -0.99663727    0.04360942
```

Using the standard deviation of the parameters one can then determine the confidence intervals for a chosen level of significance $\alpha = 5\% = 0.05$.

```
alpha = 0.05 ; % level of significance
p95_n = p + norminv(1-alpha/2)*[-sigma +sigma]    % normal distribution
% Student-t distribution
p95_t = p + tinv(1-alpha/2,length(x_d)-3)*[-sigma +sigma]
-->
p95_n =   -0.016453    0.015125
           0.905105    1.074535
          -1.082110   -0.911164

p95_t =   -0.017660    0.016333
           0.898628    1.081012
          -1.088645   -0.904629
```

The numerical result for the Student-t distribution implies that with a level of confidence of 95% the parameters p_i in $y = p_1\,1 + p_2\,x + p_3\,x^2$ satisfy

$$-0.018 \;\leq\; p_1 \;\leq\; +0.016$$
$$+0.90 \;\leq\; p_2 \;\leq\; +1.08$$
$$-1.09 \;\leq\; p_3 \;\leq\; -0.90 \,.$$

This confirms the "exact" values of $p_1 = 0$, $p_2 = +1$ and $p_3 = -1$. The best fit parabola in Figure 3.14(a) is rather close to the "true" parabola.

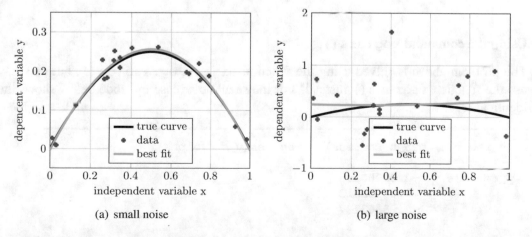

(a) small noise

(b) large noise

Figure 3.14: Linear regression for a parabola, with a small or a large noise contribution

The effect of large noise contributions

The above can be repeated with a considerably large noise contribution.

```
noise_big  = 0.5*randn(N,1);  % the large (random) noise
...
parameters = [p sigma]
...
p95 = p + norminv(1-alpha/2)*[-sigma +sigma]
p95 = p + tinv(1-alpha/2,length(x_d)-3)*[-sigma +sigma]
-->
parameters =  -0.23680    0.25600
               3.54557    1.37354
              -3.74329    1.38583
p95_n = -0.73856    0.26496
         0.85348    6.23766
        -6.45947   -1.02710
p95_t = -0.77692    0.30332
         0.64765    6.44348
        -6.66714   -0.81943
```

These numbers imply

$$p_1 \approx -0.24 \quad , \quad p_2 \approx 3.55 \quad \text{and} \quad p_3 \approx -3.74 .$$

This is obviously far from the "true" values and confirmed by the very wide confidence intervals.

$$-0.78 \leq p_1 \leq +0.30$$
$$0.65 \leq p_2 \leq +6.44$$
$$-6.67 \leq p_3 \leq -0.82$$

The best fit parabola in Figure 3.14(b) is poor approximation of the "true" parabola.

Using the command regress()

The problem above is solved using the function LinearRegression(), but one may use the function regress() instead. For the small noise case the code below shows the identical results.

```
[p_regress, p95_t_regress] = regress(y_d,F,alpha)
-->
p_regress =  -0.00066395
              0.98982013
             -0.99663727

p95_t_regress =  -0.017660    0.016333
                  0.898628    1.081012
                 -1.088645   -0.904629
```

and for the large noise case obtain again

```
[p_regress, p95_t_regress] = regress(y_d,F,alpha)
-->
p_regress =   -0.23680
               3.54557
              -3.74329

p95_t_regress =  -0.77692    0.30332
                  0.64765    6.44348
                 -6.66714   -0.81943
```

Improving the confidence interval by using more data points

In the above Figure 3.14(b) the size of the noise was large and as a consequence the results for the parameters p_i was rather unreliable, i.e. a large standard variation and wide confidence intervals. This can be improved by using more data points, assuming the the noise is given by a normal distribution. By multiplying the number of data points by 100 (N $\rightarrow$ 100*N) theory predicts that the stanadard devition σ should be $\sqrt{100} = 10$ times smaller. The results confirm this expectation.

```
noise_big   = 0.5*randn(100*N,1);
x_d = rand(100*N,1) ; y_d = x_d.*(1-x_d) + noise_big;

F = [ones(size(x_d)) x_d x_d.^2];  % construct the regression matrix
[p, e_var, r, p_var, fit_var] = LinearRegression(F,y_d);
sigma = sqrt(p_var);
parameters = [p sigma]  % show the parameters and standard deviations
% Student-t distribution
p95_t=p+tinv(1-alpha/2,length(x_d)-3)*[-sigma +sigma]
-->
parameters =   -0.0086227    0.0334205
                1.1139283    0.1555626
               -1.1457702    0.1514078

p95_t =   -0.074165    0.056920
           0.808846    1.419010
          -1.442704   -0.848836
```

Using more data points will increase the reliability of the result, but only if the residuals are normally distributed. If you are fitting a straight line to data on a parabola the results can not improve. Thus one has to make a visual check of the fitting (Figure 3.15(a)), to realize the the result is impossibly correct, even if the standard deviations of the parameters p_i are small. The problem is also visible in a histogram of the residuals (Figure 3.15(b)). Obviously the distribution of the residuals is far from a normal distribution, and thus the essential assumption is violated.

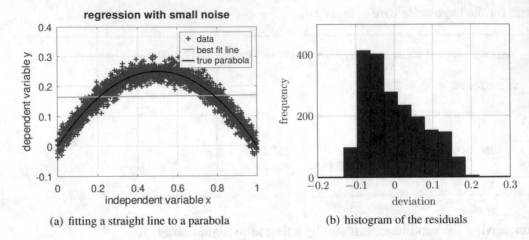

(a) fitting a straight line to a parabola   (b) histogram of the residuals

Figure 3.15: Fitting a straight line to data close to a parabola

```
noise_small = 0.02*randn(100*N,1); y_d = x_d.*(1-x_d) + noise_small;
F = [ones(size(x_d)) x_d]; % regression matrix for a straight line
[p, e_var, r, p_var, fit_var] = LinearRegression(F,y_d);
sigma = sqrt(p_var);
parameters = [p sigma]    % show the parameters and standard deviations
% Student-t distribution
p95_t =p+tinv(1-alpha/2,length(x_d)-3)*[-sigma +sigma]

y_fit = p(1) + p(2)*x; % compute the fitted values
figure(3); plot(x,y,'k',x_d,y_d,'b+',x,y_fit,'g')
           xlabel('independent variable x')
           ylabel('dependent variable y')
           title('regression with small noise')
           legend('true parabola','data','best fit line')

residuals = F*p-y_d;
figure(4); hist(residuals)
           xlabel('deviation'); ylabel('frequency')
-->
parameters =  0.1639191    0.0034668
              0.0035860    0.0059815


p95_t =   0.1571200    0.1707181
         -0.0081447    0.0153166
```

3.2.6 Example 1: Intensity of light of an LED depending on the angle of observation

The intensity of the light emitted by an LED will depend on the angle α of observation. The data sheets of the supplier should show this information. A sample of real data is stored in the file `LEDdata.m`. In Section 3.9.3 you find the information on how to import the data from the data sheet into *Octave*. Then Figure 3.16(b) is generated by simple code.

```
LEDdata; % load the data
figure(1); plot(angle,intensity,'*');
        xlabel('angle'); ylabel('intensity')
```

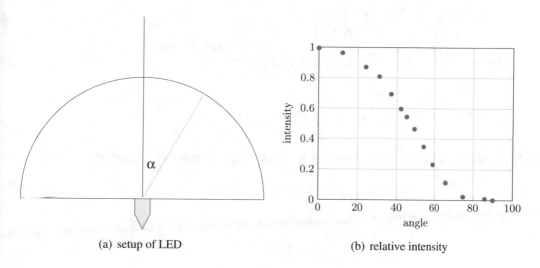

(a) setup of LED   (b) relative intensity

Figure 3.16: Intensity of light as function of the angle

To do further analysis it can be useful to have a formula for the intensity as function of the angle and linear regression is one of the options on how to obtain such a formula. The following code will fit a polynomial of degree 5 through those points and then display the result in Figure 3.17.

```
LEDdata; % load the data
n = 6;    % try with a polynomial of degree 5
F = ones(length(angle),n);
for k = 1:n
  F(:,k) = angle.^(k-1);
end

[p,int_var,r,p_var] = LinearRegression1(F,intensity);
[p,sqrt(p_var)]   % display the estimated values for the parameters

al = (0:1:90)'; % consider angles from 0 to 90 degree
```

```
Fnew = ones(length(al),n);
for k = 1:n
   Fnew(:,k) = al.^(k-1);
end%for

Inew = Fnew*p;
plot(angle,intensity,'*',al,Inew)
grid on; xlabel('angle');ylabel('intensity')
-->
ans =    2.3976e-05    9.0090e-06
         4.0603e-04    1.5255e-04
         4.8620e-03    1.8263e-03
        -1.9866e-04    9.0627e-05
         2.6932e-06    1.4458e-06
        -1.2086e-08    7.4267e-09
```

The resulting parameters point towards an intensity function

$$
\begin{aligned}
T(\alpha) \ \approx \ & 2.40 \cdot 10^{-5} + 4.06 \cdot 10^{-4}\,\alpha + 4.86 \cdot 10^{-4}\,\alpha^2 - 1.99 \cdot 10^{-4}\,\alpha^3 + \\
& + 2.69 \cdot 10^{-6}\,\alpha^4 - 1.21 \cdot 10^{-8}\,\alpha^5 \, .
\end{aligned}
$$

The result in Figure 3.17 is obviously useless. The estimated variances of the parameters

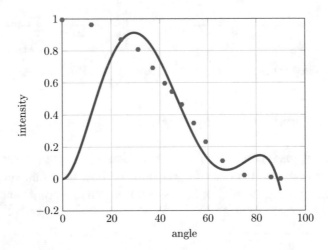

Figure 3.17: Intensity of light as function of the angle and a first regression curve

are of the same order of magnitude as the values of the parameters, or even larger. We need to find the reason for the failure and how to avoid the problem.

The above implementation of the linear regression algorithm has to solve a system of equations with the matrix $\mathbf{F}' \cdot \mathbf{F}$. With the help of

```
F'*F
-->
 1.4000e+01   6.7000e+02   4.1114e+04   2.8161e+06   2.0767e+08   1.6113e+10
 6.7000e+02   4.1114e+04   2.8161e+06   2.0767e+08   1.6113e+10   1.2949e+12
 4.1114e+04   2.8161e+06   2.0767e+08   1.6113e+10   1.2949e+12   1.0666e+14
 2.8161e+06   2.0767e+08   1.6113e+10   1.2949e+12   1.0666e+14   8.9414e+15
 2.0767e+08   1.6113e+10   1.2949e+12   1.0666e+14   8.9414e+15   7.5911e+17
 1.6113e+10   1.2949e+12   1.0666e+14   8.9414e+15   7.5911e+17   6.5056e+19
```

observe that the matrix contains numbers of the order 1 and of the order 10^{19} and one should not be surprised by trouble when solving such a system of equations. Mathematically speaking this is a very large **condition number** of 10^{16} and thus one will loose many digits of precision when performing calculations with this matrix. Entries of vastly different sizes are an indication for large condition numbers, but other effects also matter and you will have to consult specialized literature or a mathematician to obtain more information.

There are different measures to be taken to avoid the problem. For a good, reliable solution they **should all be used**.

1. **Rescaling**

   For a polynomial of degree 6 and angles of $90°$ the matrix **F** will contain numbers of the size 1 and 90^4. Thus $\mathbf{F}' \cdot \mathbf{F}$ will contain number of the size $90^8 \approx 100^8 = 10^{20}$. If we switch to radians instead of degrees this will be reduced to $\left(\frac{\pi}{2}\right)^8 \approx 100$ and thus this problem should be taken care of. The code below will generate a good solution.

```
LEDdata;
scalefactor = 180/pi;   angle = angle/scalefactor;

n = 6;
F = ones(length(angle),n);
for k = 1:n
  F(:,k) = angle.^(k-1);
end

[p,int_var,r,p_var] = LinearRegression1(F,intensity);
result = [p sqrt(p_var)]     % display the estimated values

al = ((0:1:90)')/scalefactor;% consider angles from 0 to 90 degree

Fnew = ones(length(al),n);
for k = 1:n
  Fnew(:,k) = al.^(k-1);
end

Inew = Fnew*p;
plot(angle*scalefactor,intensity,'*',al*scalefactor,Inew)
grid on
xlabel('angle');ylabel('intensity')
```

This is confirmed by the smaller condition number.

```
cond(F' *F)
-->
2.1427e+05
```

2. **Better choice of basis functions**

Since the intensity function $I(\alpha)$ has to be symmetric with respect to α, i.e. $I(-\alpha) = I(\alpha)$, there can be no contributions of the form α, α^3 or α^5. Thus we seek a good fit for a function of the type

$$I(\alpha) = p_1 + p_2\,\alpha^2 + p_3\,\alpha^4 .$$

The code below leads to the result in Figure 3.18. The condition number of $\mathbf{F}' \cdot \mathbf{F}$ is approximately 200 and thus poses no problem. The result in Figure 3.18 is now useful for further investigations and the computations indicate that the intensity is approximated by

$$I(\alpha) = 1.02951 - 0.95635\,\alpha^2 + 0.21890\,\alpha^4$$

The new code is a slight modification of the previous code.

```
LEDdata;
scalefactor = 180/pi;    angle = angle/scalefactor;
n = 3;
F = ones(length(angle),n);
for k = 1:n
  F(:,k)  = angle.^(2*(k-1));
end

[p,int_var,r,p_var] = LinearRegression1(F,intensity);
result = [p,sqrt(p_var)]     % display the estimated values

al = ((0:1:90)')/scalefactor;% consider angles from 0 to 90 degree

Fnew = ones(length(al),n);
for k = 1:n
  Fnew(:,k)  = al.^(2*(k-1));
end

Inew = Fnew*p;
plot(angle*scalefactor,intensity,'*',al*scalefactor,Inew)
grid on
xlabel('angle');ylabel('intensity')
```

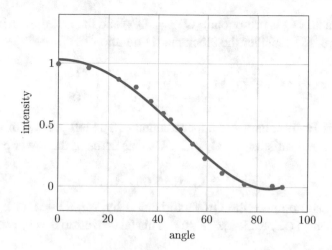

Figure 3.18: Intensity of light as function of the angle and regression with even function

This point is by far the most important aspect to consider when using the linear regression method.

> Choose your basis functions for linear regression very carefully,
> based on information about the system to be examined.

There are many software packages (*Mathematica*, MATLAB, *Octave*, Excel, ...) to perform linear regression with polynomials of high degrees. This author is not aware of **one single problem** where a polynomial of high degree leads to useful information. All software behave according to the **GIGO**[6] principle.

3. **QR factorization instead of the matrix $\mathbf{F}' \cdot \mathbf{F}$**
   Idea and consequences of this change in algorithm are based on QR factorization and are given in the next section. Any serious implementation of a linear regression method should use this modification. In all the above code the naive impementaion in function `LinearRegression1()` has to be replaced by `LinearRegression()` to take advantage of the improved algorithm, based on the QR factorization.

3.2.7 QR factorization and linear regression

For a $n \times m$ matrix $\mathbf{F}$ with more rows than columns ($n > m$) a **QR** decomposition of the matrix can be computed

$$\mathbf{F} = \mathbf{Q} \cdot \mathbf{R}$$

---

[6]Garbage In, Garbage Out

where the $n \times n$ matrix $\mathbf{Q}$ is orthogonal ($\mathbf{Q}^{-1} = \mathbf{Q}^T$) and the $n \times m$ matrix $\mathbf{R}$ has an upper triangular structure. No consider the block matrix notation

$$\mathbf{Q} = \begin{bmatrix} \mathbf{Q}_l & \mathbf{Q}_r \end{bmatrix} \quad \text{and} \quad \mathbf{R} = \begin{bmatrix} \mathbf{R}_u \\ \mathbf{0} \end{bmatrix}.$$

The $m \times m$ matrix $\mathbf{R}_u$ is square and upper triangular. The left part $\mathbf{Q}_l$ of the square matrix $\mathbf{Q}$ is of size $n \times m$ and satisfies $\mathbf{Q}_l^T \mathbf{Q}_l = \mathbb{I}_n$. Use the zeros in the lower part of $\mathbf{R}$ to verify that

$$\mathbf{F} = \mathbf{Q} \cdot \mathbf{R} = \mathbf{Q}_l \cdot \mathbf{R}_u.$$

MATLAB/*Octave* can compute the QR factorization by [Q,R]=qr(F) and the reduced form by the command [Ql,Ru]=qr(F,0). This factorization is very useful to implement linear regression.

Multiplying a vector $\vec{r} \in \mathbb{R}^n$ with the orthogonal matrix $\mathbf{Q}$ or its inverse $\mathbf{Q}^T$ corresponds to a rotation of the vector and thus will not change its length. This observation can be used to rewrite the linear regression problem from Section 3.2.2.

$$\begin{aligned}
\mathbf{F} \cdot \vec{p} - \vec{y} &= \vec{r} \quad \text{length to be minimized} \\
\mathbf{Q} \cdot \mathbf{R} \cdot \vec{p} - \vec{y} &= \vec{r} \quad \text{length to be minimized} \\
\mathbf{R} \cdot \vec{p} - \mathbf{Q}^T \cdot \vec{y} &= \mathbf{Q}^T \cdot \vec{r}
\end{aligned}$$

$$\begin{bmatrix} \mathbf{R}_u \cdot \vec{p} \\ \mathbf{0} \end{bmatrix} - \begin{bmatrix} \mathbf{Q}_l^T \cdot \vec{y} \\ \mathbf{Q}_r^T \cdot \vec{y} \end{bmatrix} = \begin{bmatrix} \mathbf{Q}_l^T \cdot \vec{r} \\ \mathbf{Q}_r^T \cdot \vec{r} \end{bmatrix}$$

Since the vector $\vec{p}$ does not change the lower part of the above system, the problem can be replaced by a smaller system of m equations for m unknowns, namely the upper part only of the above system.

$$\mathbf{R}_u \cdot \vec{p} - \mathbf{Q}_l^T \cdot \vec{y} = \mathbf{Q}_l^T \cdot \vec{r} \quad \text{length to be minimized}$$

Obviously this length is minimized if $\mathbf{Q}_l^T \cdot \vec{r} = \vec{0}$ and thus we find the reduced equations for the vector $\vec{p}$.

$$\begin{aligned}
\mathbf{R}_u \cdot \vec{p} &= \mathbf{Q}_l^T \cdot \vec{y} \\
\vec{p} &= \mathbf{R}_u^{-1} \cdot \mathbf{Q}_l^T \cdot \vec{y}
\end{aligned}$$

In *Octave* the above algorithm can be implemented with two commands only.

```
[Q,R] = qr(F,0);
p = R\(Q'*y);
```

It can be shown that the condition number for the $\mathbf{QR}$ algorithm is much smaller than the condition number for the algorithm based on $\mathbf{F}^T \cdot \mathbf{F} \cdot \vec{p} = \mathbf{F}^T \cdot \vec{y}$. Thus there are fewer accuracy problems to be expected and we obtain results with higher reliability[7].

---

[7]A careful computation shows that using the QR factorization $\mathbf{F} = \mathbf{Q}\,\mathbf{R}$ in $\mathbf{F}^T \mathbf{F} \vec{p} = \mathbf{F}^T \vec{y}$ also leads to $\mathbf{R}_u \vec{p} = \mathbf{Q}_l^T \vec{y}$.

3.2.8 Weighted linear regression

The general method

So far we minimized the length of the residual vector

$$\vec{r} = \mathbf{F} \cdot \vec{p} - \vec{y}$$

using the standard length $\|\vec{r}\|^2 = \sum_{i=1}^{n} r_i^2$. There are situations where not all errors have equal weight and thus one has to minimize a weighted length

$$\|\vec{r}\|_W^2 = \sum_{i=1}^{n} w_i^2 r_i^2 = \langle \mathbf{W} \cdot \vec{r}, \mathbf{W} \cdot \vec{r} \rangle .$$

If the estimated standard deviation for each measurement y_i is given by σ_i, use $w_i = 1/\sigma_i$. Thus points measured with a high accuracy obtain a larger weight. The weight matrix $\mathbf{W}$ is given by

$$\mathbf{W} = \mathrm{diag}(\vec{w}) = \begin{bmatrix} w_1 & & & \\ & w_2 & & \\ & & \ddots & \\ & & & w_n \end{bmatrix}.$$

A large value of the weight w_i implies that an error r_i in that component has large weight. Thus the algorithm will try to keep r_i small.

Now an algorithm similar to the previous section can be applied to estimate the optimal values for the parameters $\vec{p}$.

$$
\begin{aligned}
\mathbf{F} \cdot \vec{p} - \vec{y} &= \vec{r} \quad \text{weighted length to be minimized} \\
\mathbf{W} \cdot \mathbf{F} \cdot \vec{p} - \mathbf{W} \cdot \vec{y} &= \mathbf{W} \cdot \vec{r} \quad \text{standard length to be minimized} \\
\mathbf{Q} \cdot \mathbf{R} \cdot \vec{p} - \mathbf{W} \cdot \vec{y} &= \mathbf{W} \cdot \vec{r} \quad \text{standard length to be minimized} \\
\mathbf{R} \cdot \vec{p} - \mathbf{Q}^T \cdot \mathbf{W} \cdot \vec{y} &= \mathbf{Q}^T \cdot \mathbf{W} \cdot \vec{r} \\
\begin{bmatrix} \mathbf{R}_u \cdot \vec{p} \\ 0 \end{bmatrix} - \begin{bmatrix} \mathbf{Q}_l^T \cdot \mathbf{W} \cdot \vec{y} \\ \mathbf{Q}_r^T \cdot \mathbf{W} \cdot \vec{y} \end{bmatrix} &= \begin{bmatrix} \mathbf{Q}_l^T \cdot \mathbf{W} \cdot \vec{r} \\ \mathbf{Q}_r^T \cdot \mathbf{W} \cdot \vec{r} \end{bmatrix} \\
\mathbf{R}_u \cdot \vec{p} &= \mathbf{Q}_l^T \cdot \mathbf{W} \cdot \vec{y} \\
\vec{p} &= \mathbf{R}_u^{-1} \cdot \mathbf{Q}_l^T \cdot \mathbf{W} \cdot \vec{y}
\end{aligned}
$$

This algorithm is implemented in Figure 3.20 (see page 244).

To estimate the variances of the parameters $\vec{p}$ use assumptions on the variances σ_i of the y_i values. A heuristic argument in the next section motivates the estimate

$$\sigma_j^2 \approx \frac{1}{w_j^2} \frac{1}{n-m} \sum_{i=1}^{n} r_i^2 w_i^2$$

and then use

$$\vec{p} = \mathbf{R}_u^{-1} \cdot \mathbf{Q}_l^T \cdot \mathbf{W} \cdot \vec{y} = \mathbf{M} \cdot \vec{y}$$

to conclude

$$p_j \;=\; \sum_{i=1}^{n} m_{j,i}\, y_i \quad \text{for} \quad 1 \le j \le m$$

$$V(p_j) \;=\; \sum_{i=1}^{n} m_{j,i}^2\, \sigma_i^2 \quad \text{for} \quad 1 \le j \le m\,.$$

Observe that this calculation is only correct if the σ_i are not correlated.

Uniformly distributed absolute errors

If we set all weights to $w_i = 1$ this leads back to the results in Section 3.2.3. The expression to be minimized is

$$\chi^2 = \sum_{i=1}^{n} (y_i - f(x_i))^2 \quad \text{with} \quad f(x) = \sum_{j=1}^{m} p_j\, f_j(x)\,.$$

Uniformly distributed relative errors

In this case we expect the standard deviations to be proportional to the absolute value of y_i and thus we choose the weights $w_1 = 1/|y_i|$. The expression to be minimized is

$$\chi^2 = \sum_{i=1}^{n} \frac{(y_i - f(x_i))^2}{y_i^2}\,.$$

A priori known error distributions

If we have good estimates σ_i for the standard deviations of the values y_i we choose the weights $w_i = 1/\sigma_i$ and the expression to be minimized is

$$\chi^2 = \sum_{i=1}^{n} \frac{(y_i - f(x_i))^2}{\sigma_i^2}\,.$$

In this case the standard deviation of $w_i\, y_i$ is expected to be constant 1. The estimation of the standard deviations has to respect this fact and this leads to the estimates based on the data points and the given weights.

$$\sigma_j^2\, w_j^2 \approx \frac{1}{n} \sum_{i=1}^{n} w_i^2\, r_i^2 \qquad \Longrightarrow \qquad \sigma_j^2 \approx \frac{\sum_{i=1}^{n} w_i^2\, r_i^2}{n\, w_j^2}\,.$$

This leads to the estimate for σ_j in the previous section. The method is implemented in the code of `LinearRegression()` in Figure 3.20.

Command	Properties
`LinearRegression()`	standard and weighted linear regression
	returns standard deviations for parameters
`regress()`	standard linear regression
	returns confidence intervals for parameters
`lscov()`	gerneralized least square estimation, with weights
`ols()`	ordinary least square estimation
`gls()`	gerneralized least square estimation
`polyfit()`	regression with for polynomials only
`lsqnonneg()`	regression with positivity constraint

Table 3.4: Commands for linear regression

3.2.9 More commands for regression with *Octave* or MATLAB

In these notes I mainly use the command `LinearRegression()`, but MATLAB/*Octave* provide many more commands, some shown in Table 3.4. Consult the manuals for more information.

For nonlinear regression there are special commands too, see Table 3.5. Observe that the syntax and algorithm of these commands might differ between MATLAB and *Octave*. You definitely have to consult the manuals and examine example applications.

Command	Properties
`leasqr()`	standard non linear regression, Levenberg-Marquardt
	see section 3.2.15
`fsolve()`	can be used for nonlinear regression too
`nlinfit()`	nonlinear regression
`lsqcurvefit()`	nonlinear curve fitting
`nonlin_curvefit()`	frontend, *Octave* only
`curvefit_stat()`	statistics for the above, *Octave* only
`lsqnonlin()`	nonlinear minimization of residue
`nonlin_residmin()`	frontend, *Octave* only
`nlparci()`	determine confidence intervals of parameters, MATLAB only
`expfit()`	regression with exponential functions

Table 3.5: Commands for nonlinear regression

3.2.10   Code for the function `LinearRegression()`

The structure of the function file has the typical structure of a *Octave* function file.

- The first few lines contain the copyright.

- The first section in the file `LinearRegression.m` is the documentation. This text will be displayed by the command `help LinearRegression`. A description of the parameters and the return values is given. Find the result in Figure 3.19.

- Then the function is defined, showing all possible parameters and return values.

- The function verifies that the correct number of arguments (2 or 3) is given, otherwise returns with a message.

- The correct size of the arguments (F and y) is verified. An error message is displayed if the size of matrix and vector do not match.

- Finally the necessary computations are carried out.

- The estimated variances of the parameters and the predicted values of y are only computed if the output is requested. This is implemented by counting the return arguments of the call of the function (`nargout`).

The resulting function can be called with 1 to 5 return arguments. The function will return only the requested values.

- `p = LinearRegression(F,y)` will return the estimated value of the parameters $\vec{p}$ only .

- `[p,e_var] = LinearRegression(F,y)` will also return the variance of the y–error.

- `[p,e_var,r,p_var,fit_var] = LinearRegression(F,y)` will return all 5 results.

Find the documentation in Figure 3.19 and the code in Figure 3.20.

3.2.11   Example 2: Performance of a linear motor

In his diploma thesis in 2005 Aloïs Pfenniger examined the forces of a linear magnetic motor as function of length and diameter of the coils used to construct the motor. A typical configuration is displayed in Figure 3.21. With a lengthy computation (approximately 4 hours per configuration) he computed the forces for 25 different configurations. The result is shown in Figure 3.22.

```
help LinearRegression
-->
Function File LinearRegression (F, y, w)
  [p, e_var, r, p_var, fit_var] = LinearRegression (...)

  general linear regression

  determine the parameters p_j  (j=1,2,...,m) such that the function
  f(x) = sum_(j=1,...,m) p_j*f_j(x) is the best fit to the given values
  y_i by f(x_i) for i=1,...,n,
  i.e. minimize sum_(i=1,...,n)(y_i-sum_(j=1,...,m) p_j*f_j(x_i))^2
  with respect to p_j

  parameters:
  F is an n*m matrix with the values of the basis functions at
    the support points. In column j give the values of f_j
    at the points x_i  (i=1,2,...,n)
  y is a column vector of length n with the given values
  w is an optional column vector of length n with the weights of
    the data points. 1/w_i is expected to be proportional to the
    estimated uncertainty in the y values. Then the weighted
    expression sum_(i=1,...,n)(w_i^2*(y_i-f(x_i))^2) is minimized.

  return values:
  p     is the vector of length m with the estimated values
        of the parameters
  e_var is the vector of estimated variances of the residuals,
        i.e. the difference between the provided y values and the
        fitted function.
        If weights are provided, then the product e_var_i * w^2_i
        is assumed to be constant.
  r     is the weighted norm of the residual
  p_var is the vector of estimated variances of the parameters p_j
  fit_var is the vector of estimated variances of the fitted function
        values f(x_i)

  To estimate the variance of the difference between future y values
  and fitted y values use the sum of e_var and fit_var

  Caution: do NOT request fit_varfor large data sets, as a n*n matrix
           is generated

  see also: ols, gls, regress, leasqr, nonlin_curvefit, polyfit,
            wpolyfit, expfit

  Copyright (C) 2007-2018 Andreas Stahel <Andreas.Stahel@bfh.ch>
```

Figure 3.19: Documentation of the command `LinearRegression()`

```
┌──────────────────────── LinearRegression.m ────────────────────────┐
  function [p, e_var, r, p_var, fit_var] = LinearRegression(F, y, weight)

    if (nargin < 2 || nargin >= 4)  print_usage (); end%if

    [rF, cF] = size (F);    [ry, cy] = size (y);
    if (rF ~= ry || cy > 1)
       error ('LinearRegression: incorrect matrix dimensions');
    end%if

    if (nargin == 2)  % set uniform weights if not provided
       weight = ones (size (y));
    else
       weight = weight(:);
    end%if

    wF = diag (weight) * F;     % this efficent with the diagonal matrix
    [Q, R] = qr (wF, 0);        % estimate the values of the parameters
    p = R \ (Q' * (weight .* y));

    %% Compute the residual vector and its weighted norm
    residual = F * p - y;
    r = norm (weight .* residual);
    weight2 = weight.^2;
    %% If the variance of data y is sigma^2 * weight.^2, var is an
    %% unbiased estimator of sigma^2
    var = residual.^2' * weight2 / (rF - cF);
    %% Estimated variance of residuals
    e_var = var ./ weight2;

    %% Compute variance of parameters, only if requested
    if (nargout > 3)
      M = R \ (Q' * diag (weight));
      %% compute variance of the fitted values, only if requested
      if (nargout > 4)
      %% WARNING the nonsparse matrix M2 is of size rF by rF,
      %% wehre rF = number of data points
        M2 = (F * M).^2;
        fit_var = M2 * e_var;  % variance of the function values
      end%if
      p_var = M.^2 * e_var;   % variance of the parameters
    end%if

  end%function
└─────────────────────────────────────────────────────────────────────┘
```

Figure 3.20: Code for the command LinearRegression()

Figure 3.21: A magnetic linear motor

```
PfennigerData;
figure(1); plot(long(:,1),force(:,1),long(:,2),force(:,2),...
                long(:,3),force(:,3),long(:,4),force(:,4),...
                long(:,5),force(:,5));
            xlabel('length of coil'); ylabel('force');

figure(2); mesh(diam,long,force)
            xlabel('diameter'); ylabel('length'); zlabel('force');
            view(-30,30);
```

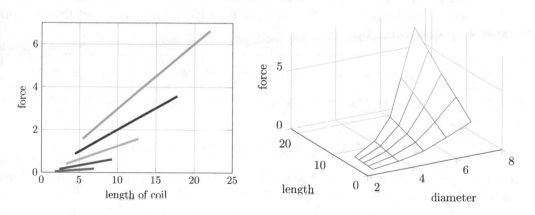

Figure 3.22: Force as function of length of coils, for 5 different diameters

General function

These graphs indicate that the force f might depend linearly on the length l and quadratically on the diameter d.

$$f(l,d) = p_1 + p_2\, l + p_3\, l\, d + p_4\, l\, d^2$$

A call of `LinearRegression()` and `mesh()`

```
diam1 = diam(:); long1 = long(:);force1=force(:);

F = [ones(size(long1)) long1 long1.*diam1 long1.*(diam1.^2)];
coef = LinearRegression(F,force1)

[L,DIA] = meshgrid(2:30,2:0.5:8);
forceA = coef(1)+L.*(coef(2)+coef(3)*DIA+coef(4)*DIA.^2);
figure(2); mesh(DIA,L,forceA)
          xlabel('diameter of coil'); ylabel('length of coil');
          zlabel('force');  view(-10,30);
```

leads to the approximate function

$$f(l, d) = -0.0252 + 0.0193\, l - 0.0114\, l\, d + 0.0065\, l\, d^2$$

and the residual of $r \approx 0.065$ gives an indication on the size of the error. The results generated by the code

```
forceA2  = coef(1)+long.*(coef(4)*diam.^2+coef(3)*diam+coef(2));
maxerror = max(max(abs(forceA2-force)))
maxrelerror = max(max(abs(forceA2-force)./force))
```

show the maximal error of 0.04 and a relative error of 10%.

If we seek to minimize the relative error we have to replace the call of the function LinearRegression() by

```
[coef,f_var,r,coef_var] = LinearRegression(F,force1,1./sqrt(force1))
```

and will find a larger maximal error of 0.05, but a smaller relative error of only 3%. The approximate function is

$$f(l, d) = -0.00639 + 0.00662\, l - 0.00730\, l\, d + 0.00617\, l\, d^2\ .$$

The contour plot in Figure 3.23 can be generated. The level curves are 0.5 apart, with values from 0.5 to 8.

```
contour(DIA,L,forceA,[0.5:0.5:8])
xlabel('diameter of coil'); ylabel('length of coil');
```

Adapted function

Physical reasoning might make believe that the form of the function should be simpler than in the previous section. Search for a solution of the form

$$f(l, d) = p_1\, l + p_2\, l\, d^2$$

and apply a weighted linear regression to keep the relative errors small.

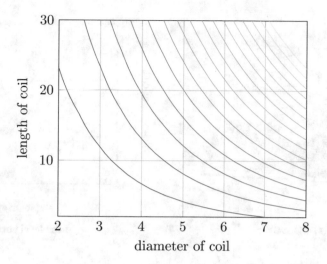

Figure 3.23: Level curves for force as function of length and diameter of coil

```
F = [long1 long1.*(diam1.^2)];
[coef,f_var,r,coef_var] = LinearRegression(F,force1,1./sqrt(force1))

forceB = L.*(coef(1)+coef(2)*DIA.^2);

figure(1); mesh(DIA,L,forceB)
          xlabel('diameter of coil'); ylabel('length of coil');
          zlabel('force');   view(-10,30);

figure(2); contour(DIA,L,forceA,[0.5:0.5:8])
          xlabel('diameter of coil'); ylabel('length of coil');

forceB2 = long.*(coef(1)+coef(2)*diam.^2);
maxrelerror = max(max(abs(forceB2-force)./force))
maxerror = max(max(abs(forceB2-force)))
```

The graphical result can be seen in Figure 3.24 and the numerical results indicates a solution

$$f(l,d) = -0.00990\, l + 0.00543975\, l\, d^2 \,.$$

The maximal error is 0.1 and the maximal relative error is 5%. With the above function further computations can be carried out quite easily.

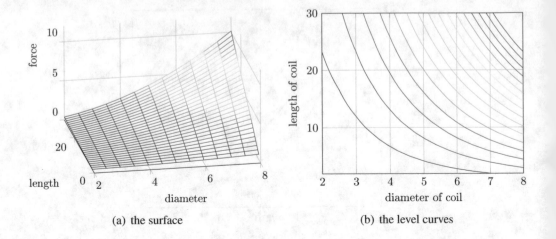

(a) the surface   (b) the level curves

Figure 3.24: Computations with a simplified function

3.2.12   Example 3: Calibration of an orientation sensor

Description of the problem

With the help of two accelerations sensors one can determine the vertical (x) and horizontal (y) components of the gravitational field. Under perfect conditions we would find

$$x = g \cos(\alpha) \quad \text{and} \quad y = g \sin(\alpha)$$

where α is the angle by which the device was rotated, clockwise. Typical sensor yield a

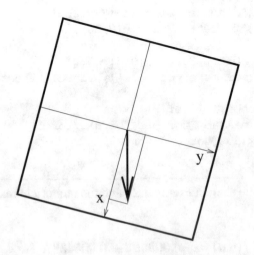

Figure 3.25: A slightly rotated direction sensor

signal proportional to the applied field, but there might be an offset. The sensor might not be perfectly orthogonal and not be mounted perfectly. Thus we actually receive signals of the type

$$x = x_0 + r_x \cos(\alpha - \phi_x) \quad \text{and} \quad y = y_0 + r_y \sin(\alpha - \phi_y).$$

If the orientation of the device is to be determined we have to compute α, given the values of x and y. Use

$$\frac{x - x_0}{r_x} = \cos(\alpha - \phi_x) \quad \text{and} \quad \frac{y - y_0}{r_y} = \sin(\alpha - \phi_y)$$

where the parameters x_0, y_0, r_x, r_y, ϕ_x and ϕ_y might be different for each sensor.

Solution with the help of linear regression

Assume that the x direction sensor is mounted in the direction given by a vector $\vec{m}_x$. Its sensitivity is given by $\|\vec{m}_x\|$. Then the signal on the x–sensor is given by

$$s_x = \langle \vec{g}, \vec{m}_x \rangle + c_x.$$

The constant c_x corresponds to the offset of the sensor, i.e. the sensors output signal for a zero input. For a number of given vectors $\vec{g}_i$ ($1 \leq i \leq n$) and the resulting signals s_i we have to minimize the residual vector $\vec{r}$ determined by

$$\begin{bmatrix} g_{x,1} & g_{y,1} & 1 \\ g_{x,2} & g_{y,2} & 1 \\ g_{x,3} & g_{y,3} & 1 \\ & \vdots & \\ g_{x,n} & g_{y,n} & 1 \end{bmatrix} \cdot \begin{pmatrix} m_{x,1} \\ m_{x,2} \\ c_x \end{pmatrix} - \begin{pmatrix} s_1 \\ s_2 \\ s_3 \\ \vdots \\ s_n \end{pmatrix} = \begin{pmatrix} r_1 \\ r_2 \\ r_3 \\ \vdots \\ r_n \end{pmatrix}.$$

A linear regression will give the optimal values of $\vec{m}_x$ and c_x. Similar calculations can be applied for the y–sensor, leading to the best values for $\vec{m}_y$ and c_y.

Once the parameter values are determined we can compute the signal at the sensors for a given orientation of the $\vec{g}$ vector by

$$\vec{s} = \begin{pmatrix} s_r \\ s_y \end{pmatrix} = \begin{bmatrix} m_{x,1} & m_{x,2} \\ m_{y,1} & m_{y,2} \end{bmatrix} \cdot \begin{pmatrix} g_1 \\ g_2 \end{pmatrix} + \begin{pmatrix} c_x \\ c_y \end{pmatrix} = \mathbf{M} \cdot \vec{g} + \vec{c}$$

This can be solved for the vector g by

$$\vec{g} = \mathbf{M}^{-1} (\vec{s} - \vec{c}).$$

This is the expression to determine the direction of the $\vec{g}$ vector as function of the signals $\vec{s}$ at the sensors. The angle β between the x axis and the $\vec{g}$ field the given by

$$\tan \beta = \frac{g_y}{g_x}.$$

The above algorithm is implemented in the code below, using some simulated data.

```
                                    OrientationTest.m
OrientationData; %% read the values of alpha, x and y
gx = cos(al); gy = sin(al);
F = [gx gy ones(size(al))];
[px,xvar,r,pvar] = LinearRegression(F,x);
[py,xvar,r,pvar] = LinearRegression(F,y);
mx = px(1:2); cx = px(3); my = py(1:2); cy = py(3);

m = [mx my]
c = [cx cy]
xn = F*px;   yn = F*py;
plot(x,y,'*r',xn,yn,'b')
axis('equal')
-->
m =    2.4030   -0.1966
       0.3038    2.1379
c =    2.1549    1.8899
```

This computation leads to Figure 3.26 and the numerical results are

$$\vec{s} = \begin{pmatrix} s_x \\ s_y \end{pmatrix} = \begin{bmatrix} 2.40299 & -0.19664 \\ 0.30382 & 2.13793 \end{bmatrix} \cdot \begin{pmatrix} g_1 \\ g_2 \end{pmatrix} + \begin{pmatrix} 2.1549 \\ 1.8899 \end{pmatrix} = \mathbf{M} \cdot \vec{g} + \vec{c}.$$

The diagonal dominance of this matrix indicates the two sensor have (almost) the same orientation as the coordinate axis. The numbers show that the x sensor has an offset of $x_0 \approx 2.155$ and an amplification of $r_x = \sqrt{m_{1,x}^2 + m_{x,2}^2} \approx 2.4221$. Similarly determine the offset of the y–sensor as $y_0 \approx 1.89$ and an amplification of $r_y \approx 2.15$.

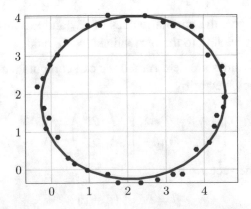

Figure 3.26: Measured data and the fitted circle

The results also allow to determine the angles of the sensors with respect to their axis. If $\vec{g}$ points in x–direction then we should obtain a maximal signal on the x sensor and if $\vec{g}$ points in y–direction we expect no signal on the x sensor. Thus the angle at which the sensor is mounted can be estimated by

$$\phi_x \approx \arctan\left(\frac{m_{x,1}}{m_{x,2}}\right)$$

and similarly for the y-sensor. The code

```octave
atan2(mx(2),mx(1))*180/pi
atan2(my(2),my(1))*180/pi-90
```

leads to deviations of $\phi_x \approx 7.21°$ and $\phi_y \approx 5.26°$. The difference of these two angles corresponds to the angle between the two sensors.

To determine the direction of $\vec{g}$ using the measurements s_x and s_y use

$$\vec{g} = \mathbf{M}^{-1} \cdot (\vec{s} - \vec{c}) = \begin{bmatrix} 0.411365 & 0.037835 \\ -0.058458 & 0.462365 \end{bmatrix} \cdot \begin{pmatrix} s_x - 2.1549 \\ s_y - 1.8899 \end{pmatrix} .$$

This formula contains all calibration data for this (simulated) sensor.

Estimation of errors

In this subsection we want to estimate the variances of the parameters. The variances of $\mathbf{M}$, x_0 and y_0 are directly given by the return parameters of the command p_var of LinearRegression(). This might be sufficient to estimate the measurement errors for the vector $\vec{g}$.

As a next step we estimate the variances or r_x, r_y and the two angles ϕ_x and ϕ_y. Use the notation $p_2 = m_{x,1}$ and $p_3 = m_{x,2}$ and

$$r_x = \sqrt{m_{x,1}^2 + m_{x,2}^2} = \sqrt{p_2^2 + p_3^2} \quad , \quad \frac{\partial r_x}{\partial p_2} = \frac{p_2}{r_x} \quad \text{and} \quad \frac{\partial r_y}{\partial p_3} = \frac{p_3}{r_x}$$

to determine the variance of r_x as

$$V(r_x) = \frac{p_2^2}{r_x^2} V(p_2) + \frac{p_3^2}{r_x^2} V(p_3).$$

If $V(p_2) \approx V(p_3)$ this simplifies to $V(r_x) \approx V(p_2)$. The similar result is valid for $V(r_y)$.
Since $\frac{\partial}{\partial x} \arctan x = \frac{1}{1+x^2}$ conclude

$$\begin{aligned}
\phi_x &= \arctan \frac{p_3}{p_2} \\
\frac{\partial}{\partial p_2} \phi_x &= \frac{1}{1 + (p_3/p_2)^2} \frac{-p_3}{p_2^2} = \frac{-p_3}{p_2^2 + p_3^2} = \frac{-p_3}{r_x^2} \\
\frac{\partial}{\partial p_3} \phi_x &= \frac{1}{1 + (p_3/p_2)^2} \frac{1}{p_2} = \frac{p_2}{p_2^2 + p_3^2} = \frac{p_2}{r_x^2} \\
V(\phi_x) &\approx \frac{1}{r_x^4} \left(p_3^2 V(p_2) + p_2^2 V(p_3) \right) \\
\sigma(\phi_x) &= \sqrt{V(\phi_x)} \approx \frac{1}{r_x^2} \sqrt{p_3^2 V(p_2) + p_2^2 V(p_3)}
\end{aligned}$$

and similarly[8] for ϕ_y. If $V(p_2) = V(p_3)$ this simplifies drastically to

$$\sigma(\phi_x) \approx \frac{\sqrt{V(p_2)}}{r_x} = \frac{\sigma(p_2)}{r_x} .$$

Repeat the simulation with the data

$$x_0 = 0.5 \quad , \quad y_0 = 0 \quad , \quad r_x = 1 \quad , \quad r_y = 1.1 \quad , \quad \phi_x = 0° \quad , \quad \phi_y = 5°$$

and add a random perturbation to x and y of the size 0.001 (variance of the simulated values). Then find with identical computations the approximated values

$$x_0 \approx 0.5 \quad , \quad y_0 \approx 0 \quad , \quad r_x \approx 1 \quad , \quad r_y \approx 1.1 \quad , \quad \phi_x \approx 0.01° \quad , \quad \phi_y \approx 4.99°$$

and all variances $V(p_i)$ are of the order 10^{-8} and thus the standard deviations of the order 10^{-4}. This is small compared to the standard deviations of x and y, i.e. $\sigma(x) = \sigma(y) = 0.001$. One can verify that the standard deviation of ϕ_x for the above simulation is approximately 0.013° and thus explains the deviation of ϕ_x from zero.

3.2.13   Example 4: Analysis of a sphere using an AFM

In his diploma thesis in 2006 Ralph Schmidhalter used an atomic force microscope (AFM) to examine the surface of ball bearing balls, produced by the local company Micro Precision Systems (MPS). The AFM yields a measured height $h(x, y)$ as function of the horizontal coordinates x and y. One can then try to determine the radius R of the ball with the given data.

Approximation of the sphere

Examine the height of a sphere with radius R and the highest point at (x_0, y_0) . Use the Taylor approximation $\sqrt{1 + z} \approx 1 + \frac{1}{2} z$ to express the height h as a linear combination of the four functions 1, x, y and $(x^2 + y^2)$.

$$
\begin{aligned}
h(x, y) &= h_0 + \sqrt{R^2 - (x - x_0)^2 - (y - y_0)^2} \\
&= h_0 + R\sqrt{1 - \frac{1}{R^2}\left((x - x_0)^2 + (y - y_0)^2\right)} \\
&\approx h_0 + R - \frac{(x - x_0)^2}{2R} - \frac{(y - y_0)^2}{2R} \\
&= h_0 + R - \frac{x_0^2 + y_0^2}{2R} + \frac{x_0}{R}x + \frac{y_0}{R}y - \frac{1}{2R}(x^2 + y^2) \\
&= p_1 + p_2\,x + p_3\,y + p_4\,(x^2 + y^2)
\end{aligned}
$$

8

$$\tan\phi_y = \frac{-p_2}{p_3} \implies \begin{array}{l} \frac{\partial}{\partial p_2}\phi_y = \frac{1}{1+(p_2/p_3)^2}\frac{-1}{p_3} = \frac{-p_3}{r_x^2} \\[2mm] \frac{\partial}{\partial p_3}\phi_y = \frac{1}{1+(p_2/p_3)^2}\frac{+p_2}{p_3^2} = \frac{p_2}{r_x^2} \end{array} \implies V(\phi_x) \approx \frac{1}{r_x^4}\left(p_3^2\,V(p_2) + p_2^2\,V(p_3)\right)$$

where

$$p_1 = h_0 + R - \frac{x_0^2 + y_0^2}{2R} \quad , \quad p_2 = \frac{x_0}{R} \quad , \quad p_3 = \frac{y_0}{R} \quad \text{and} \quad p_4 = -\frac{1}{2R} \,.$$

If all values of p_i are known, solve for the parameters of the sphere.

$$R = -\frac{1}{2p_4} \quad , \quad x_0 = R\,p_2 \quad , \quad y_0 = R\,p_3 \quad , \quad h_0 = p_1 - R + \frac{x_0^2 + y_0^2}{2R} \,.$$

In particular obtain the estimated radius R of the sphere.

Reading the data, visualize and apply linear regression

The data is measured and then stored in a file `SphereData.csv`. The first few lines of the file are shown below. For each of 5 different values of y, ranging from 0 to 14 μm, 256 values of x were examined, also ranging from 0 to 14 μm.

```
                          SphereData.csv
0,0,5.044e-006
5.491e-008,0,5.044e-006
1.098e-007,0,5.042e-006
1.647e-007,0,5.044e-006
2.196e-007,0,5.048e-006
2.745e-007,0,5.05e-006
3.294e-007,0,5.052e-006
3.844e-007,0,5.054e-006
4.393e-007,0,5.055e-006
4.942e-007,0,5.058e-006
5.491e-007,0,5.06e-006
6.040e-007,0,5.06e-006
...
```

Each row contains the values of x, y and the height z. These values have to be read into variables in *Octave*. We may use the command `dlmread()` introduced in Section 1.2.8.

```
LLL = dlmread ('SphereData.csv');
x = tt(:,1); y = tt(:,2); z = tt(:,3);
N = length(x);
```

As a next step we generate the plots with the surface and another plot with the level curves. Find the results in Figure 3.27.

```
steps = 5;
xx = reshape(x,N/steps,steps);
yy = reshape(y,N/steps,steps);
zz = reshape(z,N/steps,steps);

figure(1); contour(xx,yy,zz,5);    %% create a contour plot
```

```
                axis('equal')

 figure(2); mesh(xx,yy,zz);              %% create a surface plot
                axis('normal');
```

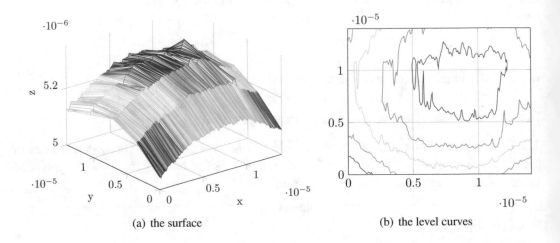

(a) the surface   (b) the level curves

Figure 3.27: The surface of a ball and the level curves

A quick look at Figure 3.27 confirms that the top of the ball is within the scanned area. This allows for a quick check of the validity of the Taylor approximation at the start of this section.

The radius of the ball is approximately $300~\mu m$ and since the top is part of the scanned area we may estimate $|x - x_0| \leq 14~\mu m$ and $|y - y_0| \leq 14~\mu m$. This leads to

$$z = \frac{1}{R^2} \left((x - x_0)^2 + (y - y_0)^2 \right) \leq 0.0044$$

and since the error of the approximation $\sqrt{1 + z} \approx 1 + \frac{1}{2}z$ is typically given by $\frac{1}{8}z^2 \leq 2.5 \cdot 10^{-6}$. This approximation error is considerably smaller than the variation in the measured data. This justifies the simplifying approximation.

Linear regression and an error analysis

The height is written as a linear combination of the functions 1, x, y and $x^2 + y^2$ and thus we have to construct the matrix

$$\mathbf{F} = \begin{bmatrix} 1 & x_1 & y_1 & x_1^2 + y_1^2 \\ 1 & x_2 & y_2 & x_2^2 + y_2^2 \\ 1 & x_3 & y_1 & x_3^2 + y_3^2 \\ & & \vdots & \\ 1 & x_n & y_n & x_n^2 + y_n^2 \end{bmatrix} .$$

Then determine the estimates for the parameters p_i and their standard deviations Δp_i and thus for the radius R and the center (x_0, y_0).

```
F = [ones(size(x)) x y x.^2+y.^2];
[p,e_var,r,p_var] = LinearRegression(F,z);
Radius = -1/(2*p(4))
x0 = Radius*p(2)
y0 = Radius*p(3)
```

To estimate the standard deviations for R, x_0 and y_0 apply the rules of error propagation and find

$$R = \frac{-1}{2\,p_4}$$

$$\Delta R \approx \left|\frac{\partial R}{\partial p_4}\right| \Delta p_4 = \frac{1}{2\,p_4^2}\,\Delta p_4 = 2\,R^2\,\Delta p_4$$

$$x_0 = R\,p_2$$

$$\Delta x_0^2 \approx \left(\frac{\partial x_0}{\partial p_2}\,\Delta p_2\right)^2 + \left(\frac{\partial x_0}{\partial R}\,\Delta R\right)^2 = (R\,\Delta p_2)^2 + (p_2\,\Delta R)^2$$

$$\Delta x_0 \approx \sqrt{(R\,\Delta p_2)^2 + (p_2\,\Delta R)^2}$$

$$\Delta y_0 \approx \sqrt{(R\,\Delta p_3)^2 + (p_3\,\Delta R)^2}\,.$$

These results are readily translated to code,

```
deltaRadius = 2*Radius^2*sqrt(p_var(4))
deltaX0 = sqrt(Radius^2*p_var(2) + p(2)^2*deltaRadius^2)
deltaY0 = sqrt(Radius^2*p_var(3) + p(3)^2*deltaRadius^2)
```

leading to the results

$$R \pm \Delta R \approx 296.4 \pm 1.7 \ \mu m$$

$$x_0 \pm \Delta x_0 \approx 8.46 \pm 0.07 \ \mu m$$

$$y_0 \pm \Delta y_0 \approx 8.77 \pm 0.07 \ \mu m\,.$$

Thus this seems to be a valid measurement of the radius R and the center (x_0, y_0) of the circle. Unfortunately different measurements of R lead to vastly different results, thus the problems requires some further analysis.

Regression with general second order surface

If we replace the approximation of a sphere by general surface of second order

$$h(x, y) = p_1 + p_2\,x + p_3\,y + p_4\,x^2 + p_5\,y^2 + p_6\,x\,y\,.$$

The radii of curvature are determined by the function

$$f(x,y) = p_4\, x^2 + p_5\, y^2 + p_6\, x\, y = \left\langle \begin{pmatrix} x \\ y \end{pmatrix}, \begin{bmatrix} p_4 & p_6/2 \\ p_6/2 & p_5 \end{bmatrix} \cdot \begin{pmatrix} x \\ y \end{pmatrix} \right\rangle.$$

If λ_i and $\vec{e}_i$ are the two eigenvalues and eigenvectors of the symmetric matrix

$$\mathbf{A} = \begin{bmatrix} p_4 & p_6/2 \\ p_6/2 & p_5 \end{bmatrix}.$$

Then any vector $(x, y)^T$ can be written in the form $t\,\vec{e}_i + s\,\vec{e}_2$ and

$$f(x,y) = \lambda_1\, t^2 + \lambda_2\, s^2$$

and consequently the two principal radii are given by

$$R_1 = \frac{-1}{2\,\lambda_1} \quad \text{and} \quad R_2 = \frac{-1}{2\,\lambda_2}.$$

This leads to *Octave* code

```
F2   = [ones(size(x)) x y x.^2 y.^2 x.*y];
[p,e_var,r,p_var] = LinearRegression(F2,z);
RadiusNew = -0.5./eig([p(4), p(6)/2;p(6)/2,p(5)])
```

and the results

$$R_1 \approx 267\,\mu m \quad \text{and} \quad R_2 \approx 316\,\mu m.$$

Observe an enormous difference between the two radii, which does certainly not correspond to reality. This was confirmed by different measurements. Thus there must be a systematic error in the measurements. A possible candidate is an inadequate calibration of the AFM microscope.

3.2.14   Example 5: A force sensor with two springs

In 2013 Remo Pfaff examined a mechanical spring system consisting of two springs with spring constants k_1 and k_2. The second spring will only effect the force if a critical x_c position is exceeded. Thus we have the following form of a force (f) distance (x) relation:

$$f(x) = \begin{cases} a + k_1\, x & \text{for} \quad x \leq x_c \\ a + (k_1 + k_2)\, x & \text{for} \quad x \geq x_c \end{cases}.$$

Using the offset $p_1 = a$ and the spring constants $p_2 = k_1$, $p_3 = k_2$ this can be written in the form

$$f(x) = p_1 + p_2\, x + p_3\, \max\{0, x - x_c\}. \tag{3.4}$$

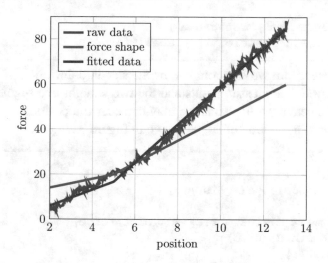

Figure 3.28: A data set for a two spring force distance system and two models

For a given data set we can try to find the optimal values for $\vec{p}$, such that measured data and (3.4) fit together. Find an example in Figure 3.28. There find the measured data, a poorly fitting model and the best possible fit for the value of $x_c = 5$.

For a fixed value of x_c we use the three basis functions

$$f_1(x) = 1 \quad , \quad f_2(x) = x \quad \text{and} \quad f_3(x) = \max\{0, x - x_c\}$$

and finding the optimal parameters p_i in equation (3.4) is a linear regression problem. The code below generated Figure 3.28 with the raw data, a poorly fitting model $f(x) = 10 + 2 \cdot x + 3 \cdot \max\{0, x - 5\}$ and the best fitting model with for the fixed value of $x_c = 5$.

```
force = load('ForceData.dat'); % read the data
dist  = load('DistanceData.dat');

xMin = 2; xMax = 13;  % select the useful domain
ind = find ((dist>=xMin).*(dist<=xMax));
dist = dist(ind); force = force(ind);

x_c = 5;  % choose a horizontal position for the break point
M = ones(length(dist),3);
M(:,2) = dist(:);
M(:,3) = max(0,dist(:)-x_c);
[p,e_var,r,p_var] = LinearRegression(M,force);
[p sqrt(p_var)]
force_fit = M*p;
force_off = 10 + 2*dist + 3*max(0,dist-x_c);

figure(1); plot(dist,force,'b',dist,force_off,'m',dist,force_fit,'r')
            legend('raw data','force shape','fitted data',...
```

```
          'location','northwest')
          xlabel('position'); ylabel('force')
```

In the above algorithm we can compute σ, the standard deviation of the residuals, i.e. the difference between the measurements and the two straight line segments. The smaller the value of σ, the better the fit. Thus we may consider the position x_c of the break as a variable and plot σ as a function of x_c, as shown in Figure 3.29(a).

```
x_list = 2:0.5:12;
sigma_list = zeros(size(x_list));

function  sigma = EvaluateBreak(x_c,dist,force)
  M = ones(length(dist),3);
  M(:,2) = dist(:);
  M(:,3) = max(0,dist(:)-x_c);
  [p,sigma] = LinearRegression(M,force);
  sigma = mean(sigma);
endfunction

for k = 1:length(x_list)
  sigma_list(k) = EvaluateBreak(x_list(k),dist,force);
endfor

figure(2); plot(x_list,sigma_list)
           xlabel('position of break'); ylabel('sigma')
```

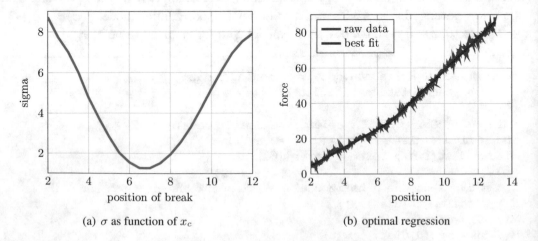

(a) σ as function of x_c　　　　　　　　(b) optimal regression

Figure 3.29: Optimal values for the two spring system

In Figure 3.29(a) it is clearly visible that there is an optimal value of x_c, such that σ is minimized. Use the command fminunc() to solve the unconstrained minimization problem.

```
[x_opt,sigma_opt,Info] = fminunc(@(x_c)EvaluateBreak(x_c,dist,force),5)
M = ones(length(dist),3);   % redo the linear regression
M(:,2) = dist(:);           % with the optimal value for x_c
M(:,3) = max(0,dist(:)-x_opt);
[p,e_var,r,p_var] = LinearRegression(M,force);
param = [p sqrt(p_var)]
force_fit = M*p;

figure(3); plot(dist, force, 'b', dist, force_fit,'r')
           legend('raw data','best fit')
           xlabel('position'); ylabel('force')
-->
x_opt =  6.7645
sigma_opt =  1.2391
Info =  3
param =  -6.048381   0.138453
          5.252886   0.026087
          3.789025   0.038254
```

Thus the best position of the break point is at $x_c \approx 6.76$ and the resulting minimal value is $\sigma \approx 1.24$. This is confimed by Figure 3.29(a). Find the optimal result in Figure 3.29(b). The numerical results

$$
\begin{pmatrix} p_1 \\ p_2 \\ p_3 \end{pmatrix} = \begin{pmatrix} -6.05 \\ 5.253 \\ 3.789 \end{pmatrix} \quad \text{and} \quad \begin{pmatrix} \sigma_1 \\ \sigma_2 \\ \sigma_s \end{pmatrix} = \begin{pmatrix} 0.14 \\ 0.026 \\ 0.038 \end{pmatrix}
$$

point towards a best fitting function in equation (3.4)

$$
f(x) = p_1 + p_2\,x + p_3\,\max\{0, x - x_c\} = -6.05 + 5.253\,x + 3.789\,\max\{0, x - 6.76\}\ .
$$

3.2.15   Nonlinear regression, introduction and a first example

The commands in the above section are well suited for linear regression problems, but there are many important **nonlinear** regression problems. Examine Table 3.6 to distinguish linear and nonlinear regression problems. Unfortunately nonlinear regression problems are considerably more delicate to work with and special algorithm have to be used. For many problems the critical point is to find good initial guesses for the parameters to be determined. Linear and nonlinear regression problems may also be treated as minimization problems. This is often not a good idea, as regression problems have special properties that one can, and should to, take advantage of.

Find a list of commands for nonlinear regression in Table 3.5 on page 241. In the next section `leasqr()` is used to illustrate the typical process when solving a nonlinear regression problem. Observe that this may be considerably more difficult than using linear regression.

function	parameters	
$y = a + m\,x$	a, m	linear
$y = a\,x^2 + b\,x + c$	a, b, c	linear
$y = a\,e^{c\,x}$	a, c	nonlinear
$y = d + a\,e^{c\,x}$	a, c, d	nonlinear
$y = a\,e^{c\,x}$	a	linear
$y = a\,\sin(\omega t + \delta)$	a, ω, δ	nonlinear
$y = a\,\cos(\omega t) + b\,\sin(\omega t)$	a, b	linear

Table 3.6: Examples for linear and nonlinear regression

Nonlinear least square fit with `leasqr()`

The optimization package of *Octave* provides the command `leasqr()`[9]. It is an excellent implementation of the Levenberg–Marquardt algorithm. The *Octave* package also provides one example as `leasqrdemo()` and you can examine its source.

As a first example we try to fit a function of the type

$$f(t) = A\,e^{-\alpha t}\,\cos(\omega\,t + \phi)$$

through a number of measured points (t_i, y_i). We search the values for the parameters A, α, ω and ϕ to minimize

$$\sum_i |f(t_i) - y_i|^2 \,.$$

Since the function is nonlinear with respect to the parameters A, α, ω and ϕ we can **not** use linear regression.

In *Octave* the command `leasqr()` will solve nonlinear regression problems. As an example we will:

1. Choose "exact" values for the parameters.

2. Generate normaly distributed random numbers as perturbation of the "exact" result.

3. Define the appropriate function and generate the data.

Find the code below and the generated data points are shown in Figure 3.30, together with the best possible approximation by a function of the above type.

---

[9]MATLAB users may use the code provided with the samples codes for these notes. Use `leasqr.m` and `dfdp.m`.

```
Ae = 1.5; ale = 0.1; omegae = 0.9 ; phie = 1.5;
noise = 0.1;
t = linspace(0,10,50)'; n = noise*randn(size(t));
function y = f(t,p)
  y = p(1)*exp(-p(2)*t).*cos(p(3)*t + p(4));
endfunction
y =  f(t,[Ae,ale,omegae,phie])+n;
plot(t,y,'+;data;')
```

You have to provide the function `leasqr()` with good initial estimates for the parameters. Examining the selection of points in Figure 3.30 we estimate

- $A \approx 1.5$: this might be the amplitude at $t = 0$.

- $\alpha \approx 0$: there seems to be very little damping.

- $\omega \approx 0.9$: the period seems to be slightly larger than 2π, thus ω slightly smaller than 1.

- $\psi \approx \pi/2$: the graph seems to start out like $-\sin(\omega t) = \cos(\omega t + \frac{\pi}{2})$.

The results of your simulation might vary slightly, caused by the random numbers involved.

```
A0 = 2; al0 = 0; omega0 = 1; phi0 = pi/2;
[fr,p] = leasqr(t,y,[A0,al0,omega0,phi0],'f',1e-10);
p'

yFit = f(t,p);
plot(t,y,'+', t,yFit)
legend('data','fit')
-->
p =  1.523957  0.098949  0.891675  1.545294
```

The above result contains the estimates for the parameters. For many problems the deviations from the true curve are randomly distributed, with a normal distribution, with small variance. In this case the parameters are also randomly distributed with a normal distribution. The diagonal of the covariance matrix contains the variances of the parameters and thus we can estimate the standard deviations by taking the square root.

```
pkg load optim  % load the optimization package in Octave
[fr,p,kvg,iter,corp,covp,covr,stdresid,Z,r2] =...
       leasqr(t,y,[A0,al0,omega0,phi0],'f',1e-10);
pDev = sqrt(diag(covp))'
-->
pDev =  0.0545981  0.0077622  0.0073468  0.0307322
```

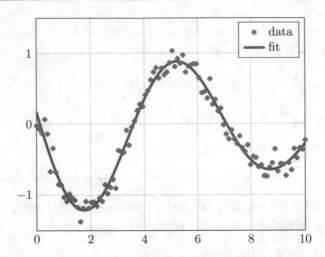

Figure 3.30: Least square approximation of a damped oscillation

With the above results obtain Table 3.7. Observe that the results are consistent, i.e. the estimated parameters are rather close to the "exact" values. To obtain even better estimates, rerun the simulation with less noise or more data points.

parameter	estimated value	standard dev.	"exact" value
A	1.52	0.055	1.5
α	0.099	0.0078	0.1
ω	0.892	0.0073	0.9
ϕ	1.54	0.031	1.5

Table 3.7: Estimated and exact values of the parameters

The above algorithm is applicable if we have only very few periods of the signal to examine. For a longer signal it typically fails miserably. Consider Fourier methods or ideas examined in Section 3.7 on a vibrating cord.

Nonlinear regression with `fsolve()`

The command `fsolve()` is used to solve systems of nonlinear equations, see Section 1.3.3. Assume that a function depends on parameters $\vec{p} \in \mathbb{R}^m$ and the actual variable x, i.e. $y = f(\vec{p}, x)$. A few (n) points are given, thus $\vec{x} \in \mathbb{R}^n$, and the same number of values of $\vec{y}_d \in \mathbb{R}^n$ are measured. For precise measurements expect $\vec{y}_d \approx \vec{y} = f(\vec{p}, \vec{x})$. Then search for the optimal parameter $\vec{p} \in \mathbb{R}^m$ such that

$$f(\vec{p}, \vec{x}) - \vec{y}_d = \vec{0}.$$

If $m < n$ this is an **over determined** system of n equation for the m unknowns $\vec{p} \in \mathbb{R}^m$. In this case the command `fsolve()` will convert the system of equations to a minimization problem

$$\|f(\vec{p}, \vec{x}) - \vec{y}_d\| \quad \text{is minimized with respect to } \vec{p} \in \mathbb{R}^m \quad .$$

It is also possible to estimate the variances of the optimal parameters, using the techniques from Section 3.2.3.

As an illustrative example some data $y = \exp(-0.2\,x) + 3$ are generated and then some noise is added. As initial parameters we use the naive guess $y(x) = \exp(0 \cdot x) + 0$. The best possible fit is determined and displayed in Figure 3.31.

```
b0 = 3; a0 = 0.2;   % chose the data
x = 0:.5:5;
noise = 0.1 * sin (100*x);
y = exp (-a0*x) + b0 + noise;

[p,fval,info,output] =  fsolve (@ (p) (exp(-p(1)*x) + p(2)-y), [0, 0]);
plot (x,y,'+', x,exp(-p(1)*x)+p(2));          xlabel('x'); ylabel('y')
```

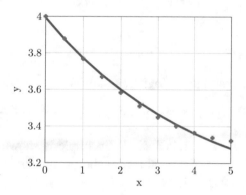

Figure 3.31: Nonlinear least square approximation with `fsolve()`

Nonlinear regression with `nonlin_curvefit()`, *Octave only*

The command `nonlin_curvefit()` does solve nonlinear regression problems too and with `curvefit_stat` some information about the results can be extracted. Assume that a function depends on parameters $\vec{p} \in \mathbb{R}^m$ and the variable x, i.e. $y = f(\vec{p}, x)$. A few points are given with some added noise. Then search for the optimal parameter $\vec{p} \in \mathbb{R}^m$ such that

$$\sum_{i=1}^{n} |f(\vec{p}, x_i) - y_i|^2 \quad \text{is minimized with respect to } \vec{p} \in \mathbb{R}^m \quad .$$

The command `curvefit_stat` does estimate the variances of the optimal parameters, using the techniques from Section 3.2.3. As an illustrative example some data points $y_i = \exp(-0.2\,x_i) + 3$ are generated and then some noise is added. As initial parameters use the

naive guess $y(x) = \exp(0 \cdot x) + 0$. The best possible fit is determined by the code below
and the graphical result is similar to Figure 3.31.

```
b0 = 3; a0 = 0.2;  % chose the data
x = 0:.1:5; noise = 0.1 * sin (100*x);
f = @(p,x)exp (-p(1)*x) + p(2);
y = f([a0;b0],x) + noise;
[p, fy, cvg] = nonlin_curvefit (f, [0;0], x, y);
settings = optimset('ret_covp', true,'objf_type','wls');
FitInfo = curvefit_stat (f, p, x, y, settings);
[p , sqrt(diag(FitInfo.covp))]
plot(x,y,'+', x, f(p,x));  xlabel('x'); ylabel('y')
```

3.2.16 Nonlinear regression with a logistic function

Many growth phenomena can be described by rescaling and shifting the basic logistic[10]
growth function $g(x) = \frac{\exp(x)}{1+\exp(x)} = \frac{1}{1+\exp(-x)}$. It is easy to see that this function is mono-
tonically increasing and

$$\lim_{x\to-\infty} g(x) = 0 \quad , \quad g(0) = \frac{1}{2} \quad \text{and} \quad \lim_{x\to+\infty} g(x) = 1 \,.$$

By shifting and rescaling examine the logistic function

$$f(x) = p_1 + p_2\, g(p_3\,(x - p_4)) = p_1 + \frac{p_2}{1 + \exp(-p_3\,(x - p_4))} \tag{3.5}$$

with the four parameters p_i, $i = 1, 2, 3, 4$. An example is shown in Figure 3.32. For the
given data points (in red) the optimal values for the parameters p_i have to be determined.
This is a nonlinear regression problem.

An essential point for a nonlinear regression problems is to find good estimates for the
values of the parameters. Thus we examine the graph of the logistic function (3.5) carefully:

- At the midpoint $x = p_4$ find $f(p_4) = p_1 + p_2\,\frac{1}{2}$.

- For the extreme values observe $\lim_{x\to-\infty} f(x) = p_1$ and $\lim_{x\to+\infty} f(x) = p_1 + p_2$.

- The maximal slope is at the midpoint and given by[11] $f'(p_4) = \frac{p_2\,p_3}{4}$.

Assuming $p_2, p_3 > 0$ we can now find good estimates for the parameter values.

- p_1 offset: minimal height of the data points

- p_2 amplitude: difference of maximal and minimal value

- p_3 slope: the maximal slope is $m = \frac{p_2\,p_3}{4}$ and thus $p_3 = \frac{4\,m}{p_2}$

- p_4 midpoint: average of x values.

Based on this use the code below to determine the estimated values.

---

[10] Also called sigmoid function

[11] For $g(x) = \frac{\exp(x)}{1+\exp(x)}$ use $g'(0) = \frac{1}{4}$ and then some rescaling to determine $f'(p_4)$.

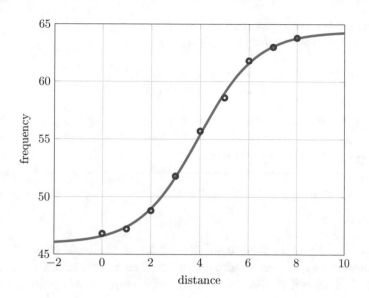

Figure 3.32: Data points and the optimal fit by a logistic function

```
x_data =   [0 1 2 3 4 5 6 7 8]';
y_data =   [46.8 47.2 48.8 51.8 55.7 58.6 61.8 63 63.8]';

p1 = min(y_data);
p2 = max(y_data)-min(y_data);
p3 = 4*max(diff(y_data)./diff(x_data))/p2;
p4 = mean(x_data);
```

This result can now be used to apply a nonlinear regression, using the functions `leasqr()`, `fsolve()` or `lsqcurvefit()`.

Solution by `leasqr()`

To determine the optimal values of the parameters:

- Define the logistic function, with the parameters p_i.

- Call `leasqr()`, returning the values and the covariance matrix. On the diagonal of the covariance matrix find the estimated variances of the parameters p_i.

Find the result in Figure 3.32. As numerical result the optimal values of p_i and their standard deviations are shown. In addition the number of required iterations and the resulting residual $(\sum_{i=1}^{n}(f(x_i) - y_i)^2)^{1/2}$ is displayed.

```
f = @(x,p) p(1) + p(2)*exp(p(3)*(x-p(4)))./(1+exp(p(3)*(x-p(4))));
[fr, p, ~, iter, ~, covp] = leasqr(x_data,y_data,[p1,p2,p3,p4],f);
```

```
optimal_values = [p';sqrt(diag(covp))']
iter_residual = [iter,norm(fr-y_data)]

figure(1); plot(x,f(x,p),x_data,y_data,'or')
          xlabel('distance'); ylabel('frequency')
-->
optimal_values =    45.931829    18.428664    0.838742    3.932786
                     0.380858     0.645210    0.062353    0.080993
iter_residual   =    4  0.64832
```

Solution by `fsolve()`

The command `fsolve()` is used to solve systems of nonlinear equations. If more data points than parameters are given (more equations than unknowns), then a nonlinear least square solution is determined. Thus we can solve the above problem using this command.

```
f2 = @(p) p(1) + p(2)*exp(p(3)*(x_data-p(4)))./(1+exp(p(3)*...
                  (x_data-p(4))))-y_data;
[p,fval] = fsolve(f2,[p1,p2,p3,p4]);
optimal_values = p
residual = norm(fval)
-->
optimal_values =    45.93183    18.42866    0.83874    3.93279
residual       =     0.64832
```

It is no surprise that the same result is found. `fsolve()` does not estimate standard deviations for the parameters. One might use `nlparci()` to determine confidence intervals.

Solution by `lsqcurvefit()`

With the command `lsqcurvefit()` the method of nonlinear least squares can be used to fit a function to data points. A solution for the above problem is given by

```
f3 = @(p,x_data) p(1) + p(2)*exp(p(3)*(x_data-p(4)))./(1+exp(p(3)*...
                  (x_data-p(4))));
[p,residual] = lsqcurvefit(f3,[p1,p2,p3,p4],x_data,y_data)
optimal_values = p'
residual = sqrt(residual)
-->
optimal_values =    45.93183    18.42866    0.83874    3.93279
residual       =     0.64832
```

It is no surprise that the same result is found. `lsqcurvefit()` does not estimate standard deviations for the parameters.

3.2.17 Nonlinear regression with an arctan runction

Similar to the previous section on can use a rescaled and shifted arctan function to describe a similar curve. The function for the regression is thus given by

$$f(x) = p_1 + p_2 \arctan(p_3 (x - p_4)) . \tag{3.6}$$

For this function observe:

- At the midpoint $x = p_4$ find $f(p_4) = 0$.

- For the extreme values $\lim_{x \to -\infty} f(x) = -\frac{\pi}{2}$ and $\lim_{x \to +\infty} f(x) = +\frac{\pi}{2}$ and

- The maximal slope is at the the midpoint and given by $f'(p_4) = p_2 p_3$.

Assuming $p_2, p_3 > 0$ we can now find good estimates for the parameter values.

- p_1 offset: average height of the data points

- p_2 amplitude: difference of maximal and minimal value, divided by π.

- p_3 slope: the maximal slope is $m = p_2 p_3$ and thus $p_3 = \frac{m}{p_2}$

- p_4 midpoint: average of x values.

Based on this use the code below to determine the estimated values.

```
p1 = mean(y_data)
p2 = (max(y_data)-min(y_data))/pi
p3 = max(diff(y_data)./diff(x_data))/p2
p4 = mean(x_data)
```

This result can now be used to apply a nonlinear regression, using one of the functions `leasqr()`, `fsolve()` or `lsqcurvefit()`.

Solution by `leasqr()`

With the code below the optimal values for the parameters p_i and the estimated standard deviations are computed. In addition the number of required iterations and the resulting residual are shown. Find the result in Figure 3.33.

```
f = @(x,p) p(1) + p(2)*atan(p(3)*(x-p(4)));
[fr, p, ˜, iter,˜, covp] = leasqr(x_data,y_data,[p1,p2,p3,p4],f);
optimal_values = [p';sqrt(diag(covp))']
iter_residum = [iter,norm(fr'-y_data)]

x = linspace(-2,10);
figure(1); plot(x,f(x,p),x_data,y_data,'or')
          xlabel('distance'); ylabel('frequency')
```

```
-->
optimal_values =    55.112684    7.731968    0.521908    3.917840
                     0.244363    0.481495    0.065494    0.106043
iter_residum  =    4  0.79932
```

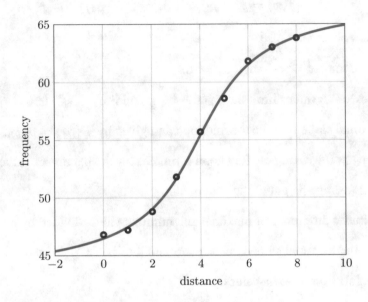

Figure 3.33: Data points and the optimal fit by an arctan function

The residual norm of 0.799 for the arctan function is larger than the residual norm of 0.648 for the approximation by a logistic function. Thus the approximation by an arctan function is slightly worse than the logistic approach.

Solution by `fsolve()`

We may again of `fsolve()` for an over determined system to apply the nonlinear regression algorithm.

```
f2 = @(p) p(1) + p(2)*atan(p(3)*(x_data-p(4)))-y_data;
[p,fval] = fsolve(f2,[p1,p2,p3,p4]);
optimal_values = p
residal = norm(fval)
-->
optimal_values =    55.11278    7.73150    0.52198    3.91788
residal       =    0.79932
```

It is no surprise that the same result is found. `fsolve()` does not estimate standard deviations for the parameters.

Solution by `lsqcurvefit()`

With the command `lsqcurvefit()` the method of nonlinear least squares can be used to fit a function to data points. A solution for the above problem is given by

```
f3 = @(p,x_data) p(1) + p(2)*atan(p(3)*(x_data-p(4)));
[p,residual] = lsqcurvefit(f3,[p1,p2,p3,p4],x_data,y_data)
optimal_values = p'
residual = sqrt(residual)
-->
optimal_values =   55.11268    7.73197    0.52191    3.91784
residual       =    0.79932
```

It is no surprise that the same result is found. `lsqcurvefit()` does not estimate standard deviations for the parameters.

3.2.18 Approximation by a Tikhonov regularization

A different approach to generate a function fitting the given data points (x_i, y_i) is given by a Tikhonov regularization. For given parameters $\lambda_1 \geq 0$ and $\lambda_2 \geq 0$ find the function $u(x)$ minimizing the functional

$$F(\lambda_1, \lambda_2) = \sum_{i=1}^{n}(u(x_i) - y_i)^2 + \lambda_1 \int_{-2}^{10} (u'(x))^2 \, dx + \lambda_2 \int_{-2}^{10} (u''(x))^2 \, dx .$$

The package `splines` in *Octave* provides a command[12] solving this problem and the code below leads to Figure 3.34. The shape of the solution can be modified by changing the values of $\lambda_1 = 0.1$ and $\lambda_2 = 0.03$. The residual of 0.413 is smaller than the residuals by the nonlinear regression approaches. An additional advantage of the regularization approach is that one does not have to choose the type of function, i.e. arctan or logistic.

```
―――――――――――――――――――――― Octave ――――――――――――――――――――――
x_data =  [0 1 2 3 4 5 6 7 8];
y_data =  [46.8 47.2 48.8 51.8 55.7 58.6 61.8 63 63.8];

F1.lambda = 1e-1;   F2.lambda = 3e-2;
pkg load splines
[x,y] = regularization([x_data',y_data'],[-2,10],100,F1,F2);

figure(1); plot(x,y,x_data,y_data,'or')
           xlabel('distance'); ylabel('frequency')

y_fit = interp1(x,y,x_data);
residual = norm(y_fit-y_data)
-->
residual =  0.41310
```

---

[12] Additional information is available at github.com/AndreasStahel/RegularizationOctave.

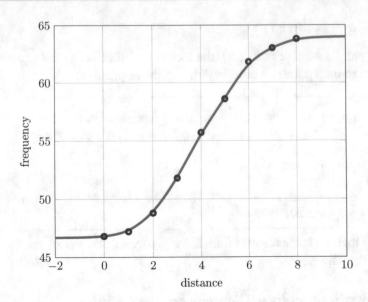

Figure 3.34: Data points and the optimal fit by a Tikhonov regularization

3.2.19 A real world nonlinear regression problem

For her Bachelor project Linda Welter had to solve a nontrivial nonlinear regression problem. The dependent variable was assumed to be the sum of a linear function and a trigonometric function with exponentially decaying amplitude. For a given set of points examine a function of the form

$$y = f(t) = p_1 \cdot \exp(-p_2 \cdot t) \cdot \cos(p_3 \cdot t + p_4) + p_5 + p_6 \cdot t$$

and one has to find the optimal values for the six parameters p_i. At first sight this is a straight forward application for the function leasqr(), presented in the previous section. Thus we run the following code[13].

```
ReadData  % read the data
f_exp_trig_lin = @(t,p)p(1)*exp(-p(2)*t).*cos(p(3)*t + p(4)) +...
                   p(5) + p(6)*t;

p_in = zeros(6,1);  % guess for initial values for parameters
[fr,p] = leasqr(t,y,p_in,f_exp_trig_lin,1e-8);

y_fit1 = f_exp_trig_lin(t,p);
figure(1); plot(t,y,t,y_fit1)
           xlabel('t'); legend('y','y_{fit1}'); grid on
```

---

[13]The presented code works with *Octave*, for MATLAB minor adaptations are required.

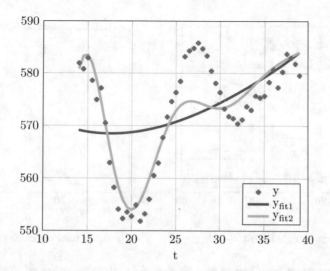

Figure 3.35: Raw data and two failed attempts of nonlinear regression

This first result (the red curve in Figure 3.35) is clearly of low quality and we try to improve by using better initial estimates for the parameters. Examine the graph carefully and estimate the values, leading to the code and second regression result (yellow) in Figure 3.35.

```
p_in = [1,0,0.5,0,570,0]; % guess for initial values for parameters
[fr,p] = leasqr(t,y,p_in,f_exp_trig_lin,1e-8);
y_fit2 = f_exp_trig_lin(t,p);
figure(1); plot(t,y,t,y_fit1,t,y_fit2)
           xlabel('t'); legend('y','y_{fit1}','y_{fit2}'); grid on
```

To improve upon the above result we need a plan on how to proceed, and then implement the plan.

1. First determine a good estimate on the linear function by fitting a straight line through those points.

2. The difference of the straight line and the given data should be a trigonometric function with exponentially decaying amplitude. Use a nonlinear regression to determine those parameters.

3. Use the above parameter results to run a full nonlinear regression, but now with good initial guesses.

Now implement and test the above, step by step.

1. Using LinearRegression() fit a straight line through the given data points.

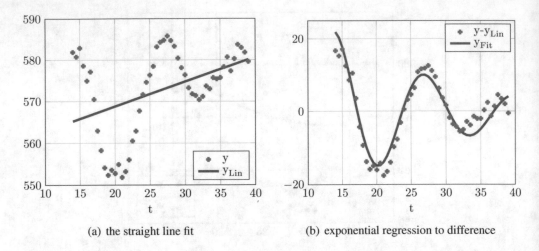

(a) the straight line fit   (b) exponential regression to difference

Figure 3.36: Regression by a linear and an exponentially decaying trigonometric function

```
%%% fitting a straight line
F = ones(length(t),2); F(:,2) = t;
pLin = LinearRegression(F,y)
yLin = F*pLin;

figure(2); plot(t,y,'+-',t,yLin)
         xlabel('t'); legend('y','yLin'); grid on
-->
pLin =    556.84180
            0.59710
```

Thus the best possible line has a slope of approximately 0.6 and a y–intercept at $y \approx 557$. This is confirmed by Figure 3.36(a).

2. Now we examine the difference of the optimal straight line and the actual data. Using a new function and `leasqr()` we find the best fitting function. The initial parameters are estimated by using Figure 3.36(b). Find the estimated standard deviations in the square roots of the diagonal elements of the covariance matrix.

```
% nonlinear regression with leasqr
AEst = 50; alphaEst = log(16/12)/14; omegaEst = 0.5 ; phiEst =-15;

f_exp_trig = @(t,p)p(1)*exp(-p(2)*t).*cos(p(3)*t + p(4));

[fr,p,kvg,iter,corp,covp] = leasqr(...
        t,y-yLin,[AEst,alphaEst,omegaEst,phiEst],f_exp_trig,1e-4);
pVal = p'
```

```
  pDev = sqrt (diag (covp))'
  -->
  pVal =   51.054390    0.060480    0.476920  -12.902044
  pDev =    8.6404054   0.0078864   0.0082692   0.1842718
```

The above implies

$$y(t) - y_{lin}(t) \approx 51 \exp(-0.06\,t) \cdot \cos(0.477\,t - 13)$$

The estimated standard deviations of the parameters are rather large, e.g. for the initial amplitude 51.1 ± 8.6. Now we can generate Figure 3.36(b).

```
  yFit = f_exp_trig(t,p);
  figure(3); plot(t,y-yLin,'+-',t,yFit)
             xlabel('t'); legend('y-yLin','yFit'); grid on
```

To verify the above result, rerun `leasqr ()` using the previously obtained parameters as starting values and asking for more accuracy. The result should not differ drastically from the above. This is confirmed by the following code and result.

```
  [fr,p,kvg,iter,corp,covp,covr,stdresid,Z,r2] =...
     leasqr(t,t-tLin,pVal,f_exp_trig,1e-8);
  pVal = p'
  pDev = sqrt (diag (covp))'
  -->
  pVal =   51.011746   0.06044   0.477038   -12.904584
  pDev =    8.630768   0.00788   0.008269     0.184299
```

3. Now we have good estimates for all parameters and are ready to rerun the original, fully nonlinear regression.

```
  pNew = [p;pLin];   % combine the two parameter sets
  [fr,p2,kvg,iter,corp,covp]=leasqr(t,y,pNew,f_exp_trig_lin,1e-8);
  p2Val = p2'
  p2Dev = sqrt (diag (covp))'
  yFit2 = f_exp_trig_lin(t,p2);
  figure(4); plot(t,y,'+-',t,yFit2)
             xlabel('t'); legend('y','yFit2'); grid on
  -->
  p2Val =   56.050   0.064194   0.48503   -13.162   550.00   0.83888
  p2Dev =    7.962   0.006757   0.00781     0.177     1.54   0.05403
```

$$y = f(t) = 56 \cdot \exp(-0.064 \cdot t) \cdot \cos(0.485 \cdot t - 13.16) + 550 + 0.84 \cdot t$$

The result in Figure 3.37 is obviously superior to the naive attempt shown in Figure 3.35.

This example clearly illustrates that one of the most important aspect of nonlinear regression problems is to have good estimates for the parameters to be determined. If you start a fishing expedition in the dark for too many parameters, you will fail.

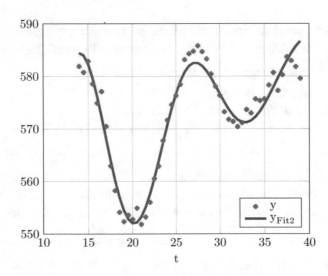

Figure 3.37: The optimal fit, using nonlinear regression

3.2.20   The functions `lsqcurvefit` and `lsqnonlin`

The above nonlinear problems could be solved by the similar functions `lsqcurvefit()` and/or `lsqnonlin()`. For a project I used `nlparci()` to determine the confidence interval, this function is not available yet in *Octave*. In *Octave* those two functions use the more general functions `nonlin_curvefit` and `nonlin_residmin`.

3.2.21   List of codes and data files

In the previous section the codes and data files in Table 3.8 were used.

3.3   Regression with Constraints

In this section a few regression problems with an additional constraint are examined.

- The first example is the fitting of a straight line through a set of given points, using the geometric distance. In the previous section the vertical distance was used.

file name	function
`LinearRegression.m`	function to perform general linear regression, *Octave*
`LinearRegression.m`	for MATLAB in subdirectory `Matlab`
`LinearRegression1.m`	temporary code for linear regression
`NSHU550ALEDwide.pdf`	data sheet for an LED
`ReadGraph.m`	script file to grab data for LED from PDF file
`LEDdata.m`	script file with the intensity data for above LED
`LinearMotorData.m`	script file with the data on the linear motor
`LinearMotor1.m`	script file for a first regression
`LinearMotor2.m`	improved script file for the regression
`OrientationTest.m`	script file for calibration
`OrientationData.m`	data set 1 for the calibration
`OrientationData2.m`	data set 2 for the calibration
`SphereRegression.m`	script file for radius of sphere
`SphereData.csv`	data file for radius of sphere
`ForceSensor.m`	code for the two spring system
`DistanceData.dat`	data the two spring system
`ForceData.dat`	data the two spring system
`EvaluateBreak.m`	function to evaluate situation
`RegressExpLinTrig.m`	script file for the real nonlinear regression
`ReadData.m`	script file to read data for the above
`Matlab`	directory with MATLAB compatible files

Table 3.8: Codes and data files for section 3.2

- The same idea and code is used to fit a plane in space $\mathbb{R}^3$ to a given set of data points.

- The idea and algorithm is used to identify straight lines in an image.

- The problem of fitting an ellipse through a given set of data points in a plane is examined next.

3.3.1 Example 1: Geometric line fit

The Hessian form of the equation of a straight line is given by

$$n_1\,x + n_2\,y + d = \langle \begin{pmatrix} n_1 \\ n_2 \end{pmatrix}, \begin{pmatrix} x \\ y \end{pmatrix} \rangle + d = \langle \vec{n}, \vec{x} \rangle + d = 0 \quad \text{where} \quad \|\vec{n}\| = 1\ .$$

For a number of points $(x_i\,,\,y_i)$ for $1 \le i \le n$ the signed distance r_i of these points from the line is given by

$$r_i = \langle \begin{pmatrix} n_1 \\ n_2 \end{pmatrix}, \begin{pmatrix} x \\ y \end{pmatrix} \rangle + d$$

Thus the best fitting straight line may be characterized by sum of the squared distances to the points being minimal. Thus examine the problem of minimizing the length of the vector $\vec{r}$. This leads to the formulation in Figure 3.38.

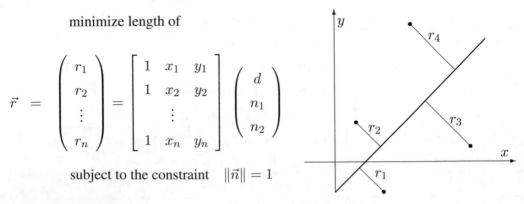

Figure 3.38: A straight line with minimal distance from a set of given points

We will first present a general approach to solve this type of problem and then come back to this example, with a solution. There are other algorithms to solve this problem, e.g. by using PCA (Principal Component Analysis), see [Stah08].

3.3.2 An algorithm for minimization problems with constraints

Most of the results in this section are based on [GandHreb95, §6].

Description of the algorithm

For a $n \times m$ matrix $\mathbf{F}$ with $n > m$ minimize the length of the vector $\vec{r}$ where

$$\mathbf{F} \cdot \vec{p} = \vec{r} \quad \text{subject to} \quad \|\vec{n}\| = 1 \quad \text{where} \quad \vec{p} = \begin{pmatrix} \vec{d} \\ \vec{n} \end{pmatrix} \in \mathbb{R}^{m_1 + m_2}. \tag{3.7}$$

The algorithm is based on a QR factorization and one might consult Section 3.2.7.

$$\mathbf{F} \cdot \vec{p} = \mathbf{Q} \cdot \mathbf{R} \cdot \vec{p}$$

Examine the matrix $\mathbf{R}$ in block form

$$\mathbf{R} = \begin{bmatrix} \mathbf{R}_u \\ \mathbf{0} \end{bmatrix} = \begin{bmatrix} \mathbf{R}_{1,1} & \mathbf{R}_{1,2} \\ \mathbf{0} & \mathbf{R}_{2,2} \\ \mathbf{0} & \mathbf{0} \end{bmatrix}$$

where $\mathbf{R}$, $\mathbf{R}_{1,1}$ and $\mathbf{R}_{2,2}$ are upper triangular matrices. This leads to a new formulation of the minimization problem. For each of the expressions below the length of the vector on the RHS has to be minimized, subject to the constraint $\|\vec{n}\| = 1$.

$$\mathbf{F} \cdot \vec{p} = \mathbf{Q} \cdot \mathbf{R} \cdot \vec{p} = \vec{r}$$
$$\mathbf{R} \cdot \vec{p} = \mathbf{Q}^T \cdot \vec{r} = \vec{z}$$

$$\begin{bmatrix} \mathbf{R}_u \\ \mathbf{0} \end{bmatrix} \cdot \begin{pmatrix} \vec{d} \\ \vec{n} \end{pmatrix} = \begin{pmatrix} \vec{z}_u \\ \vec{z}_l \\ \vec{0} \end{pmatrix}$$

$$\begin{bmatrix} \mathbf{R}_{1,1} & \mathbf{R}_{1,2} \\ \mathbf{0} & \mathbf{R}_{2,2} \end{bmatrix} \cdot \begin{pmatrix} \vec{d} \\ \vec{n} \end{pmatrix} = \begin{pmatrix} \vec{z}_u \\ \vec{z}_l \end{pmatrix}$$

For a given vector $\vec{n}$ the first set of equations in

$$\mathbf{R}_{1,1} \cdot \vec{d} = -\mathbf{R}_{1,2} \vec{n} + \vec{z}_u$$
$$\mathbf{R}_{2,2} \cdot \vec{n} = \vec{z}_l$$

can be solved such that $\vec{z}_u = \vec{0}$. Thus we have to minimize the length of $\vec{z}_l$ by finding the best vector $\vec{n}$. This subproblem can be solved with two different algorithms.

- Eigenvalue computation: Examine the gradient with respect to $\vec{n}$ of

$$\|\vec{z}_l\|^2 = \langle \mathbf{R}_{2,2} \cdot \vec{n}, \mathbf{R}_{2,2} \cdot \vec{n} \rangle = \langle \mathbf{R}_{2,2}^T \cdot \mathbf{R}_{2,2} \cdot \vec{n}, \vec{n} \rangle$$
$$\nabla_{\vec{n}} \|\vec{z}_l\|^2 = 2 \mathbf{R}_{2,2}^T \cdot \mathbf{R}_{2,2} \cdot \vec{n}$$

to realize that the smallest eigenvalue of the symmetric matrix $\mathbf{R}_{2,2}^T \cdot \mathbf{R}_{2,2}$ leads to the minimal value for $\|\vec{z}_l\|^2$ and the corresponding eigenvector equals the vector $\vec{n}$ for which the minimum is attained.

- Singular value decomposition (SVD): The matrix $\mathbf{R}_{2,2}$ can be decomposed as product of three matrices

$$
\mathbf{R}_{2,2} = \mathbf{U} \cdot \begin{bmatrix} \sigma_1 & & & \\ & \sigma_2 & & \\ & & \ddots & \\ & & & \sigma_k \end{bmatrix} \cdot \mathbf{V}^T \quad \text{where} \quad \sigma_1 \geq \sigma_2 \geq \ldots \geq \sigma_k > 0
$$

with orthogonal matrices $\mathbf{U}$ and $\mathbf{V}$. The smallest value σ_k in the diagonal matrix gives the minimal value of the function to be minimized and the last column of $\mathbf{V}$ equals the vector $\vec{n}$ for which the minimum is attained.

Use this vector $\vec{n}$ and

$$
\mathbf{R}_{1,1} \cdot \vec{d} = -\mathbf{R}_{1,2}\,\vec{n}
$$

to detremine the optimal solution of the problem in equation (3.7).

Weighted regression with constraint

The result in the previous section can be modified to take weights of the different points into account. Instead of minimizing the standard norm

$$
\|\vec{r}\|^2 = \sum_{i=1}^{n} r_i^2
$$

we want to minimize the weighted norm

$$
\|\mathbf{W} \cdot \vec{r}\|^2 = \sum_{i=1}^{n} \sqrt{w_i}\, r_i^2 \ .
$$

Using the weight matrix $\mathbf{W}$ and the QR factorization of $\mathbf{W}\cdot\mathbf{F}$ the algorithm can be modified.

$$
\begin{aligned}
\mathbf{F} \cdot \vec{p} &= \vec{r} \quad \text{weighted length to be minimized} \\
\mathbf{W} \cdot \mathbf{F} \cdot \vec{p} &= \mathbf{W} \cdot \vec{r} \quad \text{standard length to be minimized} \\
\mathbf{Q} \cdot \mathbf{R} \cdot \vec{p} &= \mathbf{W} \cdot \vec{r} \quad \text{standard length to be minimized} \\
\mathbf{R} \cdot \vec{p} &= \mathbf{Q}^T \cdot \mathbf{W} \cdot \vec{r} = \vec{z} \\
\begin{bmatrix} \mathbf{R}_{1,1} & \mathbf{R}_{1,2} \\ \mathbf{0} & \mathbf{R}_{2,2} \end{bmatrix} \cdot \begin{pmatrix} \vec{d} \\ \vec{n} \end{pmatrix} &= \begin{pmatrix} \vec{z}_u \\ \vec{z}_l \end{pmatrix}
\end{aligned}
$$

The remaining part of the algorithm is unchanged. The final code can be found in Figure 3.39.

RegressionConstraint.m

```
function [p,y_var,r] = RegressionConstraint(F,nn,weight)

% [p,y_var,r] = RegressionConstraint(F,nn)
% [p,y_var,r] = RegressionConstraint(F,nn,weight)
% regression with a constraint
%
% determine the parameters p_j  (j=1,2,...,m) such that the function
% f(x) = sum_(i=1,...,m) p_j*f_j(x) fits as good as possible to the
% given values y_i = f(x_i), subject to the constraint that the norm
% of the last nn components of p equals 1
%
% parameters
% F   n*m matrix with values of the basis functions at support points
%      in column j give the values of f_j at the points x_i
% (i=1,2,...,n)
% nn number of components to use for constraint
% weight   n column vector of given weights
%
% return values
% p       m vector with the estimated values of the parameters
% y_var estimated variance of the error
% r       residual  sqrt(sum_i (y_i- f(x_i))^2)

if ((nargin < 2)||(nargin>=4))
  help RegressionConstraint
  error('wrong number of arguments in RegressionConstraint()');
end%if
[n,m] = size(F);
if (nargin==2)  % set uniform weights if not provided
  weight = ones(n,1);
end%if

[Q,R] = qr(diag(weight)*F,0);
R11 = R(1:m-nn,1:m-nn);
R12 = R(1:m-nn,m nn+1:m);
R22 = R(m-nn+1:m,m-nn+1:m);
[u,l,v] = svd(R22);
p = [-R11\(R12*v(:,nn));v(:,nn)];

residual = F*p;                        % compute the residual vector
r = norm(diag(weight)*residual);       % and its norm
      % variance of the weighted y-errors
y_var = sum((residual.^2).*(weight.^4))/(n-m+nn);
```

Figure 3.39: Code for RegressionConstraint()

3.3.3   Example 1: continued

Now we use the presented algorithm to solve the problem of fitting a straight line through some given points.

The file `LineData.m` contains x and y values of a few points and with the code below we load the data and display the result.

```
LineData;
n = length(xi);
F1 = [ones(n,1) xi yi];
[p1,yvar,residual1orthogonal] = RegressionConstraint(F1,2);
p1    % display the optimal parameters
x = -2:0.1:2;
y = -(p1(1)+p1(2)*x)/p1(3);
plot(xi,yi,'*r',x,y,'g');
```

The equation of the line with minimal orthogonal distances is determined as

$$0.34978 + 0.70030\, x - 0.71385\, y = 0$$

or in the standard form

$$y = 0.49000 + 0.98101\, x\ .$$

A standard linear regression will minimize the sum of the squares of the **vertical** distances.

```
F2 = [ones(n,1) xi];
[p2,yvar,residual2vertical] = LinearRegression(F2,yi);
p2
x  = -2:0.1:2;
y2 = p2(1)+p2(2)*x;
plot(xi,yi,'*r',x,y,'b',x,y2,'g');
```

This optimal solution is given by the equation

$$y = 0.47547 + 0.91919\, x$$

or in the Hessian normal form

$$0.35006 + 0.67673\, x - 0.73623\, y = 0\ .$$

Thus the straight line in Figure 3.40 with the slightly smaller slope minimizes the vertical distances to the given set of points.

The above two solutions should be compared, leading to the results in the table below.

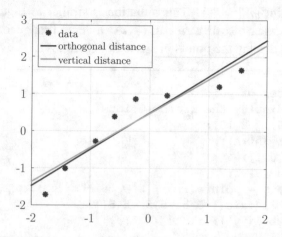

Figure 3.40: Some points with optimal vertical and orthogonal distance fit

```
y1new = -(p1(1)+p1(2)*xi)/p1(3);
residual1orthogonal
residual1vertical = sqrt(sum((yi-y1new).^2))

pp = [p2(1);p2(2);-1]/sqrt(1+p2(2)^2);
residual2vertical
residual2orthogonal = sqrt(sum((F1*pp).^2))
```

	orthogonal distance	vertical distance
optimized for orthogonal distance	0.787	1.103
optimized for vertical distance	0.799	1.085

This table confirms the results to be expected, e.g. when optimizing for orthogonal distance, then the orthogonal distance is minimal.

3.3.4 Detect the best plane through a cloud of points

Assume to work with a cloud of n points $(x_i, y_i, z_i)^T \in \mathbb{R}^3$. Then try to determine the equation of the best fitting plane. This is the context of the previous section. Minimize the length of the vector $\vec{r}$ wher

$$\vec{r} = \begin{pmatrix} r_1 \\ r_2 \\ \vdots \\ r_n \end{pmatrix} = \begin{bmatrix} 1 & x_1 & y_1 & z_1 \\ 1 & x_2 & y_2 & z_2 \\ & \vdots & & \\ 1 & x_n & y_n & z_n \end{bmatrix} \begin{pmatrix} d \\ n_1 \\ n_2 \\ n_3 \end{pmatrix}$$

subject to the constraint $\|\vec{n}\|^2 = 1$. This situation is similar to Figure 3.38 and thus use the command `RegressionConstraint()`. As a demo first generate a cloud of points almost on a plane and display the points in space in Figure 3.41.

```
nn = 100; % number of points
% generate and display the random points
A = [1 2 3;4 5 6; 7 8 10];
[V,lambda] = eig(A'*A);
T = V*diag([1 3 0.1])*V';
points = randn(nn,3)*T;
x = points(:,1); y = points(:,2); z = points(:,3);
plot3(x,y,z,'b*'); axis([-5 5 -5 5 -5 5])
xlabel('x'); ylabel('y'); zlabel('z');
```

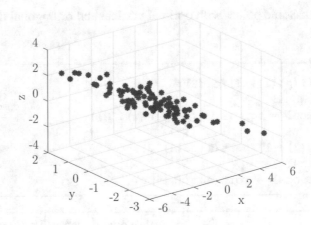

Figure 3.41: A cloud of points, almost on a plane

Then determine the normal vector $\vec{n}$ and the distance d of the optimal plane from the origin.

```
p = RegressionConstraint([ones(nn,1),points],3);
p'
-->
-0.023182   0.469395   0.548034   0.692335
```

The above result implies

$$\vec{n} \approx \begin{pmatrix} 0.469 \\ 0.548 \\ 0.692 \end{pmatrix} \quad \text{and} \quad d \approx -0.0232 .$$

There are other algorithms to solve this problem, e.g. by using PCA (Principal Component Analysis), see [Stah08].

3.3.5 Identification of a straight line in a digital image

The above method of regression with constraint and weights can be used to identify the parameters of a straight line in a digital image. The basic idea is to use a weighted linear regression where dark points have a large weight and white spots will have no weight.

In Figure 3.42 find photographs of two lines, a freehand version (3.42(a)) and one generated with a ruler (3.42(b)). The digital camera produces the images in the `jpg` format and with the command `convert` from the **ImageMagick** suite the high resolution photographs were transformed into 256×256 bitmaps, using the `bmp` format.

```
convert  Line1.jpg  -scale 256x256! Line1.bmp
convert  Line2.jpg  -scale 256x256! Line2.bmp
```

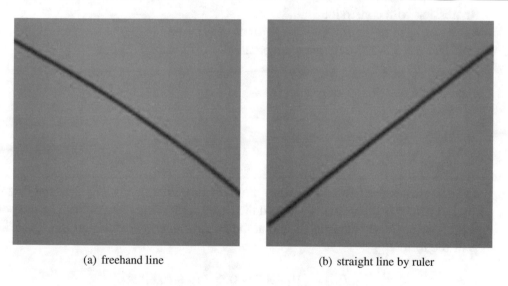

(a) freehand line   (b) straight line by ruler

Figure 3.42: Two photographs of (almost) straight lines

Now we try to determine the obvious straight lines in those images.

- Read the file and display the result
  We use the command `imread` to get the data of the picture into *Octave*.

```
aa = imread('Line1.bmp');
aa = rgb2gray(aa);
colormap(gray);
figure(1); imagesc(aa);
```

Since bright spots correspond to a high value we have to revert the scaling to obtain a high weight for the dark pixels. In addition we subtract the minimal value and ignore all points with values below a given threshold.

```
a = 255 - aa(:);      a = a - min(a(:));
pos = find(a>(0.4*max(a))); % select the points to be considered
a = double(a(pos));
numberOfPoints = length(a)   % number of points to be considered

[xx,yy] = meshgrid(1:256,1:256);
x = xx(:); x = x(pos);  y = yy(:); y = y(pos);
figure(2);   plot3(x,y,a)
-->
numberOfPoints =  1608
```

The resulting 3D graph clearly shows the points to be on a straight line.

- Regression with constraint
  Since we have to determine straight lines at all possible angles we use the method from Section 3.3.2 to determine the parameters of the straight line.

```
p = RegressionConstraint([ones(size(x)) x y],2,a)
-->
p = 16.9967
     0.5534
    -0.8329
```

Thus the distance of the origin from the line is approximately 17. Observe that the top left corner is the origin $(0, 0)$ and the lower right corner corresponds to the point $(255, 255)$. These values indicate that the straight line in the left part of Figure 3.42 is of the form

$$
\begin{aligned}
0 &= 16.9967 + 0.5534\, x - 0.8324\, y \\
y &= 20.4068 + 0.6645\, x
\end{aligned}
$$

- Estimation of the variance of the parameters
  The command RegressionConstraint() does not give any information on the variance of the parameter $\vec{p}$. To obtain this information we rotate the straight line in a horizontal position and then apply standard linear regression, including the estimation of the variance of the parameters.

```
beta = pi/2 - atan2(p(3),p(2));
rotation = [cos(beta) -sin(beta);sin(beta) cos(beta)];
newcoord = rotation*[x';y'];
xn = newcoord(1,:)';   yn = newcoord(2,:)';
```

```
[p2,d_var,r,p2_var] = LinearRegression([ones(size(x)) xn],yn,a);
p2'
p2_var'
-->
p2'      = -1.6997e+01    3.3965e-16
p2_var' =  2.5183e-02    6.6170e-07
```

The results of $\vec{p}_2 \approx (-17, 3 \cdot 10^{-16})$ confirm the distance from the origin and also show that the rotated line is in fact horizontal. The values of p2_var imply that the position of the line is determined with a standard deviation of $\sqrt{0.025} = 0.16$ and for the angle obtain a standard deviation of $\sqrt{6.62 \cdot 10^{-7}} \approx 0.0008 \approx 0.05°$.

It is worth observing that the position of the straight line is determined with a sub-pixel resolution, getting some help from statistics.

All of the above code may be rerun with the image in Figure 3.42(b). The only change is to replace the file name Line1.bmp by Line2.bmp. There are other algorithms to solve this problem, e.g. by using PCA (Principal Component Analysis), see [Stah08].

3.3.6 Example 2: Fit an ellipse through some given points in the plane

Ellipse, axes parallel to coordinate axes

The equation of an ellipse with axes parallel to the coordinate axes and semi-axes of length a and b with center at (x_0, y_0) can be given in different forms.

$$\frac{(x-x_0)^2}{a^2} + \frac{(y-y_0)^2}{b^2} = 1$$

$$\left\langle \begin{bmatrix} 1/a & 0 \\ 0 & 1/b \end{bmatrix} \begin{pmatrix} x-x_0 \\ y-y_0 \end{pmatrix}, \begin{bmatrix} 1/a & 0 \\ 0 & 1/b \end{bmatrix} \begin{pmatrix} x-x_0 \\ y-y_0 \end{pmatrix} \right\rangle = 1$$

$$\frac{1}{a^2} x^2 - \frac{2x_0}{a^2} x + \frac{1}{b^2} y^2 - \frac{2y_0}{b^2} y + \frac{x_0^2}{a^2} + \frac{y_0^2}{b^2} = 1$$

From the last form we could conclude that the search for an ellipse passing through a set of given points might be formulated as a regression problem[14]. Multiply the equation by a^2 and set $\gamma = \frac{a}{b}$ to find the equivalent equation

$$x^2 - 2x_0 x + \frac{a^2}{b^2} y^2 - \frac{2a^2 y_0}{b^2} y + x_0^2 + \frac{a^2 y_0^2}{b^2} = a^2$$

$$x^2 - 2x_0 x + \gamma^2 y^2 - 2y_0 \gamma^2 y + x_0^2 + \gamma^2 y_0^2 - a^2 = 0$$

We seek parameters $\vec{p} \in \mathbb{R}^4$ such that the length of the residual vector $\vec{r}$ is minimal, where

$$p_1 x_i + p_2 y_i^2 + p_3 y_i + p_4 + x_i^2 = r_i.$$

---

[14] A straight linear regression in the above form will fail, since there is a constant contribution $\frac{x_0^2}{a^2} + \frac{y_0^2}{b^2}$ on the left hand side. If we would know that $x_0 = y_0 = 0$ then a linear regression of the form $p_1 x^2 + p_2 y^2 = 1$ would work just fine. If a general orientation is asked for use $p_1 x^2 + p_2 y^2 + p_3 xy = 1$

With standard linear regression determine the optimal values for the parameters $\vec{p}$. Then solve for the parameters of the ellipse by solving the following system top to bottom.

$$-2\,x_0 \;=\; p_1$$
$$\frac{a^2}{b^2} = \gamma^2 \;=\; p_2$$
$$-2\,y_0\,\gamma^2 \;=\; p_3$$
$$x_0^2 + \gamma^2\,y_0^2 - a^2 \;=\; p_4$$

The *Octave* code below solves for the best ellipse, where the points of the ellipse are stored in the file `EllipseData1.m`.

```
EllipseData1;
figure(1); plot(xi,yi,'*r');
         axis([-2 2 -2 2],'equal');

F = [xi yi.^2 yi ones(size(xi))];
[p,yvar,r] = LinearRegression(F,-xi.^2);

x0 = -p(1)/2
y0 = -p(3)/(2*p(2))
a  = sqrt( x0^2 + p(2)*y0^2 - p(4))
b  = sqrt(a^2/p(2))
```

With the computed parameters

$$x_0 = 0.15032 \quad , \quad y_0 = -0.17548 \quad , \quad a = 1.7109 \quad \text{and} \quad b = 0.79109$$

we can draw the ellipse, leading to the left half of Figure 3.43.

```
phi = (0:5:360)'*pi/180;
x = x0 + a*cos(phi);
y = y0 + b*sin(phi);
plot(xi,yi,'*r',x,y,'b');
```

Ellipse with general orientation

As a starting point we consider the equation for an ellipse with semi–axes (parallel to coordinates) of length a and b in the matrix form

$$\left\langle \begin{bmatrix} 1/a & 0 \\ 0 & 1/b \end{bmatrix} \begin{pmatrix} x - x_0 \\ y - y_0 \end{pmatrix} , \begin{bmatrix} 1/a & 0 \\ 0 & 1/b \end{bmatrix} \begin{pmatrix} x - x_0 \\ y - y_0 \end{pmatrix} \right\rangle = 1 \,.$$

Rotating a vector by an angle α can be written as a matrix multiplication

$$\begin{bmatrix} \cos\alpha & -\sin\alpha \\ \sin\alpha & \cos\alpha \end{bmatrix} \begin{pmatrix} x \\ y \end{pmatrix} = \begin{bmatrix} n_1 & -n_2 \\ n_2 & n_1 \end{bmatrix} \begin{pmatrix} x \\ y \end{pmatrix}$$

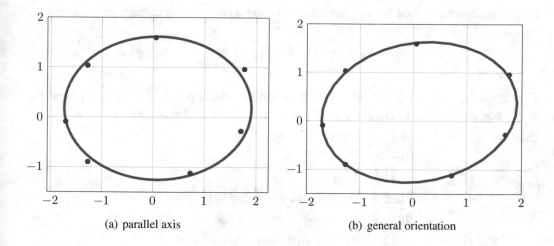

(a) parallel axis   (b) general orientation

Figure 3.43: Some points and best fit ellipses, parallel to axes and general orientation

Thus we can write down the equation of a general ellipse with the help of

$$\mathbf{M} = \left[\begin{array}{cc} 1/a & 0 \\ 0 & 1/b \end{array} \right] \cdot \left[\begin{array}{cc} n_1 & -n_2 \\ n_2 & n_1 \end{array} \right] = \left[\begin{array}{cc} \frac{n_1}{a} & \frac{-n_2}{a} \\ \frac{n_2}{b} & \frac{n_1}{b} \end{array} \right]$$

in the form

$$
\begin{aligned}
1 &= \left\langle \mathbf{M} \cdot \left(\begin{array}{c} x - x_0 \\ y - y_0 \end{array} \right) , \mathbf{M} \cdot \left(\begin{array}{c} x - x_0 \\ y - y_0 \end{array} \right) \right\rangle \\
&= \left\langle \left(\begin{array}{c} x - x_0 \\ y - y_0 \end{array} \right) , \left[\begin{array}{cc} \frac{n_1^2}{a^2} + \frac{n_2^2}{b^2} & \frac{-n_1 n_2}{a^2} + \frac{n_1 n_2}{b^2} \\ \frac{-n_1 n_2}{a^2} + \frac{n_1 n_2}{b^2} & \frac{n_2^2}{a^2} + \frac{n_1^2}{b^2} \end{array} \right] \left(\begin{array}{c} x - x_0 \\ y - y_0 \end{array} \right) \right\rangle \\
&= \left\langle \left(\begin{array}{c} x - x_0 \\ y - y_0 \end{array} \right) , \mathbf{A} \cdot \left(\begin{array}{c} x - x_0 \\ y - y_0 \end{array} \right) \right\rangle \\
&= \left\langle \left(\begin{array}{c} x \\ y \end{array} \right) , \mathbf{A} \cdot \left(\begin{array}{c} x \\ y \end{array} \right) \right\rangle \; 2 \left\langle \left(\begin{array}{c} x_0 \\ y_0 \end{array} \right) , \mathbf{A} \cdot \left(\begin{array}{c} x \\ y \end{array} \right) \right\rangle + \left\langle \left(\begin{array}{c} x_0 \\ y_0 \end{array} \right) , \mathbf{A} \cdot \left(\begin{array}{c} x_0 \\ y_0 \end{array} \right) \right\rangle .
\end{aligned}
$$

With the help of the symmetric matrix

$$\mathbf{A} = \left[\begin{array}{cc} a_{1,1} & a_{1,2} \\ a_{1,2} & a_{2,2} \end{array} \right]$$

we can compute a residual r_i for each given point (x_i , y_i).

$$
\begin{aligned}
r_i &= a_{1,1} x_i^2 + 2 a_{1,2} x_i y_i + a_{2,2} y_i^2 - 2 \left(a_{1,1} x_0 x_i + a_{1,2} \left(x_0 y_i + y_0 x_i \right) + a_{2,2} y_0 y_i \right) \\
&\quad + a_{1,1} x_0^2 + 2 a_{1,2} x_0 y_0 + a_{2,2} y_0^2 - 1 \\
&= a_{1,1} x_i^2 + 2 a_{1,2} x_i y_i + a_{2,2} y_i^2 - 2 \left(a_{1,1} x_0 + a_{1,2} y_0 \right) x_i - 2 \left(a_{2,2} y_0 + a_{1,2} x_0 \right) y_i \\
&\quad + a_{1,1} x_0^2 + 2 a_{1,2} x_0 y_0 + a_{2,2} y_0^2 - 1
\end{aligned}
$$

Dividing this expression by $a_{1,1}$ leads us to a least square problem with modified residuals.

$$\frac{r_i}{a_{1,1}} = x_i^2 + p_1\, x_i\, y_i + p_2\, y_i^2 + p_3\, x_i + p_4\, y_i + p_5$$

This is a linear regression problem for the parameter vector $\vec{p} \in \mathbb{R}^5$. With the optimal values of $\vec{p}$, compute the parameters of the ellipse with the help of the equations below.

$$
\begin{aligned}
a_{1,1}\, p_1 &= 2\, a_{1,2} \\
a_{1,1}\, p_2 &= a_{2,2} \\
a_{1,1}\, p_3 &= -2\,(a_{1,1}\, x_0 + a_{1,2}\, y_0) \\
a_{1,1}\, p_4 &= -2\,(a_{1,2}\, x_0 + a_{2,2}\, y_0) \\
a_{1,1}\, p_5 &= a_{1,1}\, x_0^2 + 2\, a_{1,2}\, x_0\, y_0 + a_{2,2}\, y_0^2 - 1
\end{aligned}
$$

Using the first two equations in the last 3, divided by $a_{1,1}$ we conclude

$$
\begin{aligned}
p_3 &= -2\, x_0 - p_1\, y_0 \\
p_4 &= -p_1\, x_0 - 2\, p_2\, y_0 \\
p_5 &= x_0^2 + p_1\, x_0\, y_0 + p_2\, y_0^2 - \frac{1}{a_{1,1}} \ .
\end{aligned}
$$

The first two equations are linear with respect to the unknowns x_0 and y_0.

$$
\begin{bmatrix} 2 & p_1 \\ p_1 & 2\,p_2 \end{bmatrix} \cdot \begin{pmatrix} x_0 \\ y_0 \end{pmatrix} = - \begin{pmatrix} p_3 \\ p_4 \end{pmatrix}
$$

Thus we know the values for x_0 and y_0. Now the last equation can be solved for the only remaining unknown $a_{1,1}$ since

$$\frac{1}{a_{1,1}} = x_0^2 + p_1\, x_0\, y_0 + p_2\, y_0^2 - p_5 \ .$$

Now we know all values in the matrix $\mathbf{A}$. The eigenvalues and eigenvectors of $\mathbf{A}$ give the lengths a and b of the semi-axis and the angle of rotation α.

$$
\begin{aligned}
\mathbf{A} &= \begin{bmatrix} n_1 & n_2 \\ -n_2 & n_1 \end{bmatrix} \cdot \begin{bmatrix} \lambda_1 & 0 \\ 0 & \lambda_2 \end{bmatrix} \cdot \begin{bmatrix} n_1 & -n_2 \\ n_2 & n_1 \end{bmatrix} \\
&= \begin{bmatrix} \cos\alpha & -\sin\alpha \\ \sin\alpha & \cos\alpha \end{bmatrix} \cdot \begin{bmatrix} \frac{1}{a^2} & 0 \\ 0 & \frac{1}{b^2} \end{bmatrix} \cdot \begin{bmatrix} \cos\alpha & -\sin\alpha \\ \sin\alpha & \cos\alpha \end{bmatrix}
\end{aligned}
$$

The above algorithm is implemented in *Octave* and the graphical result can be found in the right half of Figure 3.43.

```
EllipseData1;
plot(xi,yi,'*r');

F = [xi.*yi yi.^2  xi yi ones(size(xi))];
p = LinearRegression(F,-xi.^2);

m   = [2 p(1);p(1) 2*p(2)];
x0  = -m\[p(3);p(4)]
a11 =1/(x0(1)^2+p(1)*x0(1)*x0(2)+p(2)*x0(2)^2-p(5));

[V,la] = eig(a11*m/2);
alpha = atan(V(2,1)/V(1,1))*180/pi
a = 1/sqrt(la(1,1))
b = 1/sqrt(la(2,2))

np = 37; phi = linspace(0,2*pi,np);

xx = V*([a*cos(phi); b*sin(phi)])+x0*ones(1,np);
x = xx(1,:); y = xx(2,:);

plot(xi,yi,'*r',x,y,'b'); grid on
```

Observations about the fitting of ellipses

The algorithm in the previous section only yields good results if the points to be examined are rather close to an ellipse. If the algorithm is used for set of random points one can not expect reasonable results. Often there will be no result at all, since the values of a^2 or b^2 turn out to be negative.

Also observe that we **do not minimize the distance to the ellipse**, since the residuals r_i used in the algorithm correspond not exactly to the distance of a point (x_i, y_i) from the ellipse. The precise minimal distance problem is considerably harder to solve and leads to a nonlinear regression problem. One of the subproblems to be solved is how to determine the distance of a point from an ellipse.

We illustrate the above remarks with a simulation. First choose the parameters of an ellipse, then generate a set of points rather close to this ellipse. The values are stored in the column vectors xi and yi.

```
ain = 1.2; bin = 0.8; alphain = 15*pi/180;   x0in = 0.1 ; y0in = -0.2;
np = 15; sigma = 0.05;

phi = linspace(0,2*pi,np)';
xi  = x0in+ain*cos(phi)+sigma*randn(size(phi));
yi  = y0in+bin*sin(phi)+sigma*randn(size(phi));
```

```
xynew = [cos(alphain), -sin(alphain);...
         sin(alphain),  cos(alphain)]*[xi,yi]';
xi = xynew(1,:)'; yi = xynew(2,:)';
```

Then we fit an ellipse with axes parallel to the coordinates to those points and display the parameters of the ellipse.

```
F = [xi yi.^2 yi ones(size(xi))];
[p,yvar,r] = LinearRegression(F,-xi.^2);

x0 = -p(1)/2
y0 = -p(3)/(2*p(2))
a  = sqrt( x0^2 + p(2)*y0^2 - p(4))
b  = sqrt(a^2/p(2))

phi = (0:5:360)'*pi/180;
x = x0+a*cos(phi);   y = y0+b*sin(phi);
figure(1); plot(xi,yi,'*r',x,y,'b');
```

Then redo the fitting for a general ellipse, display the parameters and create Figure 3.44.

```
%% fit general ellipse
F = [xi.*yi yi.^2  xi yi ones(size(xi))];
p = LinearRegression(F,-xi.^2);

m  = [2 p(1);p(1) 2*p(2)];
x0 = -m\[p(3);p(4)]
a11 = 1/(x0(1)^2+p(1)*x0(1)*x0(2)+p(2)*x0(2)^2-p(5));
[V,la] = eig(a11*m/2);

alpha = atan(V(2,1)/V(1,1))*180/pi
a = 1/sqrt(la(1,1))
b = 1/sqrt(la(2,2))

np  = 37;   phi = linspace(0,2*pi,np);
xx  = V*([a*cos(phi); b*sin(phi)])+x0*ones(1,np);
xg  = xx(1,:);   yg = xx(2,:);
figure(2); plot(xi,yi,'*r',x,y,'b',xg,yg,'r'); grid on
```

If exactly 4 points are given then an ellipse parallel to the axis is uniquely determined. If 5 points are given then a general ellipse is often uniquely determined. But not all combination of points will admit solutions. The above algorithm will obtain a negative number for b^2 and thus fail to produce an ellipse. This can be illustrated with data collceted by the code EllipseClick.m.

1. Use the left button of the mouse to mark the points. Give at least 5 points, approximately on an ellipse.

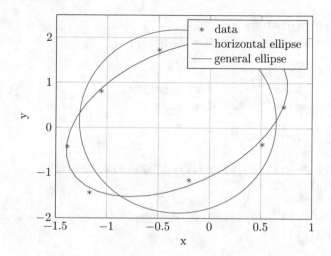

Figure 3.44: Some random points and a best fit ellipses, parallel to axes and general

2. Click on the right button to collect the last point and terminate the collection.

3. Determine the parameters of a horizontal ellipse.

4. Determine the parameters of a general ellipse.

5. Draw both ellipses and the selected points.

──────────────── **EllipseClick.m** ────────────────

```
figure(1); clf;
axislimits =[-2 2 -2 2]; axis(axislimits)
disp('Use the left mouse button to pick points.')
disp('Use the right mouse button to pick the last point.')
xi = []; yi = [];
button = 1;                         % to indicate the last point
while button == 1                   % while loop, picking up the points
    [xii,yii,button] = ginput(1); % get coodinates of one point
    xi = [xi;xii]; yi = [yi;yii];
    plot(xi,yi,'ro')                % plot all points
    axis(axislimits);               % fix the axis
end


%% fit ellipse parallel to axis
F = [xi yi.^2 yi ones(size(xi))];
[p,yvar,r] = LinearRegression(F,-xi.^2);

x0 = -p(1)/2
y0 = -p(3)/(2*p(2))
a  = sqrt( x0^2 + p(2)*y0^2 - p(4))
b  = sqrt(a^2/p(2))
```

```
phi = (0:5:360)'*pi/180;
x   = x0+a*cos(phi);  y = y0+b*sin(phi);


%% fit general ellipse
F = [xi.*yi yi.^2  xi yi ones(size(xi))];
p = LinearRegression(F,-xi.^2);


m   = [2 p(1);p(1) 2*p(2)];
x0  = -m\[p(3);p(4)]
a11 = 1/(x0(1)^2+p(1)*x0(1)*x0(2)+p(2)*x0(2)^2-p(5));
[V,la] = eig(a11*m/2);


alpha = atan(V(2,1)/V(1,1))*180/pi
a = 1/sqrt(la(1,1))
b = 1/sqrt(la(2,2))


np  = 37;  phi = linspace(0,2*pi,np);
xx = V*([a*cos(phi); b*sin(phi)])+x0*ones(1,np);
xg = xx(1,:); yg = xx(2,:);


figure(1); plot(xi,yi,'*r',x,y,'b',xg,yg,'r');
           xlabel('x'); ylabel('y')
           legend('data','horizontal ellipse','general ellipse')
```

43 Example : Another approach to generate approximate ellipse
The above problem can also be solved using a regression with constraint, using the function
`RegressionConstraint()`.

- **Ellipse with axes parallel to coordintes axes**
  The equation of an ellipse with parallel axis leads to a regression problem.

$$\frac{1}{a^2}\,x^2 - \frac{2\,x_0}{a^2}\,x + \frac{1}{b^2}\,y^2 - \frac{2\,y_0}{b^2}\,y + \frac{x_0^2}{a^2} + \frac{y_0^2}{b^2} - 1 \;=\; 0$$
$$p_1\,x^2 + p_2\,x + p_3\,y^2 + p_4\,y + p_5 \;=\; 0$$

Minimize the residual, subject to the constraint $\|\vec{p}\| = 1$. Then if $p_1 \neq 0$ divide the vector $\vec{p}$ by p_1 and you are back at the situation of the previous example.

- **Ellipse with general orientation**
  The equation of an ellipse with general leads to a regression problem.

$$\frac{1}{a^2}\,x^2 - \frac{2\,x_0}{a^2}\,x + \frac{1}{b^2}\,y^2 - \frac{2\,y_0}{b^2}\,y + \frac{x_0^2}{a^2} + \frac{y_0^2}{b^2} - 1 \;=\; 0$$
$$p_1\,x^2 + p_2\,x\,y + p_3\,y^2 + p_4\,x + p_5\,y + p_6 \;=\; 0$$

Minimize the residual, subject to the constraint $\|\vec{p}\| = 1$. Then if $p_1 \neq 0$ divide the vector $\vec{p}$ by p_1 and you are back at the situation of the previous example.

The answer will be slightly different form the previous exercise, as both approaches do not work with the geometric distance from the ellipse, but an algebraic expression. ◇

In the above example the regression problem has a special form, minimize the length of the residual vector $\vec{r} = \mathbf{F}\,\vec{p}$, subject to the constraint $\|\vec{p}\| = 1$. The solution was generated using `RegressionConstraint()`. Since all components of the parameter $\vec{p}$ are constraint, a simpler algorithm is possible, using the Lagrange multiplier theorem.

$$\begin{aligned} \|\vec{r}\|^2 &= \langle \mathbf{F}\vec{p}, \mathbf{F}\vec{p}\rangle = \langle \mathbf{F}^T\mathbf{F}\vec{p}, \vec{p}\rangle \qquad \text{to be minimized, subject to } \|vecp\| = 1 \\ \nabla\|\vec{r}\|^2 &= 2\,\mathbf{F}^T\mathbf{F}\vec{p} = \lambda\,\vec{p} \end{aligned}$$

Thus the eigenvector of the symmetric matrix $\mathbf{F}^T\mathbf{F}$, belonging to the smallest eigenvalue, is the solution of the problem.

3.3.7 List of codes and data files

In the previous section the codes and data files in Table 3.9 were used.

filename	function
`RegressionConstraint.m`	function to perform regression with constraint
`LineFitOrthogonal.m`	regression for a straight line
`LineData.m`	data file for a line fit
`ImageLine.m`	script file to determine a line in an digital image
`Line1.bmp`	image data for a first line
`Line2.bmp`	image data for a second line
`Ellipse1.m`	fit a parallel ellipse to data
`Ellipse2.m`	fit a general ellipse to data
`EllipseData1.m`	data file for an ellipse
`EllipseCompare.m`	script file to compare the two methods
`EllipseClick.m`	script file to read points with mouse and fit ellipses

Table 3.9: Codes and data files for section 3.3

3.4 Computing Angles on an Embedded Device

In this section *Octave*/MATLAB are used to design an algorithm and its implementation on a micro controller with limited hardware resources. A particular example is examined very carefully, but the methods and results are applicable to a wide variety of problems.

3.4.1 Arithmetic operations on a micro controller

On most micro controllers only the integer operations for addition, subtraction and multiplication are implemented directly. We use integer data types `int16` or `int32` to examine the results of the algorithms. The use of *Octave*/`MATLAB` to design an integer algorithm has some obvious advantages:

- `MATLAB`/*Octave* code is easier to develop than code in C, do not even think of the trouble with assembler code.

- Using the data types `int16` in will automatically take care of overflow and underflow and make rounding problems visible.

- We can compare the results of an integer computation with a similar floating point computations and thus determine approximation errors.

- We can use all the graphical power of `MATLAB`/*Octave* to visualize results and approximation errors.

- Once the integer computation algorithm is developed and tested in *Octave* we can translate to C code or even assembler.

General observations

Typical micro controllers have either 8–bit or 16–bit integer arithmetic, but lack any floating point commands. Emulating floating point operations with the help of a library leads to a huge computational overhead and should be avoided if possible. Most often software is written in C, occasionally in assembler.

Most micro controllers provide a set of arithmetic operations with a given resolution. Precise information can be found in the manuals of the micro controllers.

- 8-bit micro controller, e.g. 8051, Cygnal

  - Signed integer numbers have to be between -128 and 127.
  - Unsigned integer numbers have to be between 0 and 255.
  - 8-bit add/subtract 8-bit leads to 8-bit result.
  - 8-bit multiply 8-bit leads to 16-bit result.
  - 8-bit divide by 8-bit leads to 16-bit result, 8-bit integer part, 8-bit remainder.
  - 16-bit additions and subtracions are not very difficult to implement.

  Based on the above commands one can implement fast 16-bit additions and subtractions and also 8 bit multiplications.

- 16-bit micro controller (e.g. Cyan)

  - Signed integer numbers have to be between -2^{15} and $2^{15} - 1$.
  - Unsigned integer numbers have to be between 0 and $2^{16} - 1$.

- 16-bit add/subtract 16-bit leads to 16-bit result.

- 16-bit multiply 16-bit leads to 32-bit result.

- 32-bit divide by 16-bit leads to 32-bit result, 16-bit integer part, 16-bit remainder.

Based on the above commands one can implement fast 16-bit additions and subtractions and also 16-bit multiplications/divisions.

When writing arithmetic code for a micro controller the following facts should be kept in mind:

- Multiplication by 2^k are easy to implement as shifts of binary representation. Multiplications and division by 2^8 have not be be computed at all, since they result in byte shifts, not bit shifts.

- When implementing the algorithm one may apply shifts to make full use of the 16 bit resolution.

- When adding two numbers $z = x + y$ find the error estimations

$$z = x + y \quad \Longrightarrow \quad \Delta z \approx \Delta x + \Delta y .$$

Thus the absolute errors are to be added and the final error is dominated by the largest error of the arguments.

- Multiplication is also susceptible to loss of accuracy. Use the error analysis

$$
\begin{aligned}
z &= x \cdot y \\
\Delta z &\approx y \, \Delta x + x \, \Delta y \\
\frac{\Delta z}{z} &\approx \frac{\Delta x}{x} + \frac{\Delta y}{y} .
\end{aligned}
$$

Thus the relative errors are to be added. If one argument has a large relative error, then the result has a large relative error.

As a consequence design algorithms that use the full accuracy of the hardware.

A sample computation with `int8` **and** `int16` **data types**

As an example we want to compute $y = f(x) = 1 - 0.8 \, x^2$ for $0 \le x \le 2$ on an 16-bit processor, assuming that for x and y we need an 8-bit resolution. The code is developped with MATLAB/*Octave*. Start out by generating a graph of the function.

```
x = linspace(0,2,1001);
f = @(x)1-0.8*x.^2;
y = f(x);

figure(1); plot(x,y);
        title('original function');
        xlabel('x'); ylabel('y = 1-0.8*x*x');grid on
```

Now we want to perform the arithmetic operations for

$$y = 1 - 0.8 \cdot x^2 = 1 - 0.8 \cdot (x \cdot x)$$

keeping track of the effects of the 16-bit arithmetic.

- For the return values y we know $-3 \le y \le 1$ and this domain should be represented with the data range for int8, i.e. between -128 and +127. Thus aim for $y8 = y \cdot 2^5$ as results.

- The true values of x are $0 \le x \le 2$. To use an 8-bit resolution pass the values $x8 = x \cdot 2^6 = x \cdot 64$, thus $0 \le x8 \le 127$ and use it as an int8 data type.

```
x8 = int8(x*2^6);
```

- As a first intermediate result compute $r1 = x8 * x8 = x^2 \cdot 2^{12}$. The result will be a 16-bit integer. Verify that the data range is respected.

```
r1 = uint16(int16(x8).*int16(x8));   % r1 = x^2*2^(6+6) = x^2*2^12
% verify the limits, maximal value 2^16 (uint16)
limr1 = [min(r1),max(r1)]
-->
limr1 =   0    16129
```

- The next step is to multiply the previous result by 0.8 with an integer multiplication. Since $0.8 \cdot 256 \approx 204.8$ use the factor 205. But before multiplying $r1$ by 205, divide $r1$ by 2^{-8}, otherwise the data range is not respected. The result is rescaled, such that it may be treated as an 8-bit integer.

```
%  0.8*2^8 approximately 205
r2 = int16(205*(r1*2^-8)); % r2 = 0.8*x^2*2^12
r3 = int8(bitshift(r2,-7));% rescale, r3 = 0.8*x^2*2^5
limr3 = [min(r3),max(r3)]   % verify the limits, maximal value 2^7
-->
limr3 = 0    100
```

- Now subtract the previous result $r3$ from 1, respectively from $2^5 = 32$.

```
r4 = int8(1*2^5-r3);      % r4 =(1-0.8*x^2)*2^5
limr4 = [min(r4),max(r4)]  % verify the limits
-->
limr4 =  -68   32
```

The results equals $y \cdot 2^5$. Plot the exact function $y = 1 - 0.8 \cdot x^2$ and the 8-bit approximation.

```
r5 = single(r4)/2^5;
plot(x,y,x,r5)
xlabel('x'); ylabel('1-0.8*x*x'); grid on
```

To examine the error we plot the difference of the exact and approximate function.

```
figure(2); plot(x,r5-y)
           title('arithmetic error')
xlabel('x'); ylabel('error'); grid on
relErr = max(abs(r5-y))/max(abs(y))
bitError = log2(relErr)
bitErrorMean = log2(mean(abs(r5-y))/max(abs(y)))
-->
relErr    =  0.040962
bitError = -4.6096
bitErrorMean = -6.0771
```

3.4.2 Computing the angle based on coordinate information

A pair of sensors might give the x and y component of a point in the plane. The expression to be measured is the angle α. On a pure mathematical level the answer is given with the formulas in Figure 3.45, but for a good impementation in an actual device some further aspects have to be taken into account.

- The values of x and y are given with errors Δx and Δy. The error $\Delta \alpha$ has to be controlled and minimized.

- The evaluation of the formulas has to be implemented on a micro controller and thus should require as little computational resources as possible.

- The evaluation has to be reliable and fast.

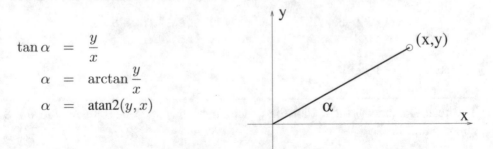

$$\tan \alpha = \frac{y}{x}$$

$$\alpha = \arctan \frac{y}{x}$$

$$\alpha = \text{atan2}(y, x)$$

Figure 3.45: The angle α as function of x and y

3.4.3   Error analysis of arctan–function

Use the derivative $\frac{\partial}{\partial u} \arctan u = \frac{1}{1+u^2}$ and a linear approximation for the function $f(x, y) = \arctan \frac{y}{x}$ to derive a formula for the error $\Delta \alpha$ generated by the deviations Δx and Δy.

$$\Delta \alpha \approx \frac{\partial f}{\partial x} \Delta x + \frac{\partial f}{\partial y} \Delta y$$

$$= \frac{1}{1 + (y/x)^2} \frac{-y}{x^2} \Delta x + \frac{1}{1 + (y/x)^2} \frac{1}{x} \Delta y$$

$$= \frac{-y}{x^2 + y^2} \Delta x + \frac{x}{x^2 + y^2} \Delta y$$

$$= \frac{1}{r^2} \left(-y \, \Delta x + x \, \Delta y \right)$$

If the errors are randomly distributed, with variance $V(x) = \sigma_x^2$ (resp. $V(y) = \sigma_y^2$), use the law of error propagation to conclude

$$V(\alpha) = \frac{1}{r^4} \left(y^2 \, V(x) + x^2 \, V(y) \right)$$

$$\sigma_\alpha = \frac{1}{r^2} \sqrt{y^2 \sigma_x^2 + x^2 \sigma_y^2} \, .$$

If the standard deviations for the angles are of equal size ($\sigma_x = \sigma_y = \sigma$) this simplifies to

$$V(\alpha) = \frac{1}{r^2} \sigma^2 textor \sigma_\alpha = \frac{\sigma}{r} \, .$$

The error contributions Δx and Δy are influenced by the hardware (e.g. resolution of the AD converters) and might be determined by statistical methods, see also Section 3.2.12.

3.4.4   Reliable evaluation of the arctan–function

The formula $\alpha = \arctan \frac{y}{x}$ might lead to an unnecessary division by zero. Thus divide the plane in 8 different sectors and use a slightly different formula for each sector in Figure 3.46 to avoid divisions by small numbers.

No	conditions			result
1	$x > 0$	$y \geq 0$	$y \leq x$	$\alpha = \arctan \frac{y}{x}$
2	$x \geq 0$	$y > 0$	$x \leq y$	$\alpha = \frac{\pi}{2} - \arctan \frac{x}{y}$
3	$x \leq 0$	$y > 0$	$|x| \leq y$	$\alpha = \frac{\pi}{2} + \arctan \frac{-x}{y}$
4	$x < 0$	$y \geq 0$	$y \leq |x|$	$\alpha = \pi - \arctan \frac{y}{-x}$
5	$x < 0$	$y \leq 0$	$|y| \leq |x|$	$\alpha = -\pi + \arctan \frac{-y}{-x}$
6	$x \leq 0$	$y < 0$	$|x| \leq |y|$	$\alpha = -\frac{\pi}{2} - \arctan \frac{-x}{-y}$
7	$x \geq 0$	$y < 0$	$|x| \leq |y|$	$\alpha = -\frac{\pi}{2} + \arctan \frac{x}{-y}$
8	$x > 0$	$y \leq 0$	$|y| \leq |x|$	$\alpha = -\arctan \frac{-y}{x}$

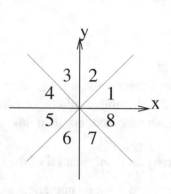

Figure 3.46: The eight sectors used to compute $\tan \alpha = \frac{y}{x}$

Using the table in Figure 3.46 we see that the arctan–function only has to be evaluated for arguments $0 \leq z \leq 1$. As additional effort we have to distinguish the eight sectors. All good mathematical packages offer this type of function. As typical example consider the result of the command `help atan2` from *Octave*/MATLAB.

Octave
```
>> help atan2
'atan2' is a built-in function from the file libinterp/corefcn/data.cc
-- atan2 (Y, X)
    Compute atan (Y / X) for corresponding elements of Y and X.
    Y and X must match in size and orientation. The signs of elements
    of Y and X are used to determine the quadrants of each resulting value.
    This function is equivalent to 'arg (complex (X, Y))'.
```

The algorithms in the next section thus only have to compute values of $y = \arctan z$ for $0 \leq z \leq 1$ leading to results $0 \leq y \leq \frac{\pi}{4}$. An approximation by polynomials can be generated with the help of a Chebyshev approximations.

3.4.5 Implementations of the arctan–function on micro controllers

Thus the remaining task is to compute the function $f(z) = \arctan z$ for arguments $0 \leq z \leq 1$. Assume that x and y are measured and then digitalized with a 10-bit AD converter. The values of x and y vary between $-r$ and $+r$. Thus expect at best a relative error of $\frac{\Delta x}{x} \approx 2^{-9} \approx \frac{1}{500} = 0.02$. Based on the result of the previous section we can not hope for a better accuracy for α. The only operations to be used on a micro controller are the basic arithmetic operations, addition, subtraction and multiplication.

Taylor series approximation

A standard Taylor series (with error estimate) leads to the formula

$$\arctan z = \sum_{k=0}^{n} \frac{(-1)^k}{2\,k+1}\, z^{2\,k+1} + R_n$$

where the approximation error is $|R_n| \approx \frac{1}{2\,k+3} z^{2\,k+3}$. For the domain to be examined $0 \leq z \leq 1$ this would require $2\,k+3 \approx 500$, that is $k \approx 250$ terms. Obviously there have to be better solutions than this.

Chebyshev polynomial of degree 2

Any good book on numerical analysis has a section on Chebyshev polynomials[15] and as an application one can verify that the polynomial

$$\begin{aligned}
\arctan z \;\; &\approx \;\; -0.003113205848 + 1.073115615\, z - 0.283642707\, z^2 \\
&= \;\; -0.003113205848 + z\,(1.073115615 - z\,0.283642707)
\end{aligned}$$

is an approximation with a maximal error of 0.003 on $0 \leq z \leq 1$. For sake of completeness a very brief explanation of Chebyshev polynomials is given in Section 3.4.6. This error is comparable to the error contribution from the 10–bit resolution of the AD converters. Using the Horner scheme simplify the expression to

$$\arctan z \approx -0.0031 + z\,(1.0731 - z\,0.2836) \quad \text{for} \quad 0 \leq z \leq 1\,.$$

This requires only 2 additions and multiplications to compute the value of $\arctan z$. The computational sequence is given by

1. Multiply z by 0.2836.

2. Subtract this result from 1.0731.

3. Multiply the result by z.

4. Subtract 0.0031 from the result.

A few plots let you recognize that all intermediate results are positive for all $0 \leq z \leq 1$. The goal is to implement these calculations on a micro controller using the following arithmetic operations with integer numbers. Since all expressions are positive we use the data type unsigned integers.

- Addition of 16-bit unsigned integers leading to a 16-bit integer result.

- Multiplication of 8-bit unsigned integers leading to a 16-bit integer result.

- Multiplications by 2^k to be implemented with arithmetic shifts.

---

[15] Another optin is to use the optimization commands `fmins()` or `fminsearc()`, see Section 1.3.4.

All of the above operations are suited for an 8-bit micro controller. For each intermediate step one has to check for possible under- and overflow. Since we also aim for accuracy we have to assure that the arguments of the multiplications are as close as possible to the maximal number 2^8.

Prepare the computations by storing the precomputed constants $0.2836 \cdot 2^8$, $1.0731 \cdot 2^{15}$ and $0.0031 \cdot 2^{16}$ and also define the Chebyshev approximation to the arctan–function for comparative purposes.

```
a0 = 0.283642707;     i0 = uint16(a0*2^8)
a1 = 1.073115615;     i1 = uint16(a1*2^15)
a2 = 0.003113205848;  i2 = uint16(a2*2^16)

myatan = @(z)-0.003113205848 + z.*( 1.073115615  - z*0.283642707);
```

Then for a given value $0 \leq z \leq 1$ compute $z \cdot 2^8$ as a 16-bit unsigned integer. The algorithm below requires 2 multiplications, 2 additions and a few shifts.

```
z = 0:0.001:1;
zi = uint16(z*2^8);
```

1. Multiply z by 0.2836.

   - Compute $(z \cdot 2^8) \cdot (0.2836 \cdot 2^8)$.
   - The intermediate result is modified by a factor of 2^{16}.

     ```
     r1 = uint16(zi.*i0);
     limits1 = [min(r1) max(r1)]
     ```

2. Subtract this result from 1.0731.

   - Divide the previous result by 2.
   - Subtract it from $1.0731 \cdot 2^{15}$.
   - The intermediate result is modified by a factor of 2^{15}.

     ```
     r2 = uint16(i1-r1/2);
     limits2 = [min(r2) max(r2)]
     ```

3. Multiply the result by z.

   - Divide the previous result by 2^7. It might be faster to divide by 2^8 and then multiply by 2.
   - Multiply it with $z \cdot 2^8$.
   - The intermediate result is modified by a factor of 2^{16}.

```
r3 = uint16(zi.*bitshift(r2,-7));
limits3 = [min(r3) max(r3)]
```

4. Subtract 0.0031 from the result.

   - Subtract $0.0031 \cdot 2^{16}$ from the result.
   - The intermediate result is modified by a factor of 2^{16}.

```
r4 = uint16(r3-i2);
limits4 = [min(r4) max(r4)]
```

This result is converted back to floats and then Figure 3.47 can be generated. The graph shows that the error is smaller than 0.005 which corresponds to $0.29°$. Considering that the possible results range from $0°$ to $45°$ we find an accuracy of 7.5 bit ($\log_2(45 * 4)$). This is quite good since we started with 8-bit accuracy for the input z. Figure 3.47 shows that the contributions from the Chebyshev approximation and the integer arithmetic are both of the same size.

```
res = single(r4)/2**16;
plot(z,res-atan(z),'r',z,myatan(z)-atan(z),'b')
legend('int16','float')
xlabel('z'); ylabel('difference'); grid on;
```

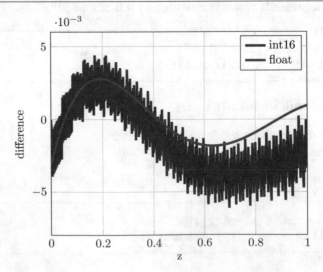

Figure 3.47: Errors for a Chebyshev approximation and its integer implementation

This can all be crammed into one lengthy formula

$$\arctan z \approx 2^{-16} \left(-0.0031 \cdot 2^{16} + (z \cdot 2^8) \cdot (2^{-7}(1.0731 \cdot 2^{15} - 2^{-1}(z \cdot 2^8) \cdot (0.2836 \cdot 2^8)))) \right),$$

but for an implementation it is wise to use the above description.

Improvements and implementation in C

The approximation error in the previous section can be improved by choosing a higher or-der approximation and by using better integer arithmetic. Example 3.4.5 shows a modified, improved version of the above solution, using a micro controller with more powerful arith-metic commands. The algorithm is implemented in C.

Look up tables, 8-bit

Another approach might be to use a look-up table for the values of the arctan–function. For easy and fast processing we choose a table of 256 equally spaced values for z. Thus we use $z_i = \frac{i-1}{255}$ as midpoints of the intervals and compute the corresponding values $y_i = \arctan z_i$. Then we scale those values to use the full range of a 8-bit resolution and we choose to round to the closest integer. We are lead to the tabulated values $T_i = $ round $\left(\frac{255 \cdot 4}{\pi} y_i\right)$. These 256 values have to be computed once and then stored on the device.

```
zc = linspace(0,1,256);
atantab = round(atan(zc)*255*4/pi);
```

For a given value of $0 \le z \le 1$ we then perform the following steps to find an approximated value of the arctan function:

- Round $255\,z$ to the closest integer. This is equivalent to the integer part of $255\,z + 0.5$.

- Add 1 to the above index, since in *Octave* and MATLAB indexes are 1–based. Use this index to acces the number on the above table of precomputed values.

```
z = 0:0.001:1;

res = zeros(size(z));
for k = 1:length(z)
   res(k) = atantab(single(255*z(k)+0.5)+1)/255*pi/4;
end%for

figure(1); plot(z,res-atan(z));
```

The resulting Figure 3.48 shows a maximal error of approximately 0.003, which corre-sponds to $0.17°$.

Piecewise linear interpolation

Lets us divide the interval $0 \le z \le 1$ into $n-1$ subintervals of length $h = \frac{1}{n}$. Then tabulate the values of the function at the n points $z_i = i\,h$ for $i = 0, 1, 2, \ldots n$. For values of z between the points of support we use piecewise linear interpolation to estimate the value of

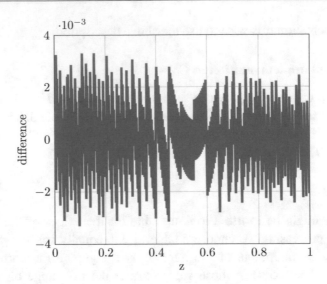

Figure 3.48: The difference for a tabulated approximation of the arctan–function

the function $\arctan z$. The linear interpolation of a function for $z_i \leq z_i + \Delta z \leq z_{i+1}$ is given by

$$f(z_i + \Delta z) \approx f(z_i) + \frac{f(z_{i+1}) - f(z_i)}{z_{i+1} - z_i} \Delta z = f(z_i) + m_i \cdot \Delta z . \qquad (3.8)$$

Using calculus one verifies that on an interval of length h the error of a linear interpolation using the left and right endpoint is estimated by

$$|\text{error}| \leq \frac{1}{8} M_2 h^2$$

where M_2 is the maximal absolute value of the second derivative of the function. For our function $f(z) = \arctan z$ with $0 \leq z \leq 1$ we have to determine M_2 by the calculations below.

$$f''(z) = \frac{-2z}{(1+z^2)^2}$$

$$f'''(z) = \frac{-2(1+z^2)^2 + 8z^2(1+z^2)}{(1+z^2)^2} = 2\frac{-1-z^2+4z^2}{(1+z^2)} = 0$$

$$z_m = \frac{1}{\sqrt{3}}$$

$$M_2 = |f''(1/\sqrt{3})| = \frac{2/\sqrt{3}}{(1+1/3)^2} = \frac{3\sqrt{3}}{8} \approx 0.64 < 1$$

If we divide the interval $0 \leq z \leq 1$ into $2^5 = 32$ subintervals of equal length we find $h = \frac{1}{32}$ and thus

$$|\text{error}| \leq \frac{1}{8} M_2 h^2 \leq \frac{1}{8 \cdot 32^2} \approx 1.3 \cdot 10^{-4} .$$

Since $\log_2 \frac{1.3 \cdot 10^{-4}}{\pi/4} \approx -12.6$ we conclude that we have at least 12-bit accuracy with this algorithm.

To understand the following computations we have to examine the binary representation of numbers. As example consider the number $z = 0.3$. The code

```
bin = dec2bin(round(0.33*2^15))
-->
bin = 010101000111101
```

implies that

$$z = 0.33 \approx \frac{1}{2^2} + \frac{1}{2^4} + \frac{1}{2^6} + \frac{1}{2^{10}} + \frac{1}{2^{11}} + \frac{1}{2^{12}} + \frac{1}{2^{13}} + \frac{1}{2^{15}}$$

and we decompose the number into the 5 leading digits of the binary representation and the remainder

$$z \approx 0. \underbrace{01010}_{\text{zint}=10} \underbrace{1000111101}_{\text{zfrac}=573} = \text{zint} \cdot 2^{-5} + \text{zfrac} \cdot 2^{-15} = z_i + \Delta z$$

Using the linear interpolation formula (3.8) conclude

$$\arctan(z) \approx \arctan(z_i) + m_i \cdot \Delta z = \arctan(\text{zint} \cdot 2^{-5}) + m_i \cdot \text{zfrac} \cdot 2^{-15}$$
$$\arctan(z) \cdot 2^{31} \approx \arctan(\text{zint} \cdot 2^{-5}) \cdot 2^{31} + m_i \cdot 2^{16} \cdot \text{zfrac}$$

The above idea leads to the following algorithm:

1. Precompute the integer parts $y_i = \arctan(\frac{i}{2^5}) \cdot 2^{15}$ for $i = 0, 1, \ldots 31$ and store these 32 values. These 16-bit values will be used as the upper half of 32-bit values. This hides a multiplication by 2^{16}.

2. Precompute the integer parts of

$$s_i = m_i \cdot 2^{16} = \frac{\arctan(\frac{i+1}{2^5}) - \arctan(\frac{i}{2^5})}{2^{-5}} \cdot 2^{16} = (\arctan(\frac{i+1}{2^5}) - \arctan(\frac{i}{2^5})) \cdot 2^{21}$$

   for $i = 0, 1, \ldots 31$ and store these 32 values. These are 16-bit values.

3. Use the 5 bits on positions 10 through 14 of $z \cdot 2^{15}$ as integer index $i = \text{zint}$ into the table entries y_i and s_i.

4. Consider bits 0 through 10 of $z \cdot 2^{15}$ as integer zfrac and compute

$$y_i \cdot 2^{16} + s_i \cdot \text{zfrac} .$$

The result is a good approximation of $\arctan(z) \cdot 2^{31}$.

The only computationally demanding task in the above algorithm are one multiplication of 16-bit numbers, with 32-bit results and one 32-bit addition[16]. The code below is one possible implementation and leads to Figure 3.49. The maximal error of $1.2 \cdot 10^{-4}$ leads to a value of `bitaccuracy`≈ -12.9 and thus we find at least 12-bit resolution of this implementation.

```
——————————————————— LinearInterpol.m ———————————————————

nn = 5; zc = linspace(0,1,2^nn+1);
atantab = int16(round(atan(zc)*(2^(15)))); % 16-bit values tabulated
%%  errorest=1/8*zc./(1+zc.^2).^2*2^(-2*nn);
%%  atantab=int16(round((atan(zc)+errorest)*(2^(15))));
atantab = int32(int32(atantab)*(2^(16))); % move to upper half of 32-bit

datantab = int32(round(diff(atan(zc))*2^(16+nn))); % 16-bit unsigned

z = 0:0.001:1-1e-10;
zint  = uint8(floor(z*2^nn));              % integer part
zfrac = int32(mod(z*2^15,2^(15-nn)));      % fractional part

res = zeros(size(z)); res2 = res;
for k = 1:length(z);
   ind = zint(k)+1;
   res(k) = int32(atantab(ind) + zfrac(k)*datantab(ind));
end%for
res = res/2^(31);

figure(1); plot(z,res-atan(z),'')
           xlabel('z'); ylabel('error'); grid on

accuracybits = log2(max(abs(res-atan(z))*4/pi))
-->
accuracybits = -12.886
```

The graph of $\arctan z$ explains why the piecewise linear interpolation is below the actual function. Thus the error in Figure 3.49 is everywhere negative. Since an estimate for the maximal error on each subinterval given by

$$\frac{1}{8} f''(z) h^2 = \frac{-2z}{8(1+z^2)^2} \frac{1}{32^2}$$

one can try to correct this error. This is implemented by the commented out section in the above code. When using this modification find a maximal absolute error of 0.00006 and `accuracybits`≈ -13.6 and thus we have a slightly improved result.

---

[16]This author developed a modification that uses only one 8-bit multiplication, with a 16-bit result, and one 16-bit addition. The resolution is slightly better than 12-bit. It requires 96 Bytes to store lookup tables. This might be a good solution for a 8-bit controller and moderate accuracy requirements. The result is given in the next section.

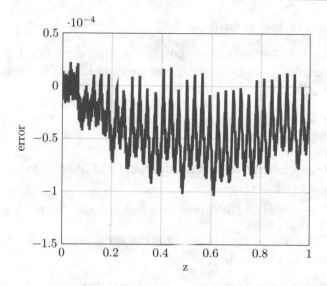

Figure 3.49: The errors for a piecewise linear approximation of the arctan–function

Standard C code for a piecewise linear interpolation

The above mentioned algorithm using piecewise linear interpolation with one 8-bit multiplication and one 16-bit addition can be implemented in the programming language **C**. The code below will, when called with an unsigned 16-bit integer z ($0 \leq z \leq 2^{16} - 1 = 65535$), return an integer value y ($0 \leq y \leq 2^{15} - 1 = 32767$), giving a good approximation of

$$y \approx (2^{15} - 1) \arctan\left(\frac{z}{2^{16} - 1}\right).$$

Observe that the computations of the integer part and the fractional part are implemented with bit operations only and thus very fast.

```
———————————————————————————————— C ————————————————————————————————
unsigned short atan16(unsigned short z){
  static unsigned short atantab[]=
  {    4, 1026, 2048, 3065, 4077, 5081, 6076, 7059, 8030, 8987, 9928,
    10852,11760,12648,13518,14367,15197,16006,16793,17560,18306,19034,
    19740,20425,21090,21734,22360,22969,23557,24129,24683,25219};
  static unsigned char datantab[]=
    {255,255,254,253,251,249,246,243,239,235,231,227,222,217,212,207,
     202,197,192,187,182,176,171,166,161,157,152,147,143,138,134,130};
  unsigned char zint;   unsigned char zfrac;
  zint  = z>>11;
  zfrac = (z&4095)>>3;
  return  atantab[zint] + (zfrac*datantab[zint])>>6;
}
```

Comparison of the previous algorithms

In Table 3.10 find some essential information on the algorithms developed in this section. If you are to choose an algorithm for a concrete application the following points should be taken into consideration.

- If 7-bit accuracy is sufficient then choose between a Chebyshev polynomial of degree 2 and a 8-bit lookup table.

  - The lookup table is the fastest algorithm and rather easy to implement, but it uses some memory.

  - The Chebyshev polynomial requires less memory, but takes longer to evaluate.

- If you need 12-bit accuracy choose between the two other options in Table 3.10 or use Example 3.4.5 and 3.4.5.

  - The Chebyshev polynomial of degree 4 in Example 3.4.5 requires very little storage, but a couple of 16-bit arithmetic operations.

  - The improved linear interpolation method in Table 3.10 needs fewer arithmetic operations, but 96 Bytes of additional memory.

  - A full 16-bit lookup table is possible and would be very fast, but requires a prohibitive amount of memory.

- If you need even higher accuracy you want to consider polynomials of higher degree or a lookup table with quadratic interpolation.

	Chebyshev degree 2	table look-up 8-bit	interpolation version 1	interpolation version 2
absolute error	0.005	0.003	0.00006	0.00001
resolution	7 bit	7 bit	13 bit	12 bit
multiplications	2 (8-bit)	0	1 (16-bit)	1 (8 -bit)
additions	2 (16-bit)	0	1 (32-bit)	1 (16 bit)
memory for lookup	6 Bytes	256 Bytes	128 Bytes	96 Bytes
table look ups	0	1	2	2

Table 3.10: Comparison of approximation algorithms for the arctan–function

44 Example : Approximation by a Chebyshev polynomial of degree 3

Use the Chebyshev approximation

$$\arctan z \quad \approx \quad -0.0011722644 + 1.038178669\, z - 0.1904775175\, z^2 - 0.06211012633\, z^3$$

$$= \quad -0.0011722644 + z\,(1.038178669 + z\,(-0.1904775175 - z\,0.06211012633$$

with the maximal error approximately 0.001 and a 16-bit micro controller to compute the arctan–function. Implement these calculations using the following arithmetic operations.

- Addition of 16-bit signed integers leading to a 16-bit integer result.

- Multiplication of 16-bit signed integers leading to a 32-bit integer result.

- Multiplications by 2^k to be implemented with arithmetic shifts.

```
                          ┌─────── Chebyshev3.m ───────
a0 = -0.06211012633; i0 = int32(a0*2^16)
a1 = -0.1904775175;  i1 = int16(a1*2^16)
a2 = +1.038178669 ;  i2 = int32(a2*2^14)
a3 = -0.0011722644;  i3 = int16(a3*2^14)


myatan = @(z)-0.0011722644 +z.*(1.038178669 + ...
              z.*(-0.1904775175-z*0.06211012633));
z   = 0:0.001:1;   zi = int32(z*2^15);
r1 = int32(zi.*i0);                %% 15+16
limits1 = [min(r1) max(r1)]
r2 = i1+int16(bitshift(r1,-15)); %% 16
limits2 = [min(r2) max(r2)]
r3 = int32(zi.*int32(bitshift(r2,-2)));   %%16-2+15=29
limits3 = [min(r3) max(r3)]
r4 = bitshift(r3,-15)+i2;          %% 29-15 =14
limits4 = [min(r4) max(r4)]
r5 = int32(zi.*int32(r4));          %% 14+15 =29
limits5=[min(r5) max(r5)]
r6 = int16(bitshift(r5,-15))+i3; %% 29-15
limits6 = [min(r4) max(r4)]
res = single(r6)*2^-14;

figure(1); plot(z,res-atan(z),'r',z,myatan(z)-atan(z),'b')
          xlabel('z'); ylabel('difference');
          legend('int32','float');
```

The graph in Figure 3.50 shows that the error of approximately 0.001 is largely dominated by the Chebyshev approximation. Thus using a polynomial of higher degree will improve the accuracy. With this solution we have 9-bit accuracy.

The above approximation can be implemented in C.

```
                          ┌─────── C ───────
//  for x = z*2^15 the value of y=2^15 * arctan(z) will be computed
int atan32(int z){

  static int i0 = -4070;    static int i1 = -12483;
  static int i2 =  17009;   static int i3 = -19;
  int r;
```

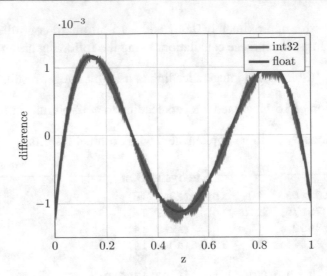

Figure 3.50: Approximation error, using a Chebyshev polynomial of degree 3

```
r = i1+((i0*x)>>15);
r = i2+((x*(r>>2))>>15);
r = i3+((x*r)>>15);
return  r<<1; }
```

A very crude measurement indicated that the above algorithm required approximately 25 CPU cycles to compute one value. of $\arctan(z)$. $\Diamond$

45 Example :   An algorithm suitable for 12-bit AD converters

Many AD converters have a resolution of 12 bit. The Chebyshev approximation

$$\arctan z \approx -0.000077171760231 + 1.00313570626204\, z - 0.0152627086731\, z^2$$
$$-0.342453819098\, z^3 + 0.140171846705\, z^4$$

shows a maximal error of 0.0001 for $0 \leq z \leq 1$. Since $\log_2 \left(\frac{0.0001 \cdot 4}{\pi} \right) \approx -13$ this approximation might allow to keep the 12-resolution of the AD converter. Below find an implementation and the graphical results.

```
                              Chebyshev4.m
a0 = +0.14017184670506358;      i0 = int32(a0*2^15)
a1 = -0.34245381909783135;      i1 = int16(a1*2^15)
a2 = -0.015262708673125458;     i2 = int16(a2*2^15)
a3 = +1.0031357062620350346;    i3 = int16(a3*2^14)
%a3 = +1.0032357062620350346;   i3 = int16(a3*2^14)%patched slope
a4 = -0.00007717176023065795;   i4 = int16(a4*2^15)
myatan = @(x)-0.00007717176023065795 + 1.0031357062620350346*x ...
```

```
              - 0.015262708673125458*x.^2 - 0.34245381909783135*x.^3 ...
              + 0.14017184670506358*x.^4;
z  = 0:0.001:1;   zi = int32(z*2^15);
r1 = int32(zi.*i0);        %% 15+16=30
limits1 = [min(r1) max(r1)]
r2 = int16(bitshift(r1,-15))+i1;   %% 30-15=15
limits2 = [min(r2) max(r2)]
r3 = int32(zi.*int32(r2));          %%15+15=30
limits3 = [min(r3) max(r3)]
r4 = int16(bitshift(r3,-15))+i2;   %% 30-15 =15
limits4 = [min(r4) max(r4)]
r5 = int32(zi.*int32(r4));          %% 15+15 =30
limits5 = [min(r5) max(r5)]
r6 = int16(bitshift(r5,-16))+i3;   %% 30-16=14
limits6 = [min(r6) max(r6)]Rabius93

r7 = int32(zi.*int32(r6));          %% 14+15 =29
limits5 = [min(r7) max(r7)]
r8 = int16(bitshift(r7,-14))+i4;   %% 29-14=15
limits6 = [min(r8) max(r8)]
res = double(r8)*2^-15;
plot(z,res-atan(z),z,myatan(z)-atan(z))
legend('int32','float'); xlabel('z'); ylabel('errors')
max_error = max(abs(res-atan(z)));
bitaccuracy = log2(max_error/pi*4)
```

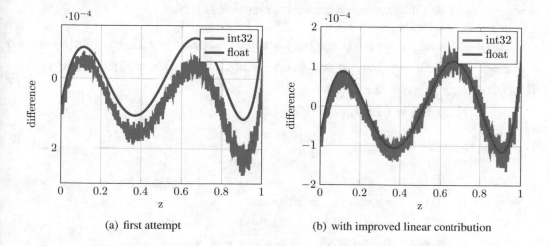

(a) first attempt   (b) with improved linear contribution

Figure 3.51: 32 bit approximation of $\arctan(z)$ by a polynomial of degree 4

Find the result in Figure 3.51(a). The value of the bit accuracy of ≈ -11.5 shows that the desired 12-bit resolution is barely achieved. The graph also shows that the maximal error of approximately 0.00027 is dominated by the Chebyshev approximation, but has a slightly negative slope. Correct the slope by By modifying the coefficient for the linear contribution, leading to Figure 3.51(b). Now the `bitaccuracy` ≈ -12.2 is good enough and the difference are clearly dominated by the Chebyshev approximation. The above code shows 4 necessary 16-bit multiplications (for `r1`, `r3`, `r5` and `r7`) and also 4 additions (for `r2`, `r4`, `r6` and `r8`). For 12-bit resolution one has to store at least $2^{12} = 4096$ numbers. Since these are 16-bit numbers one needs $2^{13} = 8192 = 8$ K bytes of memory. Adding one more bit of resolution would double the memory requirement.   ◇

3.4.6   Chebyshev approximations

The goal of this section is to present the formulas necessary to determine the values of the optimal coefficients c_n for the approximation by Chebyshev polynomials.

$$f(x) \approx \frac{c_0}{2} + \sum_{n=1}^{N} c_n T_n(x)$$

Determine the coefficient of the Chebyshev polynomials

The Chebyshev polynomials on the interval $[-1, 1]$ are defined by

$$T_n(x) = \cos(n \arccos(x)) .$$

Using the trigonometric identity $\cos(\alpha + \beta) = \cos(\alpha) \cos(\beta) - \sin(\alpha) \sin(\beta)$ and with $\alpha = \arccos(x)$ we find a recursion formula for the polynomials.

$$
\begin{aligned}
\cos(+\alpha + n\,\alpha) &= \cos(\alpha) \cos(n\,\alpha) - \sin(\alpha) \sin(n\,\alpha) = x\,T_n(x) - \sin(\alpha) \sin(n\,\alpha) \\
\cos(-\alpha + n\,\alpha) &= \cos(-\alpha) \cos(n\,\alpha) + \sin(\alpha) \sin(n\,\alpha) = x\,T_n(x) + \sin(\alpha) \sin(n\,\alpha) \\
T_{n+1}(x) + T_{n-1}(x) &= 2\,x\,T_n(x) \\
T_{n+1}(x) &= 2\,x\,T_n(x) - T_{n-1}(x)
\end{aligned}
$$

This leads to

$$
\begin{aligned}
T_0(x) &= \cos(0) = 1 \\
T_1(x) &= \cos(\arccos(x)) = x \\
T_2(x) &= 2\,x\,(x) - 1 = 2\,x^2 - 1 \\
T_3(x) &= 2\,x\,(2\,x^2 - 1) + x = 4\,x^2 - 3\,x \\
T_4(x) &= 2\,x\,(4\,x^2 - 3\,x) - 2\,x^2 + 1 = 8\,x^4 - 8\,x^2 + 1
\end{aligned}
$$

$$\vdots$$

The above recursive algorithm can be used to write *Octave* code to compute the coefficients of these Chebyshev polynomials.

--- **Chebyshev.m** ---

```
function pn = Chebyshev(n)
% compute coefficients of the n'th order Chebyshev polynomial
pA = [1];   pB = [1 0];
if n == 0          pn = [1];
  elseif n == 1;   pn = [1,0];
  else
    for i=2:n
      pn = (2*[pB,0] - [0,0,pA]);
      pA = pB;    pB = pn;
    end%for
end%if
```

The code below will generate the graphs of the first few Chebyshev polynomials in Figure 3.52.

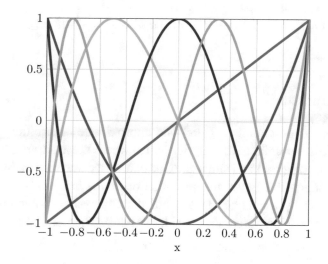

Figure 3.52: Graphs of the first 5 Chebyshev polynomials

```
nn = 5;   x = -1:0.02:1; y = zeros(nn,length(x));
for n = 1:nn
  y(n,:) = polyval(Chebyshev(n),x);
end%for
plot(x,y); legend('1','2','3','4','5')
```

Orthogonality of the Chebyshev polynomials

These polynomials are orthogonal on the interval $[-1, 1]$ with respect to the integration weight $1/\sqrt{1 - x^2}$. This is verified by an integration.

$$
\begin{aligned}
\langle T_n(x), T_m(x) \rangle &= \int_{-1}^{1} T_n(x) \, T_m(x) \, \frac{1}{\sqrt{1 - x^2}} \, dx \\
&= \int_{-1}^{1} \cos(n \cdot \arccos(x)) \, \cos(m \cdot \arccos(x)) \, \frac{1}{\sqrt{1 - x^2}} \, dx
\end{aligned}
$$

$$\text{substitution} \qquad \cos\phi = x \quad , \quad -\sin(\phi) \frac{d\phi}{dx} = 1 \quad ,$$

$$\sqrt{1 - x^2} = \sqrt{1 - \cos^2(\phi)} = \sin(\phi)$$

$$
\begin{aligned}
&= \int_{\pi}^{0} \cos(n \, \phi) \, \cos(m \, \phi) \, \frac{-\sin(\phi)}{\sin(\phi)} \, d\phi \\
&= \int_{0}^{\pi} \cos(n \, \phi) \, \cos(m \, \phi) \, d\phi = 0 \quad \text{if } n \neq m \\
\langle T_n(x), T_n(x) \rangle &= \int_{0}^{\pi} \cos(n \, \phi) \, \cos(n \, \phi) \, d\phi = \frac{\pi}{2} \quad \text{if } n \geq 1
\end{aligned}
$$

Compute the coefficients

The idea is to use the polynomials $T_n(x)$ and approximate an arbitrary function $f(x)$ in terms of $T_n(x)$. This is similar to the Fourier series, where the arbitrary function is rewritten in terms of functions $\cos(n\,x)$ and $\sin(n\,x)$. One can verify that the Chebyshev polynomials give almost the best possible uniform approximation on the interval $[-1, 1]$. The result can at most be improved by a constant factor, but the Chebyshev polynomial can easily be determined. As one possible reference consider [Rivl69, Theorem 2.2].

For a function $f(x)$ defined on the interval $[-1, 1]$ compute the coefficients

$$
\begin{aligned}
\frac{\pi}{2} \, c_n &= \int_{-1}^{1} f(x) \, T_n(x) \, \frac{1}{\sqrt{1 - x^2}} \, dx \\
&= \int_{-1}^{1} f(x) \, \cos(n \cdot \arccos(x)) \, \frac{1}{\sqrt{1 - x^2}} \, dx
\end{aligned}
$$

$$\text{substitution} \qquad \cos\phi = x \quad , \quad -\sin(\phi) \frac{d\phi}{dx} = 1 \quad , \quad \sqrt{1 - x^2} = \sqrt{1 - \cos^2(\phi)} = \sin(\phi$$

$$
\begin{aligned}
&= \int_{\pi}^{0} f(\cos(\phi)) \, \cos(n \, \phi) \, \frac{-\sin(\phi)}{\sin(\phi)} \, d\phi \\
&= \int_{0}^{\pi} f(\cos(\phi)) \, \cos(n \, \phi) \, d\phi \, .
\end{aligned}
$$

Then the Chebyshev approximation is given by

$$f(x) \approx \frac{c_0}{2} + \sum_{n=1}^{N} c_n T_n(x) = \frac{c_0}{2} + \sum_{n=1}^{N} c_n \cos(n \arccos(x)) \, .$$

A function $g(z)$ defined on an interval $[a, b]$ has to be transformed onto the interval $[-1, 1]$ by the transformations

$$z = -1 + 2\frac{x - a}{b - a} \qquad a \le x \le b$$

$$x = a + \frac{1}{2}(z + 1)(b - a) = \frac{a + b}{2} + z\frac{b - a}{2}$$

$$g(z) = f(\frac{a + b}{2} + z\frac{b - a}{2}) \qquad -1 \le z \le 1$$

and then the coefficients for this new function $g(z)$ have to be computed. We find

$$f(x) = g(z) = \frac{c_0}{2} + \sum_{n=1}^{N} c_n T_n(z) = \frac{c_0}{2} + \sum_{n=1}^{N} c_n T_n(-1 + 2\frac{x - a}{b - a}) \, . \qquad (3.9)$$

MATLAB/*Octave* code

The above results are readily implemented in *Octave* for the exemplary function $f(x) = \arctan(x)$ on the interval $A = 0 \le x \le 1 = B$.

```
%% compute the Chebyshev approximation of order n of the function fun
n = 2;
accuracy = 1e-8; % accuracy for the numerical integration
A = 0;     % left endpoint
B = 1;     % right endpoint
fun = @(x)atan(x);   % function to be approximated
```

The remaining parts of the code remain unchanged if we were to examine another function. First the function to be integrated over the standard interval $[-1, 1]$ has to be defined. Then we compute the coefficients using an integration with `quad()`. With `Chebyshev()` we then determine the coefficients of the Chebyshev approximation.

```
% redefine function on standard interval [-1,1]
newFun = @(x,A,B) fun(A+0.5*(x+1)*(B-A));

% function to be integrated
intFun = @(p,A,B,k)newFun(cos(p),A,B).*cos(k*p)

c = zeros(n+1,1);
c(1)   = quad(@(p)intFun(p,A,B,0),0,pi,accuracy)*1/pi;
for k = 1:n
```

```
  c(k+1) = quad(@(p)intFun(p,A,B,k),0,pi,accuracy)*2/pi;
end%for


coeff = zeros(n+1,n+1);
for i = 1:n+1;
  coeff(i,n-i+2:n+1) = Chebyshev(i-1);
end%for


newPol = coeff'*c
```

The vector `newPol` contains the coefficients for the modified function $g(z)$ in the previous section. To apply the transformation we use equation (3.9) and seek the coefficients p_k such that

$$y_i = g(-1 + i\,\frac{2}{n}) = p(x_i) = p(A + i\,\frac{B-A}{n}) = \sum_{k=0}^{n} p_k\,x_i^k \quad \text{for} \quad i = 0, 1, 2, 3, \ldots, n \,.$$

This can be regarded as a system of $n+1$ linear equations with a Vandermonde matrix.

$$\begin{bmatrix} x_0^n & x_0^{n-1} & x_0^{n-2} & & 1 \\ x_1^n & x_1^{n-1} & x_1^{n-2} & \cdots & 1 \\ x_2^n & x_2^{n-1} & x_2^{n-2} & & 1 \\ & \vdots & & \ddots & \vdots \\ x_n^n & x_n^{n-1} & x_n^{n-2} & \cdots & 1 \end{bmatrix} \cdot \begin{pmatrix} p_n \\ p_{n-1} \\ p_{n-2} \\ \vdots \\ p_0 \end{pmatrix} = \begin{pmatrix} y_0 \\ y_1 \\ y_2 \\ \vdots \\ p_n \end{pmatrix}$$

This is easily implemented in *Octave*.

```
y = polyval(newPol,linspace(-1,1,n+1))';
F = vander(linspace(A,B,n+1)');
p = F\y
```

Then the results can be visualized. A part of the result is shown in Figure 3.47.

```
x  = linspace(A,B,101);   y1 = fun(x);   y2 = polyval(p,x);
figure(1); plot(x,y2-y1)
           xlabel('x'); ylabel('difference');
figure(2); plot(x,y1,x,y2)
           xlabel('x'); ylabel('values');
```

3.4.7   List of codes and data files

In the previous section the codes and data files in Table 3.11 were used.

filename	function
`Chebyshev2.m`	Chebyshev approximation of degree 2
`Lookup8bit.m`	Approximation by an 8-bit look-up table
`LinearInterpol.m`	Approximation by a piecewise linear interpolation
`Chebyshev.m`	function the determine the coefficients of $T_n(x)$
`ChebyshevApproximation.m`	general Chebyshev approximation
`Chebyshev3.m`	script for Example 3.4.5
`atan32.c`	C code for Example 3.4.5
`Chebyshev4.m`	script for Example 3.4.5

Table 3.11: Codes and data files for section 3.4

3.5 Analysis of Stock Performance, Value of a Stock Option

In many situation one needs to extract information from a file generated by another code. In this section we illustrate a flexible method by analyzing the value of a given stock over an extended time. The file `IBM.csv` contains data for the stock price of IBM from 1990 through 1999[17].

3.5.1 Reading the data from the file, using `dlmread()`

The data retrieved from the internet is stored in a file, whose first few lines are shown below in a file `IBM.csv`.

```
─────────────────── IBM.csv ───────────────────
Date,Open,High,Low,Close,Volume
31-Dec-99,108.671,108.982,106.121,107.365,2870300
30-Dec-99,109.169,109.977,108.049,108.236,3435100
29-Dec-99,109.915,109.977,108.236,108.485,2683300
28-Dec-99,109.044,110.226,108.547,109.293,4083100
27-Dec-99,109.169,109.48,107.614,109.231,3740700
...
```

The easy way to go is to use the command `dlmread()` to extract the needed information. In this example we only want the second column, showing the value of the stock at the opening of each trading day. Find the result in Figure 3.53.

---

[17]The results were found at http://finance.yahoo.com trough http://finance.yahoo.com/stock-center/. The package financial of *Octave* has functions to read this type of data from the web site, but lately yahoo does not accept the commands. You obtain the data by selecting your stock, the dates, then use APPLY and download the data. Values of Swiss stock is available on Yahoo too. Similar information is available at https://www.macrotrends.net/ or at https://marktdaten.fuw.ch/overview/stocks

```
┌──────────────────────── IBMscriptDLM.m ────────────────────────────┐
% read all the data, starting at column 2 and row 2
data = dlmread('IBM.csv',',',1,1);

indata = data(:,1)';                    % use second column only
k = length(data) ;
indata =  fliplr(indata);
disp(sprintf('Number of trading days from 1990 to 1999 is %i',k))

figure(1); plot(indata)
          xlabel('days'); ylabel('value of stock');
          axis([0, k, 0, max(indata)]);  grid on
└─────────────────────────────────────────────────────────────────────┘
```

3.5.2   Reading the data from the file, using formatted reading

Instead of the above short code we can also use formatted reading. We use this simple example to illustrate the general procedure:

1. Open the file for reading.

2. Read one item of information at a time and store the useful items.

3. Close the file.

Due to the structure of the file the following operations have to be performed:

- Open the file for reading.

- Read the title line and ignore it.

- Allocate memory for all the data to be read.

- Read a first line.

- For each line in the file:

    – Determine the location of the first comma and then only use the trailing string.

    – Read the first number in the string and store it properly.

    – Read the next line.

- Close the file.

- Adjust the size of the resulting matrix and rearrange it to have early values first. Then display the number of trading days and plot the value of the stock. Find the result in Figure 3.53.

--- **IBMscript.m** ---

```
indata = zeros(1,365*10);          % allocate storage for the data
infile = fopen('IBM.csv','rt');    % open the file text for reading
tline  = fgetl(infile);            % read the title line

k = 0;                             % a counter
inline = fgetl(infile)             % read a line
while ischar(inline)               % test for end of input file
  counter = find(inline==',');     % find the first ','
                                   % then use only the rest of the line
  newline = inline(counter(1)+1:length(inline));
                                   % read the numbers
  A = sscanf(newline,'%f%c%f%c%f%c%f%c%f');
  k = k+1;
  indata(k) = A(1);                % store only the first number
  inline = fgetl(infile);          % get the next input line
end%while
fclose(infile);                    % close the file
disp(sprintf('Number of trading days from 1990 to 1999 is %i',k))

indata = fliplr(indata(1:k));      % reverse the order
figure(1); plot(indata)
         xlabel('days'); ylabel('value of stock');
         axis([0, k, 0, max(indata)]);  grid on
```

Another option would be to use the command `textread()` and then can line by line with a well constructed format string.

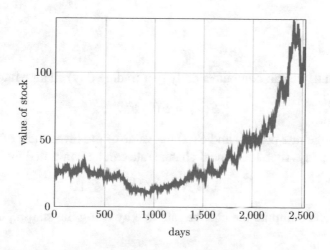

Figure 3.53: The price of IBM stock from 1990 to 1999

3.5.3   Analysis of the data

Moving averages

The value of the stock has a rather high volatility. To visualize this we might compare the actual value of the stock with the average value over a few days. The code below achieves just this and the result is shown in Figure 3.54.

```
———————————————— IBMaverage.m ————————————————
% the array indata contains the value of the stock
k = length(indata);

% find the range of trading days for each year
y1 = 1:253;    y2 = 254:506;    y3 = 507:760;

% choose the length of the averaging period
avg = 20;

% create the data
avgdata = indata;
for ii = 1:k
    avgdata(ii) = mean(indata(max(1,ii-avg):ii));
end%for
% plot results for the third year
plot(y3,indata(y3),y3,avgdata(y3))
grid on
title('Value of IBM stock and its moving average in 1992')
xlabel('Trading day'); ylabel('Value');  xlim([min(y3),max(y3)]);
text(510,15,'moving average over 20 days')
legend('data','moving average')
```

Daily change rates

Using the above data we can compute a daily (per trading day) change rate r by

$$S(n) = S(0)\, e^{nr}$$

where r is the change rate per day and $S(n)$ the value of the investment after n days. If one year has N trading days then the annual change rate can be computed by

$$e^{Nr} - 1 \approx Nr \quad \text{if} \quad Nr \ll 1.$$

Based of this we can compute the change rate rate by using the starting and final value of the stock

$$S(n) = S(0)\, e^{nr} \quad \Longrightarrow \quad e^{nr} = \frac{S(n)}{S(0)} \quad \Longrightarrow \quad r = \frac{1}{n}\, \ln \frac{S(n)}{S(0)}.$$

The code below implements this formula.

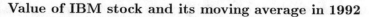

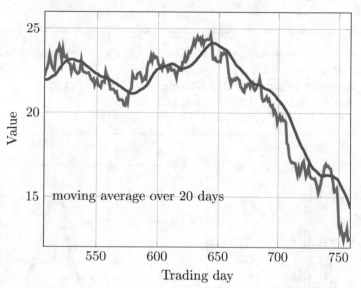

Figure 3.54: The price of IBM stock from 1992 and its average value over 20 days

```
n = length(indata);
rcomp = log(indata(n)/indata(1))/(n-1)
-->
rcomp = 6.0481e-04
```

This change rate can be compared to the average of the daily change rates, i.e. the average of the expressions

$$r(j) = \ln\left(\frac{S(j)}{S(j-1)}\right).$$

This will be a set of daily change rates and we may consider their statistical distribution of the values, i.e. determine mean and standard deviation.

```
% mean value and standard deviation of daily change rate
rates = log(indata(2:n)./indata(1:n-1));
rmean = mean(rates)
rstd  = std(rates)
-->
rmean = 6.0481e-04
rstd  = 0.019396
```

As one would expect the average of the daily change rates (computed day by day) coincides with the average daily change rate (computed by using initial and final value only).

To illustrate the distribution of the daily rates rates a histogram can be used, as shown in Figure 3.55(a).

```
dr = 0.005;   edges = [-inf,-0.1:dr:0.1,inf];  nn = length(edges);
histdata = hist(rates,edges);
figure(1); bar(edges(2:nn-1)-dr/2,histdata(2:nn-1));
          axis([-0.1,0.1,0,400]); grid on
          xlabel('rate'); ylabel('# of cases')
```

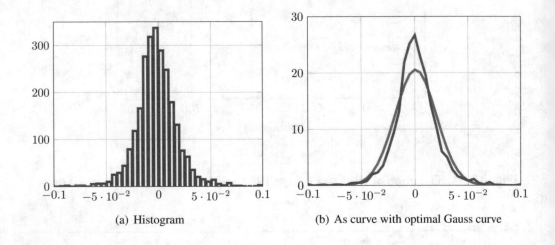

(a) Histogram   (b) As curve with optimal Gauss curve

Figure 3.55: Histogram of daily interest rate of IBM stock

It is possible to approximate the distribution of interest rates by a normal distribution with the mean value and standard deviation from above. To do this use the function `normpdf()` given by

$$\text{normpdf}(x) = \frac{1}{\sigma\sqrt{2\pi}}\, e^{-\frac{(x-\bar{x})^2}{2\sigma^2}}\ .$$

Then this function is used to generate Figure 3.55(b). The result shows the observed data and a Gauss curve with the same mean and standard deviation.

```
y = normpdf(edges(2:nn-1),rmean,rstd);
factor = sum(histdata(2:nn-1))*dr;
histnew = histdata(2:nn-1)/factor;
figure(2); plot(edges(2:nn-1),[y;histnew])
          axis([-0.1,0.1,0,max(histnew)]); grid on
          xlabel('change rate'); ylabel('percentage')
          legend('Gauss fit','data')
```

One might use the correlation coefficient of two vectors to obtain a numerical criterion on how similar the shape of the functions are. For two vectors $\vec{a}$ and $\vec{b}$ the correlation coefficient is given by

$$\cos\alpha = \frac{\langle\vec{a},\vec{b}\rangle}{\|\vec{a}\|\,\|\vec{b}\|} = \frac{\sum_i a_i b_i}{\sqrt{\sum_i a_i^2}\,\sqrt{\sum_i b_i^2}}$$

where α is the angle between the two vectors. If the vectors are generated by discretizing two functions then a correlation coefficient close to 1 implies that the graphs of the two functions are of similar shape. For the two functions (resp. vectors) in Figure 3.55 we obtain

```
y = normpdf((edges(2:nn-2)+edges(3:nn-1))/2,rmean,rstd);
correlation = histdata(2:nn-2)*y'/(norm(histdata(2:nn-2))*norm(y))
-->
correlation = 0.9812
```

Thus in this example the distribution of the daily change rates is quite close to a normal distribution.

3.5.4 A Monte Carlo simulation

Since we found an average daily change rate and its standard deviation we can use random numbers to simulate the behavior of stock values. We generate random numbers for the daily change rates with the known average value and standard deviation. Then we use these change rates to compute the behavior of the value of the stock when those change rates are applied. We can even run multiple of those simulations to extract information on an *average performance*.

Simulation of one year

We assume that the initial value of the stock is $S(0) = 1$ and one year has 250 trading days. Then we generate a vector of random numbers with the mean and standard deviation of the daily change rate of the above IBM stock. The command `randn(1,days)` will create normally distributed random numbers with average 0 and standard deviation 1. Thus we have to multiply these numbers with the desired standard deviation and then add the average value. We assume that the value of the stock is given by $S(0) = 1$ on the first trading day. For subsequent days we use

$$S(k) = S(k-1)\cdot e^{r(k)}$$

to find the value $S(k)$ on the k-th day. Then we plot the values of the stock to arrive at Figure 3.56(a). Observe that at the $(k+1)$-th day the value is given by

$$S(k+1) = S(1)\cdot e^{r(1)}\cdot e^{r(2)}\cdot e^{r(3)}\cdots e^{r(k)} = S(1)\,\exp(r(1)+r(2)+r(3)+\cdots+r(k))$$

and thus we can use the command `cumsum()` (cumulative summation) to determine the values at all days, without using a loop.

```
                              ── IBMsimulation.m ──
days = 250;      % number of trading days to be simulated
rates = randn(1,days-1)*rstd + rmean;   % daily change rates
%% solution with a loop, slow
% values = ones(1,days);                  % value of stock
% for k = 2:days
%    values(k) = values(k-1)*exp(rates(k-1));
% end

%% solution without a loop, thus fast
values = [1,exp(cumsum(rates))];
plot(values)
xlabel('trading day'); ylabel('relative value'); grid on
```

Observe that repeated runs of the same code will **not** produce identical results, due to the random nature of the simulation. Running the script a few times will convince you that the value of the stock can either move up or move down. The result of one or multiple runs are shown in Figure 3.56. Observe that most final results are rather close together, but individual runs may show a large deviation. Thus it is a good idea to examine the statistical behavior of the final value of the stock.

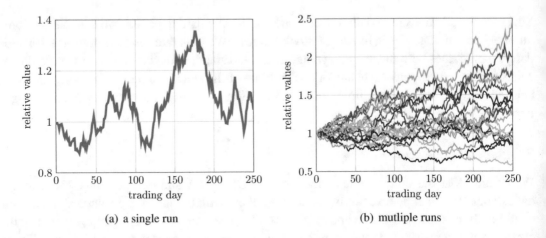

(a) a single run   (b) mutliple runs

Figure 3.56: Simulation of the annual performance of IBM stock

Multiple runs of the simulation

The above simulation can be run many times and the final value of the stock can be regarded as the outcome of the simulation. If the simulation is run many times the outcome can be drastically different, as illustrated by Figure 3.56(b). Thus we can run this simulation many times and consider the final value after one year as the result. We will obtain a probability distribution of values of the stock after one year. This function can be shown as a histogram. The code below does just this.

- Generate the random data. Observe that we removed another loop from the previous code by applying the command `cumsum()` directly to the matrix of all daily change rates. Since we only need the final values we can even use `sum()` instead of `cumsum()`. This will lead to a large speed gain, compared to the original code with two nested loops.

```
days = 250;        % number of trading days to be simulated
runs = 1000;       % number of trial runs
rates  = randn(runs,days-1)*rstd + rmean;     % daily change rates
finalvalues = exp(sum(rates,2));

MeanValue = mean(finalvalues)
StandardDeviation = std(finalvalues)
LogMeanValue = mean(log(finalvalues))
LogStandardDeviation = std(log(finalvalues))
```

One specific run of the above code leads to the numerical values shown below. Be aware that the numbers change from one run to the next, as they depend on the random simulation.

```
MeanValue = 1.2167
StandardDeviation = 0.37752
LogMeanValue = 0.14898
LogStandardDeviation = 0.30962
```

- Create the histogram of the final values, as function of the value of the stock. Find the result in Figure 3.57(a).

```
dr = 0.1;   edges = [-inf,0:dr:3,inf];
histdata = histc(finalvalues,edges)/runs;

nn = length(edges);
figure(1); bar(edges(2:nn-1)-dr/2,histdata(2:nn-1));
           title('Histogram of probability')
           xlabel('value of stock')
           axis([0 3 0 0.15]);   grid on
```

- Create the histogram of the final values, as function of the logarithm of value of the stock. Find the result in Figure 3.57(b).

```
dr = 0.1;   edges = [-inf,-1:dr:1,inf];
histdata = histc(log(finalvalues),edges)/runs;

nn = length(edges);
figure(2);  bar(edges(2:nn-1)-dr/2,histdata(2:nn-1));
            title('Histogram of probability')
            xlabel('logarithm of value of stock');
            axis([-1 1 0 0.15]);   grid on
```

The numerical results for one simulation are shown above and the graphs are given in Figure 3.57. One should realize that the logarithm of the final values are given by a normal distribution, whereas the distribution of the values is not even symmetric. Observe that the shape of this figure changes slightly from run to run, since the result is based on a Monte Carlo simulation.

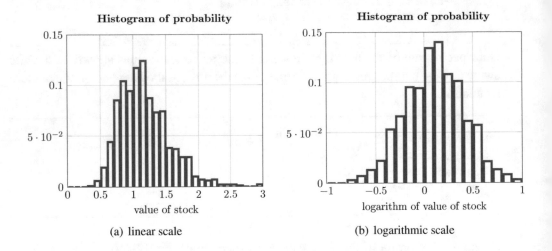

(a) linear scale   (b) logarithmic scale

Figure 3.57: Histograms of the final values of IBM stock after one year

If the daily change rate has a mean value of r and a standard deviation of σ then the logarithm of the values of the stock after N days should be given by a normal distribution with mean value $N r$ and standard deviation $\sigma \sqrt{N}$. Thus the probability density function of the logarithm of the value of the stock is given by

$$\text{PDF}(z) = \frac{1}{\sigma \sqrt{N} \sqrt{2\pi}} \exp\left(-\frac{(z - N r)^2}{2\sigma^2 N}\right).$$

Thus the probability that $\ln S(N)$ is between z and $z + \Delta z$ is given by $\text{PDF}(z)\, \Delta z$, as long as Δz is small.

$$\text{P}(z \leq \ln S(N) \leq z + \Delta z) \approx \text{PDF}(z) \cdot \Delta z$$

With the above numbers $r = 6.0481 \cdot 10^{-4}$, $\sigma = 0.0194$ and $N = 250$ we obtain

$$r N = 0.1512 \quad \text{and} \quad \sigma \sqrt{N} = 0.3067 .$$

The predicted values are very close to the simulation results LogMeanValue = 0.1490 and LogStandardDeviation = 0.3096 of the above simulation.

3.5.5 Value of a stock option : Black–Scholes–Merton

The question

Assume that today's value of IBM stock is $S_0 = 1$. A trader is offering the option to buy this stock one year from today for the fixed strike price of $C = 1.05$. Assume you acquire a few of these options. Your action taken one year from now will depend on the value S_1 of the stock at the end of the year.

- If $S_1 \leq C$ you will not use your right to buy, since it would be cheaper to buy at the stock market.

- If $S_1 > C$ you will certainly use the option and buy, as you will make a profit of $S_1 - C$.

This option has some value to you. You can not loose on the option, but you might win, if the actual value after one year is larger than the strike C.

What is a fair value (price) for this option?

Assumptions

To determine the value of this option the following assumptions can be used:

- The value of the stock is a random process, as simulated by the computations in the previous section.

- The probability for the value S_1 to satisfy $z \leq \ln S_1 \leq z + \Delta z$ is given by

$$\text{PDF}(z) \cdot \Delta z = \frac{1}{\sigma \sqrt{N} \sqrt{2\pi}} \exp\left(-\frac{(z - N r)^2}{2 \sigma^2 N} \right) \cdot \Delta z$$

  with $r = 6.0481 \cdot 10^{-4}$, $\sigma = 0.0194$ and $N = 250$.

- The fair value p of the option is determined by the condition that the expected value of the payoff should equal the value of the option.

The answer

The probability for the value S_1 of the stock after one year to satisfy $\ln C \leq z \leq \ln S_1 \leq z + \Delta z$ for Δz small is given by

$$\text{PDF}(z) \cdot \Delta z = \frac{1}{\sigma \sqrt{N} \sqrt{2\pi}} \exp\left(-\frac{(z - N r)^2}{2 \sigma^2 N}\right) \cdot \Delta z \,.$$

The graph of this function is shown in Figure 3.58. This figure has to be compared with the left part in Figure 3.57.

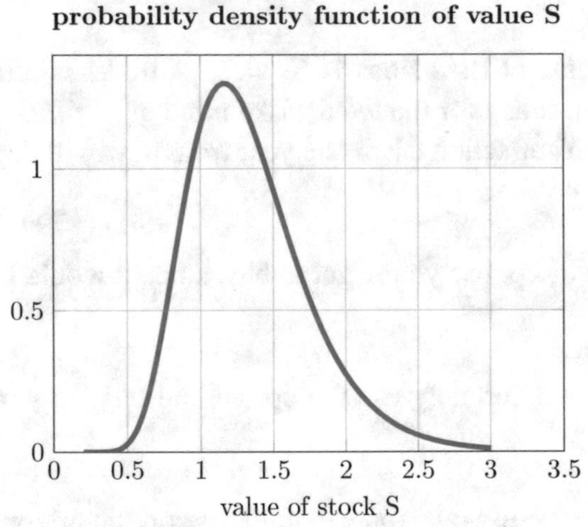

probability density function of value S

Figure 3.58: Probability density function of final values

With the help of this probability density function we can compute the probability for certain events. To find the probability that the value of the stock is larger than twice its original value we compute

$$\int_{\ln 2}^{\infty} \text{PDF}(z) \, dz = \frac{1}{\sigma \sqrt{N} \sqrt{2\pi}} \int_{\ln 2}^{\infty} \exp\left(-\frac{(z - N r)^2}{2 \sigma^2 N}\right) dz \approx 0.039 \,.$$

Thus there is only a 4% chance to double the value within one year. The integral

$$\int_{\ln 1}^{\infty} \text{PDF}(z) \, dz = \frac{1}{\sigma \sqrt{N} \sqrt{2\pi}} \int_{\ln 1}^{\infty} \exp\left(-\frac{(z - N r)^2}{2 \sigma^2 N}\right) dz \approx 0.69$$

indicates that there is a 69% chance of the value of the stock to increase. These probabilities have to be taken into account when estimating the value of the option.

If the value of the stock after one year is given by S_1 then the payoff is $\max\{0, S_1 - C\}$.

- If $S_1 \leq C$ then there is no payoff

- If $\ln C \le z \le \ln S_1 \le z + \Delta z$ then the payoff is approximately $S_1 - C = e^z - C$. Since the probability for this is given by $\text{PDF}(z) \cdot \Delta z$ we find payoff of $e^z - C$ with probability

$$\text{PDF}(z) \cdot \Delta z = \frac{1}{\sigma \sqrt{N} \sqrt{2\pi}} \exp\left(-\frac{(z - N\, r)^2}{2\,\sigma^2\, N}\right) \cdot \Delta z\,.$$

To examine the expected payoff plot the product of the payoff with the probability density function. The result is shown in Figure 3.59. Observe the following:

- You can not expect any payoff if the value of the stock will fall below $C = 1.05$.

- Values of S_1 slightly larger than C are very likely to occur, but the payoff $S_1 - C$ will be small.

- Very high values of S_1 are unlikely to happen. Thus the large payoff $S_1 - C$ is unlikely to occur.

- Most of the return from this option will occur for values of S_1 between 1.3 and 2.0.

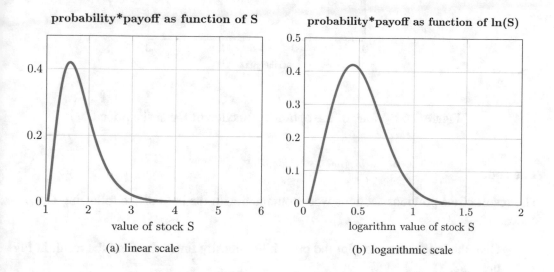

(a) linear scale   (b) logarithmic scale

Figure 3.59: Product of payoff with probability density function

All those possible contributions to the payoff have to be taken into account. The possible values for S_1 are $0 < S_1 < \infty$ and thus $-\infty < z = \ln S_1 < \infty$. By adding up, resp. integrating the above payoff we arrive at an expected value of the payoff (and thus the price of the option) of

$$
\begin{aligned}
p &= \lim_{\Delta z_i \to 0} \left(\sum_{z_i = \ln C}^{\infty} (e^{z_i} - C)\ \mathrm{PDF}(z_i)\ \Delta z_i \right) \\
&= \int_{\ln C}^{\infty} (e^z - C)\ \mathrm{PDF}(z)\ dz \\
&= \frac{1}{\sigma \sqrt{N} \sqrt{2\pi}} \int_{\ln C}^{\infty} (e^z - C)\ \exp\left(-\frac{(z - N r)^2}{2\sigma^2 N} \right) dz \approx 0.23887 .
\end{aligned}
$$

Thus the fair value of the option is $p \approx 0.24$ for a strike of $C = 1.05$.

The above computations can be repeated for multiple values of the strike price C, leading to Figure 3.60.

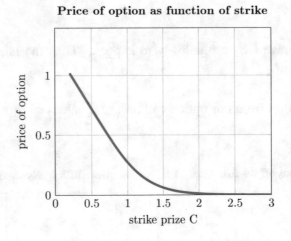

Price of option as function of strike

Figure 3.60: Value of the option as function of the strike price C

The code

The result of the previous sections were computed with the help of the following *Octave–* codes.

- Give the basic data and plot the probability density function. Find the result in Figure 3.58.

```
rmean = 6.0481e-4  % mean of the daily change rate
rstd  = 0.0194     % standard deviation of the daily change rate
N  = 250           % number of trading days in a year
C  = 1.05          % strike price
NN = 100;
```

```
zval = linspace(log(0.2),log(3),NN);
Prob = normpdf(zval,N*rmean,rstd*sqrt(N));
figure(1); plot(exp(zval),Prob)
           title('probability density function of value S');
```

- Compute the probabilities for the value of the stock to double or at least increase.

```
zval  = linspace(log(1),log(6),NN);
Prob  = normpdf(zval,N*rmean,rstd*sqrt(N));
prob1 = trapz(zval,Prob)
zval  = linspace(log(2),log(6),NN);
Prob  = normpdf(zval,N*rmean,rstd*sqrt(N));
prob2 = trapz(zval,Prob)
-->
prob1 =   0.68892
prob2 =   0.038648
```

The result show that with a probability of 69% the value of the stock will increase and with a probability of 3.8% the value will at least double.

- Compute the value of the stock option.

```
maxVal = 5*C;
zval = linspace(log(C),log(maxVal),NN);
Prob = normpdf(zval,N*rmean,rstd*sqrt(N));
payoffProb = (exp(zval)-C).*Prob;

figure(2); plot(zval,payoffProb)
           xlabel('logarithm of value of stock S')
           title('probability * payoff as function of ln(S)');

figure(3); plot(exp(zval),payoffProb)
           xlabel('value of stock S')
           title('probability * payoff as function of value S');
           % use trapezoidal integration
           OptionValue = trapz(zval,payoffProb)
-->
OptionValue =   0.23884
```

The result states that the option with a strike prize of $C = 1.05$ has a value of 0.24.

- Now examine different values for the strike price C and plot the resulting values of the option. Find the result in Figure 3.60.

```
NN   = 100;
cval = linspace(0.2,3,100);
```

```
price = zeros(size(cval));
for k = 1:length(cval)
  zval = linspace(log(cval(k)),log(6),NN);
  Prob = normpdf(zval,N*rmean,rstd*sqrt(N));
  payoffProb = (exp(zval)-cval(k)).*Prob;
  price(k) = trapz(zval,payoffProb);
end%for
figure(4); plot(cval,price)
            title('Value of option as function of strike')
            xlabel('strike prize C'); ylabel('value of option');
```

Nobel Price in Economics in 1997

This observation is used as a foundation for the famous Black–Scholes formula to find the value of stock options. Further effects have to be taken into account, e.g. changing rates and other types of options. The theory was developed by Fischer Black, Myron Scholes and Robert Merton in 1973. Since then their methods are used extensively by the financial "industry". The 1997 Nobel Prize in Economics was awarded to Merton and Scholes. Fisher Black died in 1995 and thus did not obtain the Nobel prize. Further information on the Black–Scholes–Merton method and its applications in finance can be found in many books, e.g. [Seyd00] or [Wilm98].

3.5.6   List of codes and data files

In the previous sections the codes and data files in Table 3.12 were used.

Script file	task to perform
IBMscriptDLM.m	read the data and create basic plot
IBMscript.m	formatted scanning of the data and create basic plot
IBM.csv	data file with the value of IBM stock
IBMaverage.m	compute daily interest rates
IBMhistogram.m	create histogram with interest rates
IBMsimulation.m	simulation for value of stock during one year
IBMsimulationMultiple.m	multiple runs of above simulation
IBMsimulationHist.m	histogram for multiple runs of above simulation
IBMBlackScholes.m	find the value of a stock option

Table 3.12: Codes and data files for section 3.5

3.6 Motion Analysis of a Circular Disk

3.6.1 Description of problem

A circular disk (watch) is submitted to a shock acceleration and thus will start to move. The vertical displacement is measured at several points along the perimeter. The resulting movement should be visualized. The typical situation at a given time is shown in Figure 3.61.

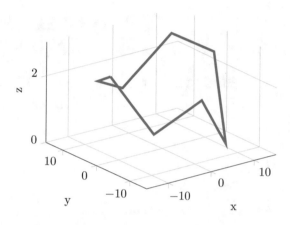

Figure 3.61: Deformed circle for a given time

3.6.2 Reading the data

The first task is to read all the data files. Each data file contains about 4460 data points. The plan is to read one out of 5 points and ignore the other measurements. Thus we read only 345 points. On the circle 8 different points were examined.

- First we create matrices big enough to contain all data.

```
nt = 345;      %% nt number of points to be read
skip = 5;      %% skip 5 frames, before the next image is created
lineskip = 1;  %% ignore one line for the header
%% nt*skip < number of points to be measured
%% Npts number of measurement points on caliber
Npts = 8;

%% define the matrices for the coordinate data of points
x = zeros(Npts+1,nt);      y = zeros(Npts+1,nt);
t = zeros(Npts+1,nt);      h = zeros(Npts+1,nt);
```

- First we define the horizontal position of each point, given by the x and y coordinates. We copy the first point to the 9th, to close the circle.

```
x(1,:) = -12.0; %% first line of x
x(2,:) = -8.4;
x(3,:) = 0;
x(4,:) = 8.4;
x(5,:) = 12;
x(6,:) = 8.4;
x(7,:) = 0;
x(8,:) = -8.4;
x(9,:) = -12.0; %% copy of the first line

y(1,:) = 0;    %% first line of y
y(2,:) = -8.4;
y(3,:) = -12;
y(4,:) = -8.4;
y(5,:) = 0;
y(6,:) = 8.4;
y(7,:) = 12;
y(8,:) = 8.4;
y(9,:) = 0; %% copy of the first line
```

- Each of the files cg*a.txt contains two columns of data. The first number indicates the time and the second the vertical displacement.

```
――――――――――――― cg1a.txt ―――――――――――――
   temps         chemin
-4.02E-7          6.66173E-3
1.551E-6         -7.02712E-2
3.504E-6         -5.89382E-2
5.457E-6          1.47049E-2
7.41E-6           9.04202E-3
9.363E-6          2.95391E-2
1.1316E-5         3.40761E-2
1.3269E-5         5.05111E-2
...
```

The entries on each line are separated by a TAB character. We use the command dlmread() to read the data files. This allows to skip the first row[18], use help dlmread.

```
――――――――――――― ReadDataDLM.m ―――――――――――――
row = 1;
data = dlmread('cg1a.txt','\t',1,0);
t(row,1:nt) = data(skip*[1:nt],1);
```

---

[18]With very recent versions of MATLAB this works too.

```
h(row,1:nt) = data(skip*[1:nt],2);

row = 2;
data = dlmread('cg10a.txt','\t',1,0);
t(row,1:nt) = data(skip*[1:nt],1);
h(row,1:nt) = data(skip*[1:nt],2);

%%%%% followed by a few similar sections of code
```

- Another option is to first open the file for reading (`fopen()`), then read each line by `fgets()` and use formatted scaning (`sscanf()`) to extract the two numbers.

ReadData.m

```
row = 1;
fid = fopen('cg1a.txt','r');
for ii = 1:lineskip
  tline = fgets(fid);
end
for ii = 1:nt
  for s = 1:skip tline = fgets(fid);end
  res = sscanf(tline,'%e %e');
  t(row,ii) = res(1);
  h(row,ii) = res(2);
end
fclose(fid);

row = 2;
fid = fopen('cg10a.txt','r');
for ii = 1:lineskip
  tline = fgets(fid);
end
for ii = 1:nt
  for s = 1:skip tline = fgets(fid);end
  res = sscanf(tline,'%e %e');
  t(row,ii) = res(1);
  h(row,ii) = res(2);
end
fclose(fid);

%%%%% followed by a few similar sections of code
```

- The new last point has to be created as a copy of the first point. This will close the cricle.

```
% copy first point to last point
t(9,:) = t(1,:);    h(9,:) = h(1,:);
```

With the above preparation one can now create the picture in Figure 3.61.

```
k = 20;
plot3(x(:,k),y(:,k),h(:,k));
     xlabel('x'); ylabel('y'); zlabel('z'); grid on
```

3.6.3   Creation of movie

By creating pictures similar to Figure 3.61 for each time slice we can now create a movie and display it on the screen. The code below does just this with the switch movie=0. If we set movie=1 then the images are written to the sub-directory pngmovie in a bitmap format (png). Then an external command (mencoder) is used to generate a movie file Circle.avi to be used without *Octave* or MATLAB. The command is composed of two strings, for them to fit on one display line. Subsequently the directory is cleaned up.

```
                        ──── MovieAVI.m ────
movie = 0; %% switch to generate movie 0: no movie, 1: movie generated
cd pngmovie
k = 1;
plot3(x(:,k),y(:,k),h(:,k));
xlabel('x'); ylabel('y'); zlabel('h');
axis([-14 14 -14 14 -360 80])
grid on
for k = 1:nt
   plot3(x(:,k),y(:,k),h(:,k));
   xlabel('x'); ylabel('y'); zlabel('h');
   axis([-14 14 -14 14 -360 80])
   grid on
   drawnow();
   if movie
     filename = ['movie',sprintf('%03i',k),'.png'];
     print(filename,'-dpng')
   end%if
end%for

if movie
   c1 = 'mencoder mf://*.png -mf fps=5 -ovc lavc -lavcopts vcodec=mpeg4';
   c2 = ' -o Circle.avi';
   system([c1,c2]);
   system('rm -f *.png');
end%if
cd ..
```

- To generate the movieone can use the program mencoder on a Linux system. To generate WMV files use

```
mencoder mf://*.png -mf fps=5 -ovc lavc ...
            -lavcopts vcodec=wmv1 -o movie.avi
```

and for MPEG files accordingly

```
mencoder mf://*.png -mf fps=5 -ovc lavc...
            -lavcopts vcodec=mpeg4 -o movie.avi
```

- Instead of the command mencoder one might use the FFmpeg suite of codes, to be found at the site http://www.ffmpeg.org/. A possible call to create a movie with 5 frames per second is given by

```
ffmpeg -r 5 -i movie%03d.png -c:v libx264 -r 30 ...
        -pix_fmt yuv420p out.mp4
```

- Similarly the command ffmpeg might be used (www.libav.org)

```
ffmpeg   -i ./*movie%03d.png out.mp4
```

or

```
ffmpeg -r 10 -i ./movie%03d.png -c:v libx264...
        -r 10 -pix_fmt yuv420p out.mp4
```

To play the movie you may use any movie player, e.g. vlc or xine.

3.6.4 Decompose the motion into displacement and deformation

The movement of the circle can be decomposed into four different actions:

1. moving the center of the circle up and down

2. rotating the circle about the y axis

3. rotating the circle about the x axis

4. internal deformation of the circle

Using linear regression we want to visualize the above movements and deformation. To determine the movement and rotations we search for coefficients p_1, p_2 and p_3 such that for any given time t the plane

$$z(t, x, y) = p_1(t) + p_2(t) \cdot x + p_3(t) \cdot y$$

describes the location of the circle a good as possible, in the least square sense. Thus for a given time t we have the following problem:

- Given:

    - location of points (x_i, y_i) for $1 \leq i \leq m$

    - measured height z_i for $1 \leq i \leq m$

- Search parameters $\vec{p}$ such that $p_1 \cdot 1 + p_2 \cdot x_i + p_3 \cdot y_i$ is as close as possible to z_i. The value of p_1 corresponds to the height of the center and p_2, p_2 show the slopes in x and y direction.

Use linear regression (see Section 3.2) to find the optimal parameters. Thus introduce a matrix notation

$$\mathbf{X} = \begin{bmatrix} 1 & x_1 & y_1 \\ 1 & x_2 & y_2 \\ 1 & x_3 & y_3 \\ & \vdots & \\ 1 & x_n & y_n \end{bmatrix} \quad \text{and} \quad \vec{z} = \begin{pmatrix} z_1 \\ z_2 \\ z_3 \\ \vdots \\ z_n \end{pmatrix} .$$

Now use the command `LinearRegression()` to determine the parameters and then create Figure 3.62. In Figure 3.62(a) find the height as function of time. The up and down movement of the circle is clearly recognizable. In Figure 3.62(b) the two slopes of the circle in x and y direction are displayed.

---
regression.m
```
X = [ones(1,8);x(1:8,1)';y(1:8,1)']';
% nt=345;
par = zeros(3,nt);

for kk = 1:nt
  p = LinearRegression(X,h(1:8,kk));
  par(:,kk) = p;
end%for

figure(3); plot(t(1,:),par(1,:));
           grid on; axis([0 0.0036, -300, 50])
           xlabel('time'); ylabel('height'); grid on

figure(4); plot(t(1,:),par(2:3,:));
           grid on; axis([t(1,1), max(t(1,:)), -10, 10])
           xlabel('time'); ylabel('slopes');
           legend('x-slope','y-slope')
           axis([0, 0.0036 -10 10])
```

As a next step we create a movie with the movement and rotations of the plane only. First we compute the position of the planes and then reuse the code from the previous section to generate a movie.

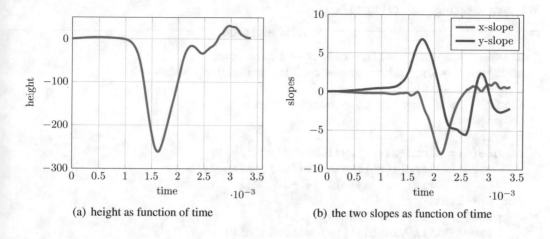

(a) height as function of time   (b) the two slopes as function of time

Figure 3.62: Height and slopes of the moving circle

```
MovieLinear.m
hlinear = X*par; hlinear(9,:) = hlinear(1,:);
horiginal = h;   hdeform = h-hlinear;  hdisp = hlinear;

movie = 0; %% switch to generate movie
cd pngmovie
k = 1;
figure(5); plot3(x(:,k),y(:,k),hdisp(:,k));
           xlabel('x'); ylabel('y'); zlabel('h');
           axis([-14 14 -14 14 -360 80]); grid on
for k = 1:nt
   plot3(x(:,k),y(:,k),hdisp(:,k));
   xlabel('x'); ylabel('y'); zlabel('h');
   axis([-14 14 -14 14 -360 80]); grid on
   drawnow()
   if movie
     filename = ['movie',sprintf('%03i',k),'.png'];
     print(filename,'-dpng')
   end%if
end%for

if movie
 c1 ='mencoder mf://*.png -mf fps=25 -ovc lavc -lavcopts vcodec=mpeg4';
 c2 =' -o movieLinear.avi';
 system([c1,c2]);
 system('rm -f *.png');
end%if
cd ..
```

Then the internal deformations can be displayed too. Since the amplitudes are smaller we have to change the scaling to be used.

MovieDeform.m

```
hlinear = X*par;   hlinear(9,:) = hlinear(1,:);
horiginal = h;     hdeform = h-hlinear;   hdisp = hdeform;

movie = 0; %% switch to generate movie
cd pngmovie
k = 1;
figure(5); plot3(x(:,k),y(:,k),hdisp(:,k));
           xlabel('x'); ylabel('y'); zlabel('h');
           axis([-14 14 -14 14 -50 50]); grid on
for k = 1:nt
   plot3(x(:,k),y(:,k),hdisp(:,k));
   xlabel('x'); ylabel('y'); zlabel('h');
   axis([-14 14 -14 14 -50 50]);  grid on
   drawnow()
   if movie
      filename = ['movie',sprintf('%03i',k),'.png'];
      print(filename,'-dpng')
   end%if
end%for

if movie
   c1 = 'mencoder mf://*.png -mf fps=25 -ovc lavc -lavcopts vcodec=mpeg4'
   c2 = ' -o movieDeform.avi';
   system([c1,c2]);
   system('rm -f *.png');
end%if
cd ..
```

3.6.5   List of codes and data files

In the previous sections the codes and data files in Table 3.13 were used.

3.7   Analysis of a Vibrating Cord

The diploma thesis of Andrea Schüpbach examined vibrating cord sensors, produced by DIGI SENS Switzerland AG. A sample is shown in Figure 3.63. An external force will lead to an increase in tension on a vibrating string, whose frequency is used to determine the force. For further developments DIGI SENS Switzerland AG needs to examine the frequency and quality factor of this mechanical resonance system. An electronic measurement system was developed and tested. In this section the data analysis for this project will be presented.

Script file	task to perform
`cg*a.txt`	data files for the moving disk
`ReadDataDLM.m`	read all data files, using `dlmread()`
`ReadData.m`	read all data files, using `sscanf()`
`MovieAVI.m`	display movie and create `Circle.avi`
`regression.m`	do the linear regression and plot the graphs
`MovieLinear.m`	display and create movie of the linear movements
`MovieDeform.m`	display create movie of the internal deformations

Table 3.13: Codes and data files for section 3.6

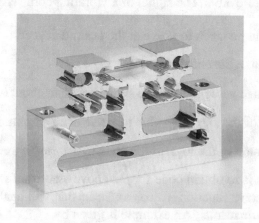

Figure 3.63: A vibrating cord sensor. Image courtesy of **DIGI SENS Switzerland AG**

3.7.1   Design of the basic algorithm

The raw signal of a sensor is measured with `LabView` and the data then written to a file. A typical result is shown in Figure 3.64. At first the cord is vibrating with a constant amplitude and then the amplitude seems to converge to zero, exponentially. Thus expect functions of the form

$$\text{first:}\quad y(t) = A\,\cos(\omega\,t) \qquad\qquad \text{then:}\quad y(t) = A\,e^{-\alpha(t-t_0)}\cos(\omega\,t)\,.$$

To characterize the behavior of the sensor the initial amplitude A and the decay exponent α have to be determined reliably.

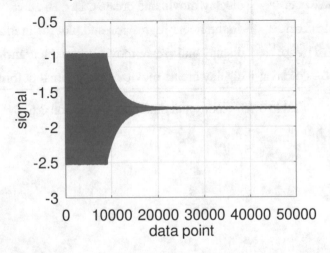

Figure 3.64: The signal of a vibrating cord sensor

The essential steps to be taken to arrive at the desired data are:

- Read the data from a file and display.

- Subtract the average value to have the signal oscillate about 0.

- Take the average of the absolute value to arrive at an amplitude signal.

- Use linear regression of the logarithmic data to determine the decay exponent.

The goal of this section is to obtain code for the above algorithm to be applied to data in a file. The result is an *Octave* function `automatic.m` that will analyze a data set, generate the graphs and display the results. An example is given by

```
[amplitude,factor] = automatic('m4')
-->
Slope: -18.3789, Variance of slope: 0.00337408,
relative error:0.000183584,  amplitude = 0.49640
factor = -0.054410
```

Thus the body of the function is given by

```
─────────────────────────── automatic.m ───────────────────────────
function [amp,fact] = automatic(filename)

write your code here

end%function
```

Reading the data

The sensor is examined with the help of a DAQ card and LabView. The result are files with the data below.

```
────────────────────────────── m4 ──────────────────────────────
waveform          [0]
t0        03.10.2007   15:36:19.
delta t 2.080000E-5

time     Y[0]
03.10.2007   15:36:19.    -1.275252E+0
03.10.2007   15:36:19.    -1.294272E+0
03.10.2007   15:36:19.    -2.469335E+0
03.10.2007   15:36:19.    -1.826837E+0
03.10.2007   15:36:19.    -9.730251E-1
03.10.2007   15:36:19.    -2.076841E+0
...
```

The data consists of 5 header lines and then the actual data lines, in this case 50000 lines. Since there are many data lines the scanning of the data will take time. The same data set will have to be analyzed many times, to determine the optimal parameters. Thus once the data is read from the file (e.g. from m4) a new binary file (e.g. m4.mat) is generated. On subsequent calls of this function the binary file will be read, leading to sizable time savings.

- We use the fact that the sampling rate is 48 kHz.

- If a file with binary data exist, read it with the help of the command load().

- Otherwise use the tools from Section 1.2.8 to read and scan the data file. After reading generate the file with the binary data.

- Generate a graph with the raw data for visual inspection.

```
──────────────────────────── Readm4.m ────────────────────────────
%% script file to read one data set
filename = 'm4';
dt   = 1/48000; %% set the sampling frequency
Nmax = 50000;    %% create arrays large enough to contain all data
x = zeros (Nmax, 1 ) ; y = x;
```

```
%% read the binary file, if it exists
%% otherwise read the original file and write binary file
if (exist([filename,'.mat'],'file')==2)
  eval(['load ',filename,'.mat'])
else
  inFile = fopen(filename ,'rt' ); % read the information from file
  for k = 1:5
    inLine = fgetl(inFile);
  end%for
  k = 1;
  for j = 1:Nmax
    inLine = fgetl(inFile);
    counter = find(inLine=='E');
    y(j) = sscanf(inLine(counter-9:length(inLine)) ,'%f');
  end%for
  fclose(inFile) ;
  eval(['save -mat ',filename,'.mat y'])
end%if

figure(1); plot(y)   %% plot the raw data
          xlabel('dada point'); ylabel('signal'); grid on
```

Determine the amplitude as function of time

The result of the above code is shown in Figure 3.64. Now we aim for the amplitude as a function of time and first subtract the average value of the result. Thus we arrive at oscillations about 0. Then we take the absolute value of the result. With the measurements we sample a function $y = \mathrm{abs}(\sin(\omega\, t))$ at equidistant times, as shown in Figure 3.65. This figure indicates that the mean value of the sampled amplitudes should be equal to the average height h of the function $\sin(t)$ on the interval $[0, \pi]$, i.e.

$$h = \frac{1}{\pi} \int_0^\pi \sin(t)\, dt = \frac{\cos(t)}{\pi}\bigg|_{y=0}^\pi = \frac{2}{\pi} \, .$$

Thus the average value of a larger sample with constant amplitude has to be multiplied by $\pi/2$ to obtain the correct amplitude. Since the instrument has to be calibrated we might as well ignore the factor $\pi/2$. One has to be careful though. The cords are vibrating with frequencies of 14-18 kHz and the sampling frequency is only 48 kHz. Thus we only find about 3 points in one period. This may cause serious discretization problems. By using 100 points the algorithm proved to be robust. A first implementation of the above is given by

```
y2 = abs(y-mean(y));   % subtract average value
FilterLength = 100+1; % average over some points
y3 = zeros(Nmax-FilterLength,1);
for k = 1:Nmax-FilterLength
  y3(k) = mean(y2(k:k+FilterLength));
end%for
```

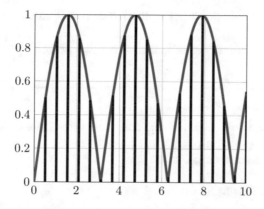

Figure 3.65: A function $y = \text{abs}(\sin(\omega\, t))$ and a sampling

leading to an excessive computation time of 35 seconds. The MATLAB/*Octave* command
conv(), short for convolution, leads to a much faster implementation with a computation
time of less than one tenth of a second. The algorithm used in conv() is based on FFT.

```
y2 = abs(y-mean(y));   % subtract average value
FilterLength = 100+1;  % average over some points
Filter = ones(FilterLength,1)/FilterLength;
y3 = conv(Filter,y2(1:end-FilterLength+1))';
y3 = y3(FilterLength:end);
N = length(y3);
```

The results are shown in Figure 3.66(a), generated by

```
t = linspace(0,(N-1)*dt,N)';
figure(2); plot(t,log(y3));  grid on
           xlabel('time t');  ylabel('log of amplitude')
```

In Figure 3.66(a) we clearly recognize the constant amplitude in the first sector and then a
straight line segment. This is caused by

$$A = A_0\, e^{-\alpha t} \qquad \Longrightarrow \qquad \ln A = \ln A_0 - \alpha\, t\,.$$

The wild variations on the right part in Figure 3.66(a) are caused by the almost 0 values. The
resolution of the involved instruments do not allow to determine small amplitudes reliably.
They will have to be ignored.

Compute the initial amplitude and the decay exponent

As full amplitude A_0 use the average of the first 200 values. Then choose an upper and a
lower cutoff value.

- Determine the first data point below the upper cutoff value.

- Determine the first data point below the lower cutoff value.

- Examine only data points between the above two points.

```
amplitude = mean(y3(1:200));
topcut = 0.8; lowcut = 0.4; % choose the cut levels on top and bottom
Nlow  = find(y3<topcut*amplitude,1);
Nhigh = find(y3<lowcut*amplitude,1);

y4 = log(y3(Nlow:Nhigh));
N  = length(y4);
t  = linspace(0,(N-1)*dt,N)';
```

On this reduced data set we use linear regression (see Section 3.2) to determine the slope of the straight line and the estimated standard deviation of the slope.

```
%% run the linear regression
F = ones(N,2); F(:,2) = t;
[p,y_var,r,p_var] = LinearRegression(F,y4');
yFit = F*p;
% display the real data and the regression line
figure(3); plot(t,y4,t,yFit)
          xlabel('time t');   ylabel('log of amplitude');    grid on
fprintf('Slope: %g, Variance of slope: %g, relative error:%g\n',...
        p(2), sqrt(p_var(2)), -sqrt(p_var(2))/p(2));
factor = 1/p(2);
```

The numerical result and Figure 3.66(b) clearly indicate that we have **one exponential function** on the decaying section of the signal.

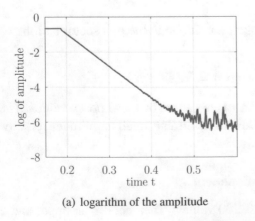

(a) logarithm of the amplitude

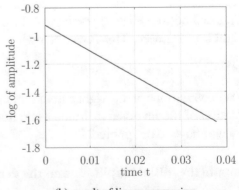

(b) result of linear regression

Figure 3.66: The logarithm of the amplitude and the regression result

Based on the above results the algorithm was implemented in `LabView` and graphs of the amplitude and quality factor as function of the frequency are generated on screen.

3.7.2 Analyzing one data set

The above basic algorithm is used to analyze one data set for a single measurement. For each frequency multiple measurements should be made and analyzed, to estimate the variance of the results. These measurements have to be repeated for many different frequencies and a usefull graph has to be generated, to be included in reports on the development of new sensors. A possible result is given in Figure 3.67. Graphs of this type could not be generated directly by *Octave*[19].

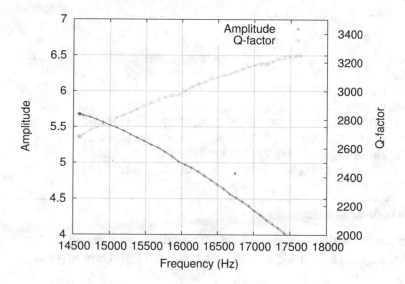

Figure 3.67: Amplitude and Q-factor as function of frequency, including error

We expect the following features for the final graphics:

- For each frequency the average value of all the measurements is shown and error lines at a distance of one standard deviation are drawn.

- The labeled axis for the amplitude is shown on the left edge of the graph.

- The labeled axis for the Q-factor is shown on the right edge of the graph.

- The scales shall not change from one data set to another. Thus it is easier to compare two different graphs.

---

[19]Up to date versions of *Octave* have the command `plotyy()` and can now not generate graphs with two different vertical axis. Thus one of the reasons to use an external program disappeared.

- The graphics should be generated in a format that can be used by most text processing software. We choose PNG. The resolution and size of this bitmap format have to be well chosen, since rescaling bitmaps is a very bad idea.

- A simpler graph without the error estimates should be generated too.

- Create a graph with the temperature as a function of the frequency.

The `LabView` program generates data files in a specified format, shown in the example below. The header lines contain information on the total number of measurements and the number of repetitions for one fixed frequency.

```
─────────────────────────── Elektro22.data ───────────────────────────
01.12.2007 12:13
42 Messungen
22 Wiederholungen

Amplitude Frequenz Q-Faktor Strom Temperatur

5.684363 14571.520000 2676.159581 0.000000 24.750000
5.685096 14572.430000 2675.312131 0.000000 24.750000
5.684854 14573.010000 2677.640868 0.000000 24.750000
5.684326 14573.550000 2678.034266 0.000000 24.750000
5.684079 14573.900000 2677.439599 0.000000 24.750000
5.684086 14574.140000 2679.226273 0.000000 24.750000
...
```

The *Octave* code to be written has to:

- Open the file with the measured data for reading and open a data file `gnu.dat` for writing the data to be used by *Gnuplot*, a powerful program to generate graphs.

- Read all the data for one frequency f.

- Compute the average amplitude A and estimate the variance ΔA.

- Compute the average Q-factor Q and estimate the variance ΔQ.

- Write one data line with the data to be displayed, i.e. write f, A, $A - \Delta A, A + \Delta A$, $Q, Q - \Delta Q$ and $Q + \Delta Q$.

- Within *Octave* a system call to *Gnuplot* has to generate the graphs.

Examine a sample call of the function `WriteData()`.

```
[freq,freqS,amp] = WriteData('Elektro22.data');
-->
mean STDEV frequency 3.804698e+00, maximal STDEV frequency 6.416919e+01
mean STDEV amplitude 5.145861e-03, maximal STDEV amplitude 1.269707e-02
mean STDEV Q-factor 4.76711, maximal STDEV Q-factor 31.0231
mimimal temperature 24.8455, maximal temperature 42.1364
```

If a data file `gnu.dat` contains the data in the above format *Gnuplot* can generate Figure 3.67 with the commands below. For most of the lines one can guess the effect of the command. Otherwise reading the manual of *Gnuplot* might be necessary, or consult the web page `http://www.gnuplot.info/`.

- The line `set terminal` allows to chose the type of output format and the size of the resulting graphics.

- `Set output` sets the name of the output file.

- The labels for the different axis have to be set.

- The range for the two vertical scales have to be set with `set yrange` and `set y2range`.

- Finally **one** plot command will use different columns in the data file `gnu.dat` to create the 6 different graphs in one image.

―――――――――― **AmpQ.gnu** ――――――――――
```
set terminal png large size 800,600
set output 'AmpQ2.png'
set y2tics border
set xlabel "Frequency (Hz)"
set ylabel "Amplitude"
set y2label "Q-factor"
set grid
set xrange [14500:18000]
set yrange [4:7]
set y2range [2000:3500]
plot 'gnu.dat' using 1:2 with points lt 1 title 'Amplitude' axes x1y1,\
     'gnu.dat' using 1:3 with lines lt 1 notitle axes x1y1,\
     'gnu.dat' using 1:4 with lines lt 1 notitle axes x1y1,\
     'gnu.dat' using 1:5 with points lt 2 title 'Q-factor' axes x1y2,\
     'gnu.dat' using 1:6 with lines lt 2 notitle axes x1y2,\
     'gnu.dat' using 1:7 with lines lt 2 notitle axes x1y2
```

The *Octave* file `WriteData.m` fullfills the above requirements[20]. The tools used are again found in Section 1.2.8. The last line of code uses a system call to use *Gnuplot* and the above command file `AmpQ.gnu` to generate the graphic files in the current directory.

―――――――――― **WriteData.m** ――――――――――
```
function [freq,freqS,amp,ampS,quali,qualiS,curr,currS,temp,tempS]...
    = WriteData(filename)
% function to write data for the DigiSens sensor
%
% WriteData(filename)
% [freq,freqS,amp,ampS,quali,qualiS,curr,currS,temp,tempS]
```

---

[20]On Win* system the last line might have to be replaced by `system('pgnuplot AmpQ.gnu');`

```
%                  = WriteData(filename)
%
% when used without return arguments  WriteData(filename) will analyze
% data  in filename and then write to the new file 'gnu.dat'.
% Then a system call 'gnuplot AmpQ.gnu' is made to generate graphs in
% in the files AmpQ.png AmpQ2.png and Temp.png
%
% when used with return arguments the consolidated data will be
% returned and may be used to generate graphs, e.g.
% [freq,freqS,amp] = WriteData('test1.dat');
% plot(freq,amp)

if ((nargin !=1))
  error('usage: give filename in WriteData(filename)');
end

calibrationFactor = 11.2; % factor for amplitudes, ideal value is 1.0
calibrationFactor = 1.0;

infile = fopen(filename,'rt');

tline = fgetl(infile); % dump top line
tline = fgetl(infile); % read number of measurements
meas  = sscanf(tline,'%i');
tline = fgetl(infile); % read number of repetitions
rep   = sscanf(tline,'%i');

freq = zeros(meas,1); freqS=freq; freqT=zeros(rep,1);
curr = freq; currS = freq; currT = freqT;
quali = freq; qualiS=freq; qualiT = freqT;
amp = freq; ampS = freq; ampT = freqT; temp = freq;

for k = 1:8;    % read the headerlines
  tline = fgetl(infile);
end%for

for im = 1:meas;
  for ir = 1:rep
    tline = fgetl(infile);
    t = sscanf(tline,'%g %g %g %g %g');
    ampT(ir) = calibrationFactor*t(1);
    freqT(ir)  = t(2);
    qualiT(ir) = t(3);
    currT(ir)  = t(4);
    tempT(ir)  = t(5);
  end%for
  amp(im)   = mean(ampT);    ampS(im)   = sqrt(var(ampT));
  freq(im)  = mean(freqT);  freqS(im)  = sqrt(var(freqT));
  quali(im) = mean(qualiT); qualiS(im) = sqrt(var(qualiT));
```

```
   curr(im)   = mean(currT);   currS(im)   = sqrt(var(currT));
   temp(im)   = mean(tempT);   tempS(im)   = sqrt(var(tempT));
 end%for
 fclose(infile);

 printf(
 'mean STDEV frequency %3e, maximal STDEV frequency %3e\n',...
               mean(freqS),max(freqS));
 printf(
 'mean STDEV amplitude %3e, maximal STDEV amplitude %3e\n',...
               mean(ampS),max(ampS));
 printf(
 'mean STDEV Q-factor %3g, maximal STDEV Q-factor %3g\n',...
               mean(qualiS),max(qualiS));
 printf(
 'mimimal temperature %3g, maximal temperature %3g\n',...
               min(temp),max(temp))

 outfile = fopen('gnu.dat',"wt");
 fprintf(outfile,...
    '# Freq Amp amp-STDEV amp+STDEV Q Q-STDEV Q+STDEV temp\n');
 for im = 1:meas;
   fprintf(outfile,"%g %g %g %g %g %g %g %g\n",...
      freq(im),amp(im),amp(im)-ampS(im),amp(im)+ampS(im),...
      quali(im),quali(im)-qualiS(im),quali(im)+qualiS(im),temp(im));
 end%for
 fclose(outfile);

 system('gnuplot AmpQ.gnu');
 end%function
```

With very similar tools the results in Figure 3.68 are generated. The file AmpQ.gnu will create all three graphs in this section.

3.7.3 Analyzing multiple data sets

To compare different sensors or the influence of external parameters it is necessary to display the results of multiple measurements in one graph. As example consider the Figures 3.69 and 3.70, where 5 measurements are displayed. These figures are created by the following code.

```
WriteDataAll('Elektro',[1:5],'.data')
-->
estimated temperature dependence of amplitude: -0.047083 um/C
estimated temperature dependence of Q-factor: -10.0248 /C
```

In this section we will examine the code carefully. The documentation of this command is contained at the top of the function file WriteDataAll.m.

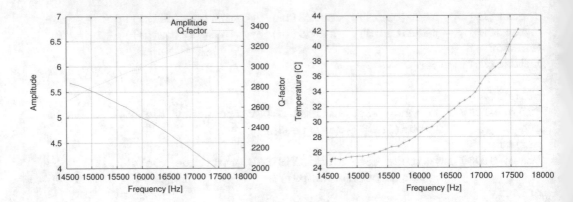

Figure 3.68: Amplitude, Q-factor and temperature as function of frequency

```
──────────────────────── WriteDataAll.m ────────────────────────
function  WriteDataAll(basename,numbers,ext)

% function to analyze data for a series of Digi Sens sensor
%
% WriteDataAll(basename,numbers,ext)
%
% will analyze data given in files and then generate graphs
% and files AmpAll.png and QAll.png
% It will also display the estimated dependencies of amplitude
% and Q-factor on the temperature
%
% sample calls:
%    WriteDataAll('Elektro",[1:5],'.data')
%    WriteDataAll('Elektro",[1, 2, 4, 5],'.data')
```

The the code in `WriteDataAll()` uses some special tricks to generate the graphs and some lines of code of require comments. You might want to read the comments shown after the code.

```
──────────────────────── WriteDataAll.m ────────────────────────
if ((nargin ~=3))
   error('usage: give filenames in WriteDataAll(basename,numbers,ext)');
end

%calibrationFactor=11.2; % factor for amplitudes, ideal value is 1.0
calibrationFactor = 1.0;

freqA = []; ampA = []; qualiA = []; tempA = [];
cmd1 = ['plot(']; cmd2 = cmd1; cmdLegend = 'legend(';
for sensor = 1:length(numbers)
   filename = [basename,num2str(numbers(sensor)),ext];
```

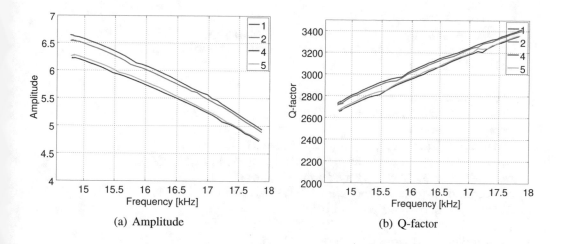

Figure 3.69: Results for multiple measurements

```
infile    = fopen(filename,'rt');

tline = fgetl(infile); % dump top line
tline = fgetl(infile); % read number of measurements
meas  = sscanf(tline,'%i');
tline = fgetl(infile); % read number of repetitions
rep   = sscanf(tline,'%i');

freq = zeros(meas,1); freqS = freq; freqT = zeros(rep,1);
curr = freq; currS = freq; currT = freqT;
quali = freq; qualiS = freq; qualiT = freqT;
amp = freq; ampS = freq; ampT = freqT; temp = freq;
for k = 1:8;    % read the headerlines
  tline = fgetl(infile);
end%for

for im = 1:meas;
  for ir = 1:rep
    tline = fgetl(infile);
    t = sscanf(tline,'%g %g %g %g %g');
    ampT(ir) = calibrationFactor*t(1);
    freqT(ir)  = t(2);
    qualiT(ir) = t(3);
    currT(ir)  = t(4);
    tempT(ir)  = t(5);
  end%for % rep
  amp(im)  = mean(ampT);    ampS(im)  = sqrt(var(ampT));
  freq(im) = mean(freqT); freqS(im)  = sqrt(var(freqT));
  quali(im)= mean(qualiT);qualiS(im)= sqrt(var(qualiT));
  curr(im) = mean(currT); currS(im)  = sqrt(var(currT));
  temp(im) = mean(tempT); tempS(im)  = sqrt(var(tempT));
```

```
  end%for  % meas
  freqA  = [freqA;freq];    ampA  = [ampA;amp];
  qualiA = [qualiA;quali]; tempA = [tempA;temp];
  fclose(infile);

  key = num2str(numbers(sensor));
  freqn = ['freq',key,'=freq/1000;']; eval(freqn);
  ampn  = ['amp',key,'=amp;'];    eval(ampn);
  qualin = ['quali',key,'=quali;'];eval(qualin);
  cmd1   = [cmd1,'freq',key,',amp',  key,','];
  cmd2   = [cmd2,'freq',key,',quali',key,','];
  cmdLegend = [cmdLegend,char(39),key,char(39),','];
end%for  % loop over all files

cmdLegend = [cmdLegend(1:end-1),')'];
cmd1 = [cmd1(1:end-1),');'];
cmd2 = [cmd2(1:end-1),');'];

figure(1); clf;
eval(cmd1)
grid('on')
axis([14500 18000 4 7]);
xlabel('Frequency [kHz]'); ylabel('Amplitude')
eval(cmdLegend)
print('AmpAll.png','-dpng');

figure(2);  clf;
eval(cmd2);
grid('on')
axis([14500 18000 2000 3500]);
xlabel('Frequency [kHz]'); ylabel('Q-factor')
eval(cmdLegend)
print('QAll.png','-dpng');

NN = length(freqA);
F = ones(NN,3);
F(:,1) = freqA;F(:,2)=tempA;
[p,y_var,r,p_var] = LinearRegression(F,ampA);
display(sprintf(...
    'estimated temperature dependence of amplitude: %g um/C\n',p(2)))
[p,y_var,r,p_var] = LinearRegression(F,qualiA);
display(sprintf(...
    'estimated temperature dependence of Q-factor: %g /C\n',p(2)))


figure(3); clf; axis();
         plot3(freqA/1000,tempA,ampA,'+')
         xlabel('frequency [kHz]'); ylabel('Temperature');
         zlabel('Amplitude')
```

- At first empty matrices are created to contain the data for all frequencies, amplitudes, quality factors and temperatures.

```
freqA = []; ampA = []; qualiA = []; tempA = [];
```

- When the function is called by `WriteDataAll('Elektro',[1,3,5],'.data')` the data files `Elektro1.data`, `Elektro3.data` and `Elektro5.data` have to be analyzed. The code uses a loop of the form

```
for sensor = 1:length(numbers)
   ...
endfor
```

to read each of the requested files and constructs the file names within the loop by

```
filename = [basename,num2str(numbers(sensor)),ext];
```

- Each data file is scanned using the tools from Section 1.2.8. Once the data is read named variables will be generated. As example consider the case `key='3'`. Then the commands

```
key    = num2str(numbers(sensor));
freqn = ['freq',key,'=freq/1000;']; eval(freqn);
```

will generate the string `'freq3=freq/1000;'` and then evaluate this command. The variable `freq3` contains the frequencies from the data file `Elektro3.data`.

- The plot command to generate Figure 3.69(a) is constructed step by step. Examine the patches of code.

```
cmd1 = ['plot('];
for sensor = 1:length(numbers) % lop over all data files
   key    = num2str(numbers(sensor));
   cmd1   = [cmd1,'freq',key,',amp',   key,','];
   cmdLegend = [cmdLegend,char(39),key,char(39),','];
end%for
cmd1 = [cmd1(1:end-1),');'];
eval(cmd1)
```

If the function is called by `WriteDataAll('Elektro',[1 3 5],'.data')` then the string `cmd1` will contain the final value

```
plot(freq1,amp1,freq3,amp3,freq5,amp5);
```

and thus `eval(cmd1)` will generate the graphics with standard *Octave* commands. With the help of `grid`, `axis()`, `legend()`, `xlabel()`, `ylabel()` the appearance of the graphics is modified.

- Finally a call of `print('AmpAll.png')` will generate the graphics in the PNG format. With recent versions of *Octave* the resolution may be given by the command `print('AmpAll.png','-S800,600');`.

- When comparing different measurements one realizes that the amplitude can not depend on the frequency only, but the temperature might be important too. This is verified by Figure 3.70, generated by

```
figure(3); clf;  axis()
          plot3(freqA/1000,tempA,ampA,'+')
          xlabel('frequency [kHz]'); ylabel('Temperature');
          zlabel('Amplitude')
```

- By rotating Figure 3.70 one might come up with the idea that all measured points are approximately on a plane, i.e. the amplitude A depends on the frequency f and the temperature T by a linear function

$$A(f,T) = c_f\, f + c_T\, T + c_0 \,.$$

The optimal values for the coefficients can be determined by linear regression.

```
NN = length(freqA);
F  = ones(NN,3);
F(:,1) = freqA;F(:,2)=tempA;
[p,y_var,r,p_var] = LinearRegression(F,ampA);
fprintf(...
 'estimated temperature dependence of amplitude: %g um/C\n',p(2))
```

As a result we find that the amplitude A decreases by 0.047 μm per degree of temperature increase.

- A similar computation shows that the Q-factor is diminished by 10 units per degree of temperature increase.

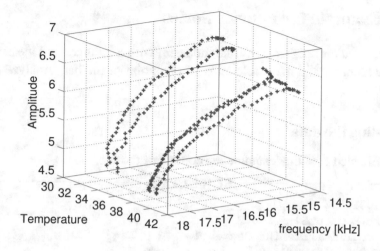

Figure 3.70: Amplitude as function of frequency and temperature

3.7.4 Calibration of the device

The electronic device to be calibrated with the help of a vibrometer, which can measure absolute amplitudes A_a of the vibrating cord. The electronic device built by Andrea Schüpbach lead to a voltage signal with amplitude A_e. Using linear regression an optimal choice of the calibration parameter α such that

$$A_a = \alpha A_e$$

was determined and then built into the Labview code. The device now used by DIGI SENS Switzerland AG yields amplitudes in micro meters.

3.7.5 List of codes and data files

In the previous section the codes in Table 3.14 were used.

filename	function
`automatic.m`	function file to examine test data
`m4`	sample data file
`Readm4.m`	script file to read the raw data
`WriteData.m`	function file to analyse one data set
`AmpQ.gnu`	command file for *Gnuplot*
`WriteDataAll.m`	function file to analyse multiple data sets
`Elektro*.data`	sample data files

Table 3.14: Codes and data files for section 3.7

3.8 An Example for Fourier Series

A beam is clamped on both sides. Strike the beam with a hammer and measure the acceleration of the hammer and the acceleration of one point of the bar. Analyze the collected data.

3.8.1 Reading the data

To read the collected data first examine the content of the file.

```
                          SC1007.TXT
LECROYLT364L,73
Segments,1,SegmentSize,10002
Segment,TrigTime,TimeSinceSegment1
#1,03-Jun-2002 15:28:33,0
Time,Ampl
-0.0060034,0.001875
-0.0059934,0.001875
-0.0059834,0.001875
-0.0059734,0.001875
-0.0059634,-0.00125
...
```

- Since the data is given in a comma separated value format use the dlmread() to read and then display the data. A possible result is shown in Figure 3.71.

```
                          ReadDataDLM.m
filename1 = 'SC1001.TXT';
indata = dlmread(filename1,',',5,0); % read data, starting row 6
k = length(indata);
disp(sprintf('Number of datapoints is  %i',k))
timedataIn = indata(:,1);  ampdataIn  = indata(:,2);
TimeIn = timedataIn(k)-timedataIn(1)
FreqIn = 1/(timedataIn(2)-timedataIn(1))
figure(1); plot(timedataIn,ampdataIn)
         title('Amplitude of input'); grid on

dom = 1070:1170;  %  choose the good domain by zooming in
figure(3); plot(timedataIn(dom),ampdataIn(dom))
         title('Amplitude of input'); grid on
%%%% similar code to read acceleration of point on vibrating bar
```

- If you want to avoid dlmread() one may read line by line and scan for the desired values. Proceed as follows

  - give the name of the files to be read
  - create an matrix of zeros large enough to store all data to be read

- open the file to be read
- read 5 lines, then ignore them
- for each of the following lines
    * read a string with the line
    * extract the first number (time), the comma and the second number (amplitude)
    * store time and amplitude in a vector
- close the file and display the number of data points read

──────────────── **ReadData.m** ────────────────

```
filename1 = 'SC1001.TXT';
indata = zeros(2,10007);         % allocate storage for the data

infile = fopen(filename1,'r');   % open the file for reading
for k = 1:5
  tline = fgetl(infile);         % read 5 lines of text
end%for

k = 0;                           % a counter
inline = fgetl(infile);          % read a line
while ischar(inline)             % test for end of input file
  A = sscanf(inline,'%f%c%f');   % read the two numbers
  k = k+1;
  indata(1,k) = A(1);            % store only the time
  indata(2,k) = A(3);            % store only the amplitude
  inline = fgetl(infile);        % get the next input line
end%while
fclose(infile);                  % close the file

disp(sprintf('Number of datapoints is  %i',k))
```

Once the information is available to *Octave* the basic data has to be extracted and displayed

- store the time and amplitudes in separate vectors
- determine the total time of the measurement and the sampling frequency
- create a graph of the amplitude as function of time for a graphical verification. A possible result is shown in Figure 3.71.
- to examine the behavior of the driving stroke by the hammer one might enlarge the section where the excitation does not vanish.

```
timedataIn = indata(1,1:k);
ampdataIn  = indata(2,1:k);
TimeIn = timedataIn(k)-timedataIn(1)
```

```
FreqIn = 1/(timedataIn(2)-timedataIn(1))

figure(1); plot(timedataIn,ampdataIn)
        title('Amplitude of input'); grid on
        xlabel('time'); ylabel('amplitude')

dom = 1070:1170;
figure(2); plot(timedataIn(dom),ampdataIn(dom))
        title('Amplitude of input'); grid on
        xlabel('time'); ylabel('amplitude')
```

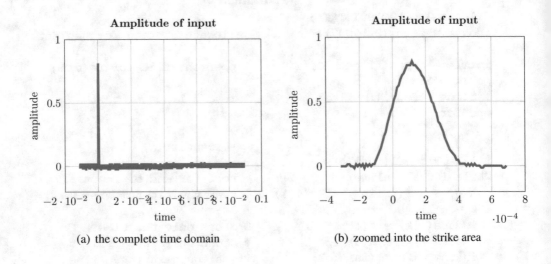

(a) the complete time domain   (b) zoomed into the strike area

Figure 3.71: Acceleration of the hammer

3.8.2   Further information

- **Time of contact**

  The result in Figure 3.71(b) shows that the time of contact can be computed. We have to choose a threshold for accelerations. If the measured acceleration is above this limit we decide that contact occurs. In this example 0.05 is used as threshold.

```
% find all points with acceleration above the threshold of 0.05
contact = ampdataIn>0.05;
ContactTimes = sum(contact) % number of timepoints above threshold
timeOfContact = sum(contact)/FreqIn % time of contact
-->
ContactTimes =   48
timeOfContact =   4.8000e-04
```

Thus detect that the hammer contacted the bar for 0.48 msec.

- **Initial speed of hammer**
  Since the signal is proportional to the acceleration $a(t)$ of the hammer we can compute the difference of the velocity of the hammer before and after contact by

$$v_2 - v_1 = \int_{t_1}^{t_2} a(t)\, dt$$

With *Octave* we can use two slightly different codes, leading to very similar results.

  – The first version uses all points with an acceleration larger than the above specified threshold of 0.05.

  – The second version integrates over the time domain specified in Figure 3.71(b).

```
speed1 = trapz(timedataIn,ampdataIn.*contact)
speed2 = trapz(timedataIn(dom),ampdataIn(dom))
-->
speed1 =   2.2515e-04
speed2 =   2.2812e-04
```

The resulting number is not equal to the actual speed, since we do not know the scale factor between the signal and the acceleration. The device would have to be calibrated to gain this information.

3.8.3   Using FFT, Fast Fourier Transform

The data file with the acceleration of the point on the bar has to be read in a similar fashion. To analyze the data proceed as follows:

- Decide on the number of data points to be analyzed. It should be a power of 2, here 2^{N2} points are used. Thus N2 decides on the artificial period T of the signal. The periodicity is introduced by the Fourier analysis. With the period T we also choose the base frequency $1/T$. Due to the Nyquist effect (aliasing) we will at best be able to analyze frequencies up to $2^{N2-1}/T$.

- Choose the number Ndisp of frequencies to be displayed. This will lead to a graph with frequencies up to Ndisp/T.

- Apply the FFT (Fast Fourier Transform) to the data.

- Plotting the absolute value of the coefficients as function of the corresponding frequencies will give a spectrum of the signal. The results are shown in Figure 3.72.

```
                                 Fourier.m
N2 = 12;        % analyze 2^N2 points
Ndisp = 200;   % display the first Ndisp frequency contributions
tdata    = timedataIn(1:2^N2); adata     = ampdataIn(1:2^N2);
PeriodIn = timedataIn(2^N2)-timedataIn(1)
frequencies = linspace(1,Ndisp,Ndisp)/PeriodIn;

fftIn = fft(adata);
figure(1); plot(frequencies,abs(fftIn(2:Ndisp+1)))
           title('Spectrum of input signal')
           xlabel('Frequency [Hz]'); ylabel('Amplitude')

tdata = timedataOut(1:2^N2);  adata = ampdataOut(1:2^N2);
PeriodOut = timedataOut(2^N2)-timedataOut(1)
fftOut = fft(adata);
figure(2); plot(frequencies,abs(fftOut(2:Ndisp+1)))
           title('Spectrum of output signal')
           xlabel('Frequency [Hz]'); ylabel('Amplitude')
```

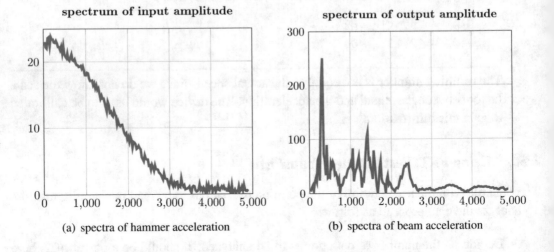

(a) spectra of hammer acceleration   (b) spectra of beam acceleration

Figure 3.72: Spectra of the accelerations of hammer and bar

The results in Figure 3.72 show that

- the input has no significant contribution for frequencies beyond 3000 Hz.

- the spectrum of the output has some significant peaks. These might correspond to eigenmodes of the vibrating beam.

3.8.4 Moving spectrum

Instead of analyzing the signal over the full time span we may also consider the spectrum on shorter sections of time. We proceed as follows:

- Examine slices of $2^{11} = 2048$ data points, thus 0.1 sec at different starting times. Here we choose starting times from 0 to 0.5 sec in steps of 0.1 sec. The starting time in the code below is chosen by setting the variable `level`. The value of `level` tells *Octave* at which point to start.

- The graphs of the above 6 computations are shown in Figure 3.73. Observe that the scales vary from one picture to the next. Obviously the spectrum changes its shape as function of time, but some features persist.

SpectrumSlice.m
```
% set level before calling this script
level = 100; %  level = 3000

N2 = 11;      % analyze 2^N2 points
Ndisp = 50;   % display the first Ndisp contributions

PeriodIn = timedataIn(2^N2)-timedataIn(1);
frequencies = linspace(1,Ndisp,Ndisp)/PeriodIn;

adata   = ampdataOut(level:level+2^N2-1);
fftOut  = fft(adata);
spectrum = abs(fftOut(2:Ndisp+1));
figure(1); plot(frequencies,spectrum)
           xlabel('frequency');  ylabel('amplitude')
```

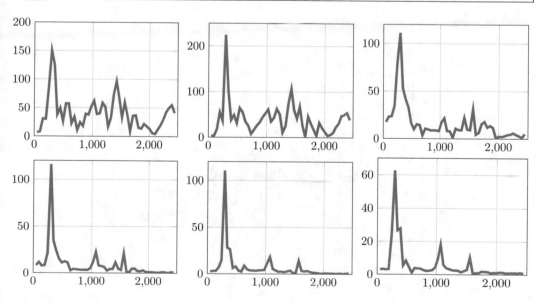

Figure 3.73: Spectra at different times. Observe the different scales for the amplitudes.

Another option would be to consider even more starting times and generate a 3D–graph.

The code below[21] does just this. The result in Figure 3.74 allows to discuss the behavior of the amplitude as function of time and frequency.

```
──────────────────────────── MovingFourier.m ────────────────────────────
N2 = 11;       % analyze 2^N2 points
Ndisp = 50;    % display the first Ndisp contributions

%levels = 1:250:5000;
levels = 1:200:8000;

spectrum    = zeros(length(levels),Ndisp);
PeriodIn    = timedataIn(2^N2)-timedataIn(1);
frequencies = linspace(1,Ndisp,Ndisp)/PeriodIn;

for kl = 1:length(levels)
  adata  = ampdataOut(levels(kl):levels(kl)+2^N2-1);
  fftOut = fft(adata);
  spectrum(kl,:) = abs(fftOut(2:Ndisp+1));
end%for

figure(3); mesh(frequencies,levels/FreqIn*1000,spectrum)
           xlabel('frequency [Hz]'); ylabel('time [ms]');
           zlabel('amplitude'); view(35,25)
           title('spectrum as function of starting time');
```

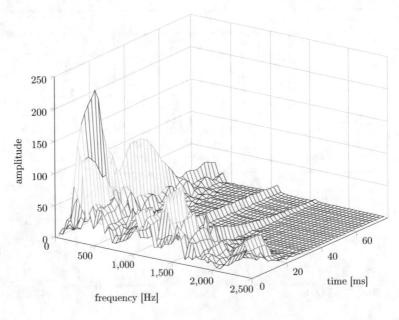

Figure 3.74: Amplitudes at different times as 3D graph, as function of starting time and frequency

---

[21]Recently your author learned about the command `specgram()` which applies a similar procedure.

3.8.5 Determine the transfer function

If we consider the acceleration of the hammer as input and the acceleration of the point on the bar as output we can examine the transfer function.

```
figure(3);
  plot(frequencies,abs(fftOut(2:Ndisp+1))./abs(fftIn(2:Ndisp+1)))
  title('Transfer Function'); grid on
  xlabel('Frequency'); ylabel('Output/Input'); grid on
```

Expect the result to be highly unreliable for frequencies above 2500 Hz. This is based on the fact that the amplitudes of input and output are small and thus minor deviations can have a drastic influence on the result of the division. Thus the large values of the transfer function for the highest frequencies in the left part of Figure 3.75 should not be taken too seriously. The right part of the figure examines only smaller frequencies. This result might be useful. It is generated by the code below.

```
nn = sum(frequencies <2500);
figure(4);
  plot(frequencies(1:nn),abs(fftOut(2:nn+1))./abs(fftIn(2:nn+1)))
  title('Transfer function'); grid on
  xlabel('Frequency [Hz]'); ylabel('Output/Input');
```

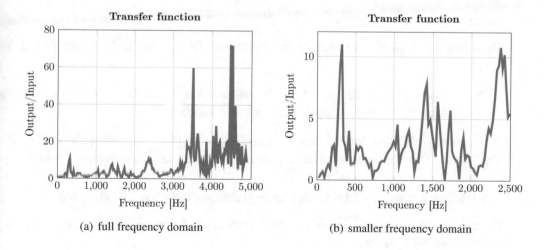

(a) full frequency domain   (b) smaller frequency domain

Figure 3.75: Transfer function for the acceleration of hammer and bar

3.8.6 List of codes and data files

In the previous section the codes and data files in Table 3.15 were used.

filename	function
ReadDataDLM.m	read the basic data from files, short version
ReadData.m	read the basic data from files, long version
Fourier.m	determine the spectra and the transfer function
SpectrumSlice.m	find the spectrum over subsection of the time interval
MovingFourier.m	create 3d graph of spectrum
SC1001.TXT	first data file with amplitude of hammer
SC2001.TXT	first data file with amplitude of point on bar
SC1002.TXT	second data file with amplitude of hammer
SC2002.TXT	second data file with amplitude of point on bar
SC1003.TXT	third data file with amplitude of hammer
SC2003.TXT	third data file with amplitude of point on bar
SC1007.TXT	fourth data file with amplitude of hammer
SC2007.TXT	fourth data file with amplitude of point on bar

Table 3.15: Codes and data files for section 3.8

3.9   Grabbing Data from the Screen and Spline Interpolation

In this section we:

- First examine how to read coordinate information from a MATLAB/*Octave* graphics window.

- Then we show to how read data from the any section of the screen with the help of mouse clicks. This only works with *Octave* on a Linux system.

- Then we examine how to read data from an *Octave* graphics window with the help of the mouse. This should work with *Octave* and MATLAB on any operations system.

3.9.1   Reading from an *Octave*/**MATLAB** graphics window by `ginput()`

With the command `ginput()` you can read information from an *Octave* or MATLAB graphics window. With the code below we

- open a graphics window and fix the axis.

- display some information for the user

- use the left mouse button to collect positions and display the corresponding points.

- exit the loop with the last point, marked by the right mouse button.

```
                      ┌──────────── GetData.m ──────────────┐
figure(1); clf                  % write to the first graphic window
axislimits = [0 2 0 1];         % gather 0<x<2 and 0<y<1
axis(axislimits)                % fixed axis for the graphs
x = []; y = [];                 % initialise the empty matrix of values
% show messages for the user
disp('Use the left mouse button to pick points.')
disp('Use the right mouse button to pick the last point.')
button = 1;                     % to indicate the last point
while button == 1               % while loop, picking up the points.
    [xi,yi,button] = ginput(1); % get coodinates of one point
    x = [x,xi]; y = [y,yi];
    plot(x,y,'ro')              % plot all points
    axis(axislimits);           % fix the axis
end%while
plot(x,y,'ro-')
xlabel('x'); ylabel('y');  axis(axislimits);
```

The command `ginput()` should work with *Octave* or MATLAB on any platform.

3.9.2 Create `xinput()` to replace `ginput()`

The command `ginput()` will read data from an *Octave*/MATLAB window only. Since *Octave* is an open source project we can use old code and generate a similar command `xinput()`, capable or reading data from any point on the screen. This will only work on Unix systems and with *Octave*. Use the following steps:

• Assure that you have copies of the files `xinput.m` and `grab.cc` in the current directory.

• Compile the code `grab.cc` using the command `mkoctfile grab.cc`. This creates the binary `grab.oct`, which is loaded when calling the function `grab()`. If you desire to do, you can also examine the source code.

• To test the function `grab()` call `[x,y]=grab()`, use the left mouse button to click on the corners of your screen, then click the middle or right button to display the resolution of your screen.

• The *Octave* function `xinput()` is a wrapper to call `grab()`. Now use `xinput()` instead of `ginput()`.

• With the files created above the commands `grab()` and `xinput()` will be available when working in this directory. To make this feature generally available you have to copy the files `xinput.m` and `grab.oct` location in the path of *Octave*. This author uses a directory `~/octave/site` and then uses the command `addpath (genpath ('~/octave/site'))`; in the startup file at `~/.octavrerc` to make the directory known to *Octave*. See also Section 1.1.1 (page 3).

The above is a good illustration of the advantages of Open Source software. On Linux systems the independent program g3data[22] might be useful for the same purpose. It does not directly read from the screen, but it works with many formats for graphics.

3.9.3   Reading an LED data sheet with *Octave*

In Figure 3.76 you find data for an LED (Light Emitting Diode). We will closely examine the intensity of the emitted light as a function of the angle. For this we want to read the data into *Octave*[23]. The command xinput() allows to read the screen coordinates of any point on the screen, even in the window covered by other applications. Thus we can launch an PDF reader (e.g. evince) on a command line or through the GUI of the operating system to display the file NSHU550ALEDwide.pdf. Locate the page shown in Figure 3.76. Then we read the data of the left part in the graph in the lower right corner.

- The first click determines the location of $x = 0$ and $y = 0$.

- The second click determines the location of $x = 90$ and $y = 1$.

- Subsequent clicks with the left mouse button specify the points to be collected.

- A click with the right button will terminate the data collection, where this last location will not be stored.

- Display the result. A possible answer is shown in Figure 3.77.

LEDread.m

```
more off
% show messages for the user
disp('Left mouse button picks points.')
disp('Right mouse button to quit.')
pause(2);  % give the user some time to get the graph in the foreground
[xi,yi] = xinput([0 90 0 1])         % read the points
figure(1); plot(xi,yi);
          grid on; axis('normal');
          xlabel('angle'); ylabel('intensity');
```

In Section 3.2.6 (page 233) the above information is used as input for linear regression to determine the intensity as a function of the angle.

With very similar code we can try to read the information from the polar plot in Figure 3.76.

- Click on the corner with angle 0° and radius 0.

- Click on the corner 90° and radius 1.

---

[22]https://github.com/pn2200/g3data

[23]With MATLAB this is unfortunately not possible, and *Octave* on Win?? systems seems to be problematic too.

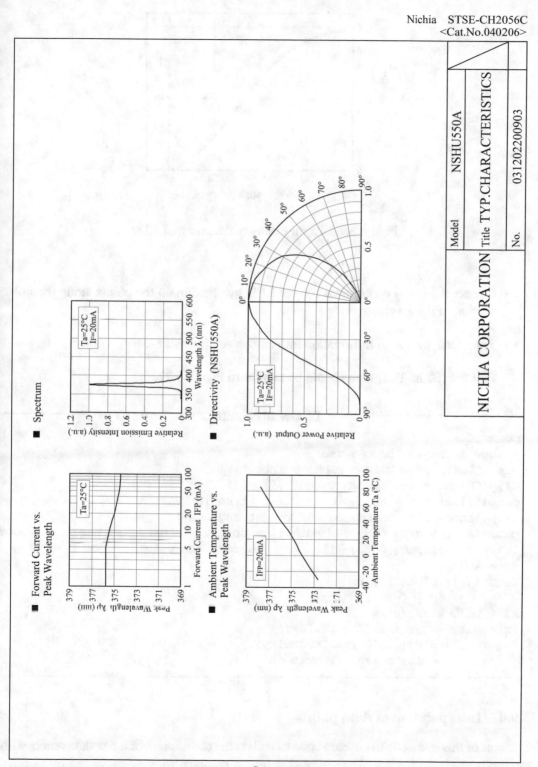

Figure 3.76: Data sheet for an LED

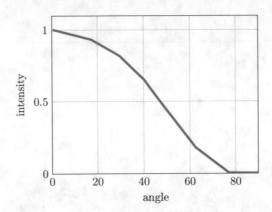

Figure 3.77: Light intensity data for an LED

- Collect the (x, y) coordinates for the LED by clicking on the points along the polar section of the graph.

- Transform the (x, y) information into angles and intensities.

- Plot the result. The graph should be similar to Figure 3.77.

──────────────── **LEDreadPolar.m** ────────────────
```
more off
% show messages for the user
disp('Left mouse button picks points.')
disp('Right mouse button to quit.')
disp('first click on the corner r=1 at angle 90')
disp('then click on the corner r=1 at angle 0')
pause(2);  % give the user some time to get the graph in the foreground
[xi,yi] = xinput([0 1 0 1]);       % read the points
xi = 1-xi;
figure(1); plot(xi,yi);

al = pi/2-atan2(yi,xi);
Intensity = sqrt(xi.^2 + yi.^2);
figure(2); plot(al*180/pi,Intensity),
         axis('normal'); grid on
```

3.9.4 Interpolation of data points

The aim of this section is to develop code to collect the coordinates of a few data points with the mouse as input device. We only read points in graphics window generated by *Octave*. Then the points should be visualized. We will use different type of interpolation to construct a curve connecting the points:

- Spline interpolation. This will lead to a smooth curve. With the help of a numerical integration, we determine the area between this curve and the horizontal axis.

- Piecewise linear interpolation. The function is tabulated at a regular set of grid points. Using the representation we compute and visualize the derivative of the function.

Getting the data, using the mouse as input device

The code below is organized as follows:

1. Initialize the graphics window and number and coordinates of the points to be collected.

2. Show a message to the user with the information on how to proceed.

3. Use the command `ginput()` (graphical input) to obtain the coordinates of the points to be used. Plot the data while it is collected.

4. Store the data in vector with the x and y components of the points

The code is shown on page 366 in a file `GetData.m`. Run this command in *Octave* or MATLAB and you will find two vectors xi and yi containing the coordinates of the points. If the data is generated by different methods, this section of the code has to be adapted.

Spline Interpolation

MATLAB and *Octave* provide the command `spline()` to compute the interpolating spline polynomial for a given set of points.

─────────────── **SplineInterpolation.m** ───────────────
```
% Interpolate with a spline curve and finer spacing.
% the code in GetData.m must be run first
n = length(x);
t  = 1:n;                        % integers from 1 to n
ts = 1: .2: n;                   % from 1 to n, stepsize 0.2
xys = spline(t,[x;y],ts);        % do the spline interpolation
xs = xys(1,:);   ys = xys(2,:);  % extract the components

% plot the interpolated curve.
figure(1); plot(xs,ys,x,y,'*-');
           xlabel('x'); ylabel('y'); grid on
% plot the two components of the spline curve separately, show labels
figure(2); plot([1:length(xs)],xs,'g-',[1:length(ys)],ys,'b-')
           legend('x values','y values')
           xlabel('numbering'); ylabel('values'); grid on
```

After having computed many points on the curve we now attempt to compute the integral of the function, i.e. for $a \leq x \leq b$ we try to determine

$$F(x) = \int_a^x f(s) \, ds .$$

Based on Figure 3.78 we use a trapezoidal integration rule, i.e.

$$\int_{x_0}^{x_5} f(x)\ dx \approx (x_1-x_0)\ \frac{y_0+y_1}{2} + (x_2-x_1)\ \frac{y_1+y_2}{2} + (x_3-x_2)\ \frac{y_2+y_3}{2} + (x_4-x_3)\ \frac{y_3+\ }{2}$$

Based on this idea we can integrate step by step, adding the area of the new rectangle at each step. This is implemented in the script file `Integration.m` shown below.

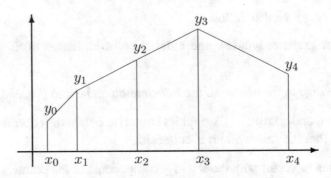

Figure 3.78: Trapezoidal integration

```
──────────────────── Integration.m ────────────────────
% use the data generated by GetData and SplineInterpolation
integral = zeros(size(xs));  % create vector of correct size
% perform a trapezoidal numerical integration by adding the contribution
for k = 2:length(xs)
    integral(k) = integral(k-1) + (xs(k)-xs(k-1))*(ys(k-1)+ys(k))/2;
end%for
figure(3); axis(axislimits)
          plot(xs,ys,xs,integral)
          legend('function','intgeral')
```

The identical result can be obtained by the command `cumtrapz()`, short for cummulative trapezoidal rule, see Section 3.1.1.

```
num_integral = cumtrapz (xs,ys);
```

Piecewise linear interpolation

Since the data points were collected with the help of the mouse, the values of x are not necessarily sorted. If y is supposed to be a explicit function of x the graph may not 'swing back'. This requires that the values of x and y be renumbered properly, as illustrated in Figure 3.79. With the sorted values we then call the function `interp1()` to determine the values of the piecewise linear interpolating function at a set of regularly spaced grid points. A plot is easily generated.

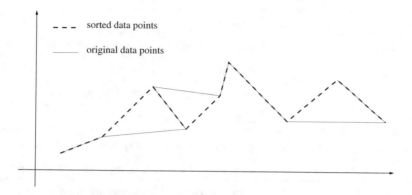

Figure 3.79: Data points in original order and sorted

```
                        LinearInterpolation.m
% Interpolate with a piecewise linear curve
% GetData must be run first
nx   = 201;                            % number of grid points
xlin = linspace(min(x),max(x),nx);     % uniformly distributed points
% sort x and y values, based on the order of the x values
xysort = sortrows([x;y]',1);

% compute the values of y at the given points xlin
ylin = interp1(xysort(:,1),xysort(:,2),xlin);

% Plot the interpolated curve.
% compute the derivatives at the midpoints
dy = diff(ylin)./diff(xlin);
% compute the midpoints of the intervals
xmid = xlin(2:length(xlin))-diff(xlin)/2;
figure(3); plot(x,y,'*', xlin,ylin,xmid,dy);
           legend('given points','interpolation','derivative');
```

A finite difference approximation was used above to find a numerical derivative of the given function.

3.9.5 List of codes and data files

In the previous section the codes and data files in Table 3.16 were used. The codes should be run in the given order.

3.10 Intersection of Circles and Spheres, GPS

In this section we build up some of the Mathematics used for the the Global Positioning System (GPS). It turns aot the the problem of finding intersection points of spheres is one of the essential tools used for the GPS, see [Thom98].

filename	function
`GetData.m`	read the basic data from screen using mouse
`LEDread.m`	read the LED data from screen, generate plot
`LEDreadPolar.m`	read the LED data from polar plot
`NSHU550ALED.pdf`	Data sheet for an LED
`SplineInterpolation.m`	perform a spline interpolation and show the results as graph of the two components and as one curve
`Integration.m`	apply a trapezoidal integration rule and show the function and its integral
`LinearInterpolation.m`	piecewise linear interpolation and graphs of the function and its derivative
`xinput.m`	script file as replacement for `ginput.m`, *Octave*
`grab.cc`	C++ source for the command `grab()`, *Octave*

Table 3.16: Codes and data files for section 3.9

- We start with the geometric problem of find the two intersection points of two circles in the plane. This can be used to determine the position of a robot in a plane, if the distence to two fixed points is known.

- A MATLAB/*Octave* function is written to determine the two intersection points.

- Similar ideas are used to determine the intersection points of three spheres in space and an application in robotics is mentioned.

- Then the problem of approximation the common intersection point of many circles in the plane is examined. The system of over-determined quadratic equations is replaced by a linear least square problem.

- The identical algorithm can be used to search for the intersection point of many spheres in space.

- By adding the additional dimension time one can use the same idea to determine the position if information of a few GPS satellites is available.

The following examples will show you how to implement the mathematical operations with MATLAB or *Octave*. We start with a geometric problem, solved by algebraic methods.

3.10.1   Intersection of two circles

As can be seen in Figure 3.80 the two points of intersection (if they exist) of two circles determine a straight line. To compute those points of intersection we may first determine

the equation of this straight line and then intersect with one of the circles. The code below
shows how this idea can be implemented.

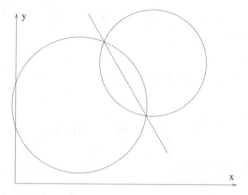

Figure 3.80: Intersection of two circles

A circle with center at $\vec{x}_m = (x_1 \, , \, y_1)^T$ and radius r_1 corresponds to the solution of the
equation

$$\|\vec{x} - \vec{x}_m\|^2 = (x - x_1)^2 + (y - y_1)^2 = r_1^2$$

and a possible parameterization of this circle is given by

$$\begin{pmatrix} x(t) \\ y(t) \end{pmatrix} = \begin{pmatrix} x_1 \\ y_1 \end{pmatrix} + r_1 \begin{pmatrix} \cos(t) \\ \sin(t) \end{pmatrix} \quad \text{for} \quad 0 \le t \le 2\pi \, .$$

First choose the parameters for the first circle

```
x1m = 2; y1m = 3;          % coordinates of the center
r1  = 1.5;                 % radius
```

then create the graph.

```
t   = linspace(0,2*pi,51);  % values of all angles, 51 steps
x1  = x1m+r1*cos(t);        % x coordinates of all points
y1  = y1m+r1*sin(t);        % y coordinates of all points
plot(x1,y1);               % create the plot
axis([0,5,0,5],"equal")     % choose a domain
```

A second circle is plotted using similar code.

```
x2m = 4; y2m = 2;          % coordinates of the center
r2 = 2;                    % radius
x2 = x2m+r2*cos(t);        % x coordinates of all points
y2 = y2m+r2*sin(t);        % y coordinates of all points
plot(x1,y1,x2,y2);         % create the plot with both circles
```

For two given circles we try to solve for the points of intersection and arrive at the system of quadratic equations

$$x^2 - 2\,x\,x_1 + x_1^2 + y^2 - 2\,y\,y_1 + y_1^2 = r_1^2$$
$$x^2 - 2\,x\,x_2 + x_2^2 + y^2 - 2\,y\,y_2 + y_2^2 = r_2^2 \ .$$

Subtracting the two equations we find the equation for a straight line on which both points of intersection have to be.

$$-2\,x\,(x_1 - x_2) + x_1^2 - x_2^2 - 2\,y\,(y_1 - y_2) + y_1^2 - y_2^2 = r_1^2 - r_2^2 \ .$$

By choosing $x = 0$ we find the point

$$\vec{x}_p = \begin{pmatrix} x_p \\ y_p \end{pmatrix} = \begin{pmatrix} 0 \\ \frac{r_1^2 - r_2^2 - y_1^2 + y_2^2 - x_1^2 + x_2^2}{-2\,(y_1 - y_2)} \end{pmatrix}$$

and

$$\vec{v} = \begin{pmatrix} v_1 \\ v_2 \end{pmatrix} = \begin{pmatrix} y_1 - y_2 \\ -x_1 + x_2 \end{pmatrix}$$

is a vector pointing in the direction of the straight line. Thus

$$\vec{x}(t) = \vec{x}_p + t\,\vec{v} \quad \text{with} \quad t \in \mathbb{R}$$

is a parameterization of this straight line.

```
xp = [0; (-r1^2+r2^2+x1m^2-x2m^2+y1m^2-y2m^2)/(2*(y1m-y2m))];
v  = [y1m-y2m; -x1m+x2m];
```

Then use this parameterization in the equation for the first circle to find

$$\begin{aligned}
r_1^2 = \|\vec{x}(t) - \vec{x}_m\|^2 &= \langle \vec{x}(t) - \vec{x}_m \,,\, \vec{x}(t) - \vec{x}_m \rangle \\
&= \langle t\,\vec{v} + \vec{x}_p - \vec{x}_m \,,\, t\,\vec{v} + \vec{x}_p - \vec{x}_m \rangle \\
&= \|\vec{v}\|^2\,t^2 + 2\,\langle \vec{v} \,,\, \vec{x}_p - \vec{x}_m \rangle\,t + \|\vec{x}_p - \vec{x}_m\|^2 \ .
\end{aligned}$$

This is a quadratic equation for the unknown parameter t with the two solutions

$$t_{1,2} = \frac{-b \pm \sqrt{b^2 - 4\,a\,c}}{2\,a} = \frac{-2\,\langle \vec{v} \,,\, \vec{x}_p - \vec{x}_m \rangle \pm \sqrt{D}}{2\,\|\vec{v}\|^2}$$

where the discriminant D is given by

$$D = 4\,(\langle \vec{v} \,,\, \vec{x}_p - \vec{x}_m \rangle)^2 - 4\,\|\vec{v}\|^2\,\|\vec{x}_p - \vec{x}_m\|^2 \ .$$

```
 a = v'*v;
 b = 2*v'*(xp-[x1m;y1m]);
 c = norm(xp-[x1m;y1m])^2-r1^2;
 D = b^2-4*a*c;   % discriminant

 % compute the two solutions
 t1 = (-b+sqrt(D))/(2*a);
 t2 = (-b-sqrt(D))/(2*a);
```

Then the two points of intersection are $\vec{x}_p + t_1\,\vec{v}$ and $\vec{x}_p + t_2\,\vec{v}$.

```
 p1 = xp + t1*v;   p2 = xp + t2*v;
```

If the discriminant $D < 0$ is negative, then there are no points of intersection.

With the above algorithm and resulting codes we have all building blocks to determine the intersection points of two general circles in a plane.

3.10.2 A function to determine the intersection points of two circles

All the above computation can be put in one function file `IntersectCircles.m`

—————————————— **IntersectCircles.m** ——————————————
```
function res = IntersectCircles(x1m,y1m,r1,x2m,y2m,r2)
% draw the graph of two circles and find the intersection points

t = linspace(0,2*pi,51);     % values of all angles, 51 steps

x1 = x1m+r1*cos(t);          % x coordinates of all points
y1 = y1m+r1*sin(t);          % y coordinates of all points
x2 = x2m+r2*cos(t);          % x coordinates of all points
y2 = y2m+r2*sin(t);          % y coordinates of all points
plot(x1,y1,x2,y2);           % create the plot with both circles
axis equal

% find the parameters for the straight line
xp = [0; (-r1^2+r2^2+x1m^2-x2m^2+y1m^2-y2m^2)/(2*(y1m-y2m))];
v  = [y1m-y2m;-x1m+x2m;];
% determine coefficients of the quadratic equation
a = v'*v;    b = 2*v'*(xp-[x1m;y1m]);   c = norm(xp-[x1m;y1m])^2-r1^2;
D = b^2-4*a*c;  % discriminant
% compute the two solutions
t1 = (-b+sqrt(D))/(2*a);
t2 = (-b-sqrt(D))/(2*a);

res = [xp + t1*v,xp + t2*v];
```

Then one command will draw the circles and determine the intersection points.

```
IntersectionPoints = IntersectCircles(2,3,1.5,4,2,2)
-->
InterPoints =   3.2368   2.0632
                3.8487   1.5013
```

3.10.3 Intersection of three spheres

When the three pairs of double beams in Figure 3.81 are moving the central point will move too. Its position is determined by the fact, that the distances of the points of attachment have known values. To determine the position of the central point in the lower part of the section as function of the position of the three guiding beams in the upper part we have to determine the intersection point of three spheres in space. A robot of this type was constructed by Sebastien Perroud in 2004, to be used as a pick and place robot. At the CSEM Sebastien did develop the concept and the results are PoketDelta and MicroDelta robots.

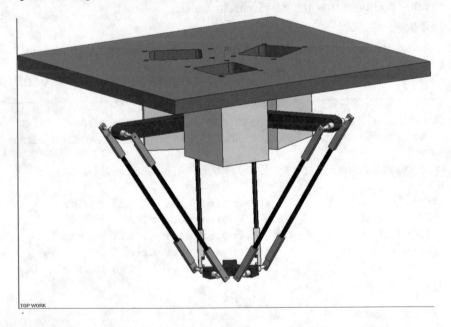

Figure 3.81: A Delta Robot

To determine the points of intersection we use the following geometric facts.

- The intersection of two spheres is typically a circle, which lies in a plane.

- The intersection of two of the above planes determines a straight line.

- With the help of this line and one of the spheres we can determine the points of intersection of the three spheres.

The code below shows how this idea can be implemented.

To find the intersection points of three spheres the following set of quadratic equations have to be solved for x, y and z.

$$
\begin{aligned}
(x - x_1)^2 + (y - y_1)^2 + (z - z_1)^2 &= r_1^2 \\
(x - x_2)^2 + (y - y_2)^2 + (z - z_2)^2 &= r_2^2 \\
(x - x_3)^2 + (y - y_3)^2 + (z - z_3)^2 &= r_3^2 \ .
\end{aligned}
$$

By subtracting these equations we find a linear system of two equations for three unknowns.

$$
-2 \left((x_1 - x_2) x + (y_1 - y_2) y + (z_1 - z_2) z \right) = r_1^2 - x_1^2 - y_1^2 - z_1^2 - r_2^2 + x_2^2 + y_2^2 + z_2^2
$$

$$
-2 \left((x_2 - x_3) x + (y_2 - y_3) y + (z_2 - z_3) z \right) = r_2^2 - x_2^2 - y_2^2 - z_2^2 - r_3^2 + x_3^2 + y_3^2 + z_3^2
$$

or using matrix notation write as

$$
\begin{bmatrix}
x_2 - x_1 & y_2 - y_1 & z_2 - z_1 \\
x_3 - x_2 & y_3 - y_2 & z_3 - z_3
\end{bmatrix}
\cdot
\begin{pmatrix} x \\ y \\ z \end{pmatrix}
= \frac{1}{2}
\begin{pmatrix}
r_1^2 - x_1^2 - y_1^2 - z_1^2 - r_2^2 + x_2^2 + y_2^2 + z_2^2 \\
r_2^2 - x_2^2 - y_2^2 - z_2^2 - r_3^2 + x_3^2 + y_3^2 + z_3^2
\end{pmatrix}
$$

or equivalently

$$
\begin{bmatrix}
\vec{M}_2 - \vec{M}_1 \\
\vec{M}_3 - \vec{M}_2
\end{bmatrix}
\vec{x} = \frac{1}{2}
\begin{pmatrix}
r_1^2 - \|\vec{M}_1\|^2 - r_2^2 + \|\vec{M}_2\|^2 \\
r_2^2 - \|\vec{M}_2\|^2 - r_3^2 + \|\vec{M}_3\|^2
\end{pmatrix} ,
$$

where $\vec{M}_i = (x_i, y_i, z_i)$. With the definitions for $\mathbf{A}$ and $\vec{b}$ this leads to an inhomogeneous system of two linear equations for three unknowns of the form $\mathbf{A} \cdot \vec{x} = \vec{b}$. The general solution of this system can be parameterized with the help of a particular solution $\vec{x}_p$ and the solution $\vec{v}$ of the homogeneous problem $\mathbf{A} \vec{v} = \vec{0}$.

$$
\vec{x}_p = \mathbf{A} \backslash \vec{b} \quad \text{and} \quad \vec{v} = \ker(\mathbf{A}) \ .
$$

All solutions of the linear system are of the form

$$
\vec{x}(t) = \vec{x}_p + t \, \vec{v}
$$

and this expression can be used with the equation of the first sphere to find a quadratic equation for the parameter $t \in \mathbb{R}$.

$$
\begin{aligned}
\|\vec{x} - \vec{M}_1\|^2 - r_1^2 &= 0 \\
\|t \vec{v} + \vec{x}_p - \vec{M}_1\|^2 - r_1^2 &= 0 \\
t^2 \|\vec{v}\|^2 + t \, 2 \langle \vec{v}, \vec{x}_p - \vec{M}_1 \rangle + \|\vec{x}_p - \vec{M}_1\|^2 - r_1^2 &= 0 \\
a t^2 + b t + c &= 0 \\
t_{1,2} &= \frac{-b \pm \sqrt{b^2 - 4 a c}}{2 a}
\end{aligned}
$$

where $\vec{M}_1 = (x_1, y_1, z_1)^T$. The two intersection points are then given by $\vec{x}_p + t_1 \, \vec{v}$ and $\vec{x}_p + t_2 \, \vec{v}$. This algorithm can be implemented in *Octave* or MATLAB.

46 Example : Intersection of spheres

Using the same ides one can generate a function `IntersectSpheres()` in *Octave* or `MATLAB` to determine the intersection points of three spheres.

──────────────────── **IntersectSpheres.m** ────────────────────

```
function res = IntersectSpheres(M1,r1,M2,r2,M3,r3)
% find the intersection points of three spheres
% Mi is a row vector with the three components of the center
% ri is the radius of the i-th sphere

% create the matrix and vector for the linear system
A = 2*[M2-M1;M3-M2];
b = [r1^2-norm(M1)^2-r2^2+norm(M2)^2;r2^2-norm(M2)^2-r3^2+norm(M3)^2];
% determine a particular solution xp and the homogeneous solution v
xp = A\b;   v  = null(A);
% determine coefficients of the quadratic equation
a = v'*v;   b = 2*v'*(xp-M1');  c = norm(xp-M1')^2-r1^2;
% solve the quadratic equation
D = b^2-4*a*c;   % discriminant
if (D<0)
   sprintf('no intersection points');    res = [];
else
   % compute the two solutions
   t1  = (-b+sqrt(D))/(2*a);     t2  = (-b-sqrt(D))/(2*a);
   res = [xp + t1*v,xp + t2*v];
end
```

As a simple test run the following commands

```
M1 = [3 -0.1  0]; r1 = 3;
M2 = [0  3    0]; r2 = 3;
M3 = [0  0.35 4]; r3 = 4;
IntersectSpheres(M1,r1,M2,r2,M3,r3)
-->
2.4652e+00    1.7077e-03
2.3841e+00    3.9699e-05
1.5948e+00    1.5339e-02
```

$\diamond$

3.10.4   Intersection of multiple circles

In Section 3.10.1 we observed that the intersection points of two circles are characterized by a system of quadratic equations

$$x^2 - 2\,x\,x_1 + x_1^2 + y^2 - 2\,y\,y_1 + y_1^2 \;=\; r_1^2$$
$$x^2 - 2\,x\,x_2 + x_2^2 + y^2 - 2\,y\,y_2 + y_2^2 \;=\; r_2^2 \;.$$

Subtraction these two leads to one linear equation

$$2\,x\,(x_2 - x_1) + (x_1^2 - x_2^2) + 2\,y\,(y_2 - y_1) + (y_1^2 - y_2^2) = r_1^2 - r_2^2$$

or

$$2\,(x_2 - x_1)\,x + 2\,(y_2 - y_1)\,y = (r_1^2 - r_2^2) - (x_1^2 - x_2^2) - (y_1^2 - y_2^2)\,.$$

If the intersection points of more than two circles are examined, then we have multiple of these equations. For N circles with centers at (x_i, y_i) and radii r_i for $i = 1, 2, 3, \ldots, N$ we have

$$2\,(x_j - x_i)\,x + 2\,(y_j - y_i)\,y = (r_i^2 - r_j^2) - (x_i^2 - x_j^2) - (y_i^2 - y_j^2) \qquad (3.10)$$

for $1 \le i \ne j \le N$. It is obvious that for $N > 3$ we have too many equations to determine the two unknown coordinates (x, y) of the point of intersection. This is mirrored by the fact that in general N circles do not have a unique point of intersection.

number of circles	2	3	4	5	N
number of equations	1	3	6	10	$\frac{1}{2}N(N-1)$

Thus it seems that we have $\frac{1}{2}N(N-1)$ equations, but this is not the full truth. Many equations are linearly dependent. The difference generated by the intersections of the pair $(1, i)$ of circles and by the pair $(1, j)$ is identical to the equation generated by the pair (i, j). Consequently we have only $N - 1$ independent equations. This confirms that three circles with a unique point of intersection lead to two equations for the two unknowns of the position of the point of intersection. The implementation below uses the pairs $(1, 2)$, $(2, 3)$, $(3, 4), \ldots (N - 1, N)$

Thus we have usually an over-determined system of $N - 1$ linear equations of the type (3.10) and this can be writen using a matrix $\mathbf{M} \in \mathbb{R}^{(N-1) \times 2}$.

$$\mathbf{M} \begin{pmatrix} x \\ y \end{pmatrix} = \vec{b} \in \mathbb{R}^{N-1}$$

When solving this over-determined system with the backslash operator \ *Octave*/MATLAB will determine the least square solution, i.e. the norm of the residual vector $\vec{r}$

$$\vec{r} = \mathbf{M} \begin{pmatrix} x \\ y \end{pmatrix} - \vec{b}$$

will be minimized.

- If the given circles happen to have a unique point of intersection, it will be determined by MATLAB/*Octave*.

- If there is not exact point of intersection, an approximation is generated.

- Even if there is no intersection point at all, the algorithm will return a result. It is the best possible result in the above least square sense.

- A major advantage of the above algorithm is that a linear regression problem is used, instead of a nonlinear system of equations.

As a first example examine the three circles with center at the origin, radius 2.3, center at $(3, 3)$, radius 2.2 and center at $(0, 3)$, radius 1.7. The approximate intersection point is at $(1.175, 1.900)$. Find the resulting graphics in Figure 3.82. Since the three straight lines have an exact point of intersection, the norm of the residual vector $\vec{r}$ is zero, but this does not imply the we have an exact point of intersection for the tree circles.

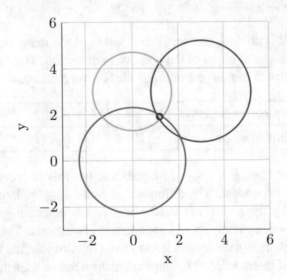

Figure 3.82: Approximate intersection point of three circles

IntersectMultipleCircle.m

```
% intersect three circles
centers = [0,0; 3,3;  0 3];
radii = [ 2.3;2.2;1.7];

% generate the plots
angle = linspace(0,2*pi,200); xr = cos(angle); yr = sin(angle);
figure(1); clf
hold on
for ii = 1:length(radii)
  plot(centers(ii,1)+radii(ii)*xr,centers(ii,2)+radii(ii)*yr)
end%for
hold off

% construct the overdetermined linear system
N = length(radii);
M = zeros(N-1,2); b = zeros(N-1,1);
```

```
for ii = 1:N-1
  M(ii,:) = 2 * (centers(ii,:) - centers(ii+1,:));
  b(ii) = radii(ii+1)^2 - radii(ii)^2 - sum(centers(ii+1,:).^2)...
                                      + sum(centers(ii,:).^2);
end%for
xy = M\b
residum = norm(M*xy-b)

hold on
plot(xy(1),xy(2),'or') % mark the point of intersection in red
hold off
axis equal; axis([-3 6 -3 6])
xlabel('x'); ylabel('y')
```

The above code can be reused to examine possible points of intersection of more circles.

circle	center	radius
1	$(0,0)$	5
2	$(8,0)$	5
3	$(4,0)$	3
4	$(3,3)$	1
5	$(4.6062, 3.35)$	0.6

Replace the first section in the code above by the data of these five circles.

```
% a set of four circles
centers = [0,0;8,0;4,0;3,3];
radii = [5;5;3;1];
% add a fifth circle
c5 = [4,3] + 0.7*[cos(pi/6),sin(pi/6)];
centers = [centers;c5]; radii = [radii;0.6];
```

The approximation for the point of intersection is $(x, y) \approx (4.0013, 3.0028)$ and clearly visible in Figure 3.83(a). But these five circles have no common point of intersection, illustrated by the zoom into the critical area in Figure 3.83(b). The length of the residual vector is $\|\vec{r}\| \approx 0.18$. If the fifth circle is dropped from the list of circles, then the exact point of intersection $(4, 3)$ is computed.

3.10.5 Intersection of multiple spheres

The problem of intersecting multiple spheres is very similar to the above question. The only additional aspect is the third coordinate z. For two spheres we have the quadratic equations

$$x^2 - 2x x_1 + x_1^2 + y^2 - 2y y_1 + y_1^2 + z^2 - 2z z_1 + z_1^2 = r_1^2$$
$$x^2 - 2x x_2 + x_2^2 + y^2 - 2y y_2 + y_2^2 + z^2 - 2z z_2 + z_2^2 = r_2^2 .$$

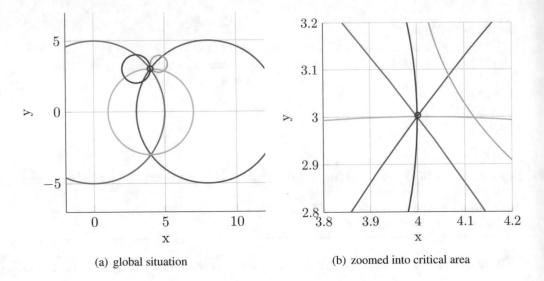

(a) global situation   (b) zoomed into critical area

Figure 3.83: Intersection of five circles

Subtraction these two leads to one linear equation

$$2\,x\,(x_2 - x_1) + (x_1^2 - x_2^2) + 2\,y\,(y_2 - y_1) + (y_1^2 - y_2^2) + 2\,z\,(z_2 - z_1) + (z_1^2 - z_2^2) = r_1^2 - r_2^2$$

or

$$2\,(x_2 - x_1)\,x + 2\,(y_2 - y_1)\,y + 2\,(z_2 - z_1)\,z = (r_1^2 - r_2^2) - (x_1^2 - x_2^2) - (y_1^2 - y_2^2) - (z_1^2 - z_2^2).$$

If the intersection points of more than two spheres are examined we have multiple of these equations. For spheres with centers at (x_i, y_i, z_i) and radii r_i for $i = 1, 2, 3, \ldots, N$ we have for $1 \leq i \neq j \leq N$

$$2\,(x_j - x_i)\,x + 2\,(y_j - y_i)\,y + 2\,(z_j - z_i)\,z = (r_i^2 - r_j^2) - (x_i^2 - x_j^2) - (y_i^2 - y_j^2) - (z_i^2 - z_j^2).$$

$$(3.11)$$

Thus the problem of intersection spheres in space is very similar to the problem of intersecting circles in the plane. To have a unique point of intersection we need at least 4 spheres, but we can use information of more and examine the corresponding least square problem.

The implementation of the algorithm is very similar to the code shown in the previous Section 3.10.4.

3.10.6 GPS

The context and some of the Mathematics of the global positioning system (GPS) is explained in the very nice article [Thom98]. The basic facts are:

- Each satellite sends accurate information on its own position and the time at which the signal was sent.

- The GPS receiver has information from N satellites and uses its own clock to determine the current distance r_i to the satellite by multiplying the measured travel time by the speed of light c.

- The (relatively inaccurate) clock of the GPS receiver might be off by ΔT, leading to a fixed error of $D = c \, \Delta T$ for the distances to each of the satellites.

With the above each satellite leads to a quadratic equation for the position (x, y, z) of the receiver.

$$(x - x_i)^2 + (y - y_i)^2 + (z - z_i)^2 = (r_i - D)^2$$
$$x^2 - 2 \, x \, x_i + x_i^2 + y^2 - 2 \, y \, y_i + y_i^2 + z^2 - 2 \, z \, z_i + z_i^2 = r_i^2 - 2 \, r_i \, D + D^2 \, .$$

Subtracting two of these equations leads to

$$2 \, (x_j - x_i) \, x + 2 \, (y_j - y_i) \, y + 2 \, (z_j - z_i) \, z - 2 \, (r_j - r_i) \, D =$$
$$= (r_i^2 - r_j^2) - (x_i^2 - x_j^2) - (y_i^2 - y_j^2) - (z_i^2 - z_j^2) \, .$$

This is one linear equation for the four unknowns z, y, z and D. With the information of at least 5 satellites determine the current position and the offset $\Delta T = \frac{D}{c}$ of the clock of the GPS receiver. Thus the accuracy of the clock in the GPS receiver does not have to be outstanding. The implementation of the algorithm is again very similar to the code shown in Section 3.10.4.

3.10.7 List of codes and data files

In the previous section the codes in Table 3.17 were used.

filename	function
TwoCircles.m	script file to determine intersection points
IntersectCircles.m	intersection points of two circle
IntersectSpheres.m	intersection points of three spheres
IntersectMultipleCircles.m	intersection points of multiple circles

Table 3.17: Codes and data files for section 3.10

3.11 Scanning a 3–D Object with a Laser

A solid is put on a plate and then scanned with a laser from a given angle. A CCD camera is placed straight above the object and is recording the laser spot on the solid. With this information one can determine the height of the solid at these points. This should lead to a 3–D picture of the solid. The aim is to show this solid in a graphic. The laser scan is performed according to the following scheme:

- For a fixed x position of the laser, move it stepwise in the y direction and detect the resulting x position of the laser spot on the solid. Compute the height z of the solid at this spot. Store the values of x, y and z.

- Advance the laser in the x direction by a fixed step size and repeat the above procedure.

Since some sections of the solid is shaded, i.e. not visible by the laser, the object is then rotated by 90° and rescanned. The two scans have to be combined.

3.11.1 Reading the data

The first task is to read the numbers in those files and store them in matrices. The code below does just this for the x values. A similar code will read the other matrices from the corresponding files.

ReadData.m

```
% read each of the three data files
x = load('Xmatnew1.txt'); y = load('Ymatnew1.txt');
z = load('Zmatnew1.txt');        [nstep,npix] = size(x);
figure(1); mesh(x,y,z);
           view(50,30); xlabel('x'); ylabel('y'); zlabel('z');

% read the rotated data
xR = load('Xmatnew2.txt'); yR = load('Ymatnew2.txt');
zR = load('Zmatnew2.txt');
```

As a result each row in the matrices x, y and z contains the values of the coordinates along one line in x direction, where the value of y is fixed. A sample is shown in Figure 3.84, generated by the code below.

```
kk = 175; plot(x(:,kk),z(:,kk))
          xlabel('x'); ylabel('z');
```

In Figure 3.85 find a visualization of this fact. Obviously the x values will not be uniformly spaced. It is in fact this nonuniform distribution of points that allows to compute the height of the solid.

With the commands mesh(x(1:25,1:30)) and mesh(y(1:25,1:30)) generate the results in Figure 3.86. This figure shows that the y values are very regularly spaced, while the x coordinate of the laser spot varies, due to the changing height of the solid, as illustrated in Figure 3.85.

The command mesh(x,y,z) will create Figure 3.87(a), a first try of a picture of the solid. On the left in this figure the traces of the shadow lines are clearly visible. In this section the shape of the solid is not correctly represented, the laser beam can not "see" this part of the solid. Thus appropriate measures have to be taken. We will scan the same object from a different angle and then try to merge the two pictures.

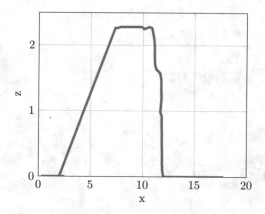

Figure 3.84: Cross section in x direction

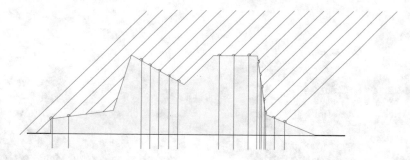

Figure 3.85: Location of laser spot, varied in x direction

3.11.2 Display on a regular mesh

For subsequent calculations it is advantageous to compute the (measured) height on a regular grid. To achieve this goal we

1. choose the number of grid points in either direction.

2. along each line in x direction (see Figure 3.84) we compute the height with the help of a piecewise linear interpolation. The command `interp1()` will perform this operation.

3. The above process has to be applied to each line in x direction.

```
nx   = 200;                         % number of grid points in x direction
xmin = 0.5; xmax = 16;              % minimal and maximal value of x
xlin = linspace(xmin,xmax,nx);      zlin = zeros(nx,npix);
for k = 1:npix
  xt = x(:,k)'; zt = z(:,k)';       % values of x and z in this row
  aa = sortrows([xt;zt]',1);        % sort with the x values as criterion
```

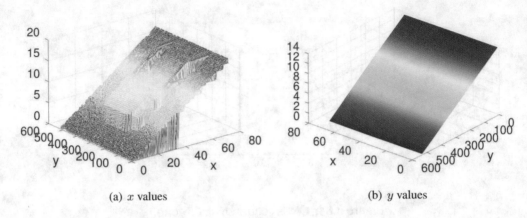

(a) x values   (b) y values

Figure 3.86: Values of the measured coordinates

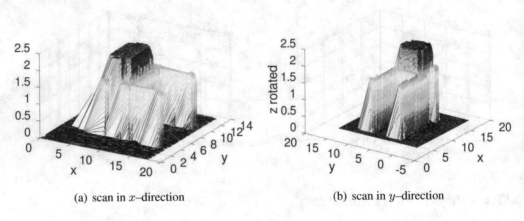

(a) scan in x–direction   (b) scan in y–direction

Figure 3.87: 3–D scan of solid from two ortogonal directions

```
xt = aa(:,1);   % add a minimal slope to prevent identical values
xt = xt+(1:length(xt))'*1e-8;
zt = aa(:,2);
t = interp1(xt,zt,xlin);      % perform a linear interpolation
zlin(:,k) = t';               % store the result in the matrix
end%for


xlin = xlin'*ones(1,npix);    % create the uniformly spaced values
ylin = ones(1,nx)'*y(1,:);

figure(2); mesh(xlin,ylin,zlin)  % plot the interpolated data
          xlabel('x'); ylabel('y'); zlabel('z');
```

As the next task we try to decide which points are on a shadow line in the plot. For the given data we know that the angle of the laser beam is $\alpha = 30° = \pi/6$ and thus the slope of the shadow is given by $\tan\alpha = 0.5$. With a comparison operator we determine all points where the slope deviates less than 0.1 from the ideal value of 0.5. The plot generated by the code below represents the shadowed area.

```
dx = xlin(2,1)-xlin(1,1);    dz = diff(zlin);

tip = abs(dz./dx-0.5)<0.1;        % mark the shadowed area
tip(nx,1:npix) = zeros(1,npix);   % no shadows on last row
figure(3); mesh(xlin,ylin,1.0*tip)
        xlabel('x'); ylabel('y'); zlabel('shadow');
```

3.11.3 Rescan from a different direction and rotate the second result onto the first result

Now the solid is rotated by $90°$ on the mounting plate and a second scan generates independent results, shown in Figure 3.87(b). The result has to be compared with Figure 3.87(a). The shadows are now falling in a different direction. The goal is to combine the two pictures by the following algorithm:

1. Rotate the second graph, such that the two pictures should coincide.

2. If a point in Figure 3.87(a) is in a shadowed area, replace the height by the result form Figure 3.87(b).

3. Plot the new, combined picture.

With the affine mapping

$$\begin{pmatrix} x \\ y \end{pmatrix} \mapsto \begin{pmatrix} +y + x_0 - y_0 \\ -x + x_0 + y_0 \end{pmatrix}$$

the direction of the two axis will be interchanged and since

$$\begin{pmatrix} x_0 \\ y_0 \end{pmatrix} \mapsto \begin{pmatrix} +y_0 + x_0 - y_0 \\ -x_0 + x_0 + y_0 \end{pmatrix} = \begin{pmatrix} x_0 \\ y_0 \end{pmatrix}$$

we have a rotation about the fixed point $(x_0, y_0)^T$. Thus the code below will create the picture of the solid in the original direction, but the laser will now throw its shadows into another direction.

```
x0 = 10.508;        y0 =  5.897;
xn = +yR -y0+x0;    yn = -xR +x0+y0;

figure(4); mesh(xn,yn,zR)
        xlabel('x'); ylabel('y'); zlabel('z rotated');
```

The next task is to compute the height of the "new" solid at the regular grid points of the first scan. This leads to an interpolation problem for a function of two variables. The algorithm is rather elaborate, implemented as `griddata()`. Then we use the matrix `tip` to construct the combined height.

- If `tip=0` then `(1-tip) * A + tip * B = A` and the first value is used.

- If `tip=1` then `(1-tip) * A + tip * B = B` and the second value is used.

The result is shown in Figure 3.88.

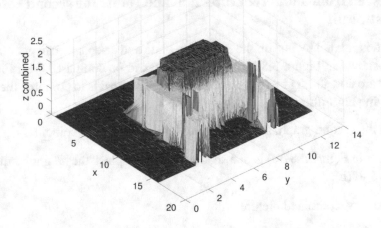

Figure 3.88: Combination of the two scans

```
zInt = griddata(xn,yn,zR,xlin,ylin,'nearest');
znew = (1-tip).*zlin + tip.*zInt;

figure(5); mesh(xlin,ylin,znew)
          xlabel('x'); ylabel('y'); zlabel('z combined'); view(50,60)
```

3.11.4 List of codes and data files

In the previous sections the codes and data files in Table 3.18 were used. The following sequence of commands in *Octave* or MATLAB should reproduce the results in this section. Use `ReadData.m` to read all data files and generate a first plot. Then use `UniformMesh.m` to examine the solid on a uniform mesh and determine the shadow areas. Finally the two scans are combined with the help of `RotateShape.m`.

filename	function
ReadData.m	read the basic data from files, including plot
UniformMesh.m	interpolate on a uniform mesh, determine shadow
RotatedShape.m	rotate the second scan and combine the two graphs
?newmat1.txt	data for the first scan
?newmat2.txt	data for the second scan

Table 3.18: Codes and data files for section 3.11

3.12 Transfer Functions, Bode and Nyquist plots

For control applications the behavior of many systems can be described by their transfer function. Bode and Nyquist plots are tools often used in connection with transfer functions. In this section we show how to create those plots

- with code of our own

- using a MATLAB toolbox

- the commands provided by *Octave*

- some operations with transfer functions

3.12.1 Create Bode and Nyquist plots, raw MATLAB/*Octave* Code

Consider the transfer function

$$G(s) = \frac{4 + 4.8\,s}{5.5 + 17.5\,s + 14.5\,s^2 + 3.5\,s^3 + s^4}\,.$$

Defining a function mytf() by

```
mytf = @(s)(4+4.8*s)./(5.5+17.5*s+14.5*s.^2 + 3.5*s.^3 + s.^4);
```

and then compute the result $G(2)$ by calling mytf(2), with the result 0.0954. Since the function uses element wise operations compute the values of the function for multiple arguments by passing a vector as argument to a single call of the function, e.g.

```
mytf([0 1 2 3 4 5])
.
ans =     0.7273     0.2095     0.0954     0.0505     0.0295     0.0184
```

By using the above code compute the values of this transfer function along the positive imaginary axis and then generate the plot. As example examine the domain $10^{-2} \leq \omega \leq 10^{5}$. The code below will generate one half of the Nyquist plot of this transfer function, as shown in Figure 3.89.

```
% generate the Nyquist plot of a transfer function
w = logspace(-2,5,200);
z = mytf(i*w);
figure(1); plot(z); hold on ; plot(z'); hold off
            axis equal; xlabel('real part'); ylabel('imaginary part')
```

A very similar code will generate the Bode plots of $G(s)$.

```
% generate the Bode plots of a transfer function, own code
figure(1); semilogx(w,20*log10(abs(z)));
            xlabel('frequency \omega'); ylabel('amplitude [dB]');
figure(2); semilogx(w,angle(z)*180/pi)
            xlabel('frequency \omega'); ylabel('phase');
```

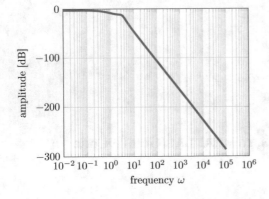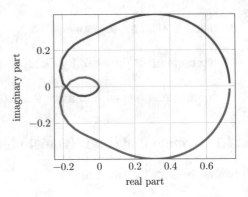

Figure 3.89: Bode and Nyquist plots of the system, programmed with *Octave*/MATLAB

With the above commands the angular velocities used for the plot were selected by `w = logspace(-2,5,200)`. By returning arguments of the command `bode()` this can be left to *Octave*/MATLAB[24], which will make a best guess for the angular velocities.

```
num = [4.8 4]; denum = [1 3.5 14.5 17.5 5.5];
[mag,phase,w] = bode(tf(num,denum));
%%mag = squeeze(mag); phase = squeeze(phase); % Matlab only
figure(1); semilogx(w,20*log10(mag));
            xlabel('frequency'); ylabel('amplitude [dB]')
```

---

[24]MATLAB returns the amplitude in a different format, aiming for systems with multiple input and output. Use `mag = squeeze(mag)` to obtain the result.

```
figure(2); semilogx(w,phase);
            xlabel('frequency'); ylabel('phase [deg]')
```

A similar feature is available for the command `nyquist()`.

When plotting the phase shift created by this transfer function by

```
mytf = @(s)(4+4.8*s)./(5.5+17.5*s+14.5*s.^2 + 3.5*s.^3 + s.^4);
w = logspace(-2,5,200); z = mytf(i*w);
figure(3); semilogx(w,angle(z)*180/pi)
            xlabel('frequency \omega'); ylabel('phase [deg]');
```

the left part in Figure 3.90 is generated, with the jump by 360° in the phase. This does not

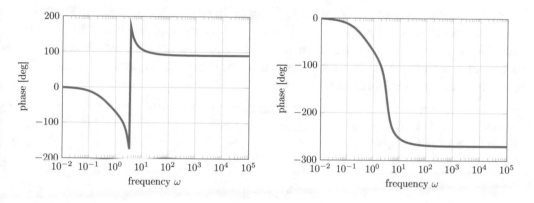

Figure 3.90: Phase of the transfer function, programmed with *Octave*/MATLAB

create a problem for this figure, but might be troublesome in another applications. Thus we seek to eliminate this artificial jump in the phase. The function `fixangles()` does take a vector of angle values as argument and returns the adjusted angles, such that no 2π jumps appear. The algorithm is rather elementary:

fixangles.m

```
function res = fixangles(angles)
%% eliminates the 2*pi jumps in a vector of angle values
n   = length(angles); res = angles;   k = 0;
da = diff(angles);
for i = 1:n-1;
  k = k - round(da(i)/(2*pi));
  res(i+1) = res(i+1) + k*2*pi;
end%for
```

- Start out with no correction k=0. Take the known angles (`angles`) and the compute the difference between subsequent angles by `da=diff(angles)`.

- For each step in the angles determine what number of 2π–steps is closest to the actual change in the angle. This is done by the command `k=k-round(da(i)/(2*pi))`. For most steps the values of k will not change.

- Then add the correct numbers of steps to the angle by adding `k*2*pi`.

Thus the code

```
figure(3); semilogx(w,fixangles(angle(z))*180/pi)
          xlabel('frequency \omega'); ylabel('phase [deg]');
```

will generate the improved figure on the right in 3.90.

3.12.2 Create Bode and Nyquist plots, using the MATLAB–toolbox

The built-in commands `bode()` and `nyquist()` in MATLAB will generate Figures 3.91 and 3.92. These functions need the coefficients of the numerator and denominator polynomial as arguments. The frequency domain will be chosen automatically. Both commands are part of a toolbox of MATLAB and thus have to be purchased separately. Consult the on-line help for more information.

```
 Matlab
num   = [4.8 4];  denum = [1 3.5 14.5 17.5 5.5];
bode(num,denum);
nyquist(num,denum);
```

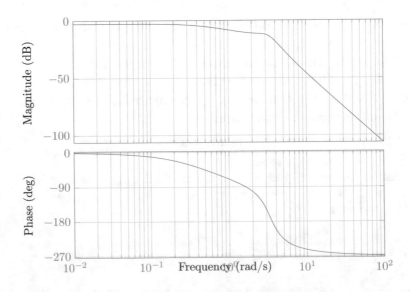

Figure 3.91: Bode plot of the system, with the control toolbox of MATLAB

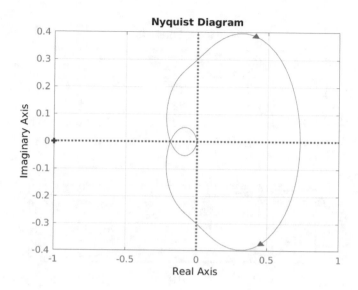

Figure 3.92: Nyquist plot of the system, with the control toolbox of MATLAB

3.12.3 Create Bode and Nyquist plots, using *Octave* commands

The Octave Forge package `control` of *Octave* provides a set of commands for control theory, including Bode[25] and Nyquist plots. The code below leads to Figures 3.93 and 3.94. You may want to consult the online help on the commands `tf()`, `bode()` and `nyquist()`.

```
Octave
pkg load control
mysys = tf([4.8 4],[1 3.5 14.5 17.5 5.5]);
figure(1);   bode(mysys);
figure(2);   nyquist(mysys);
```

3.12.4 Some commands for control theory

The control package for *Octave* has a sizable number of commands to operate on control systems. Find some documentation at https://octave.sourceforge.io/control/overview.html. The control toolbox in MATLAB provides very similar features. In these notes only continuous system with single input and output (SISO) are presented. The control package is considerably more powerful.

---

[25]The current version 6.4.0 of *Octave* seems to have a bug when trying to remove the legend. A quick fix was to use the command sequence `bode(mysys); legend off; subplot(2,1,1); legend off;`

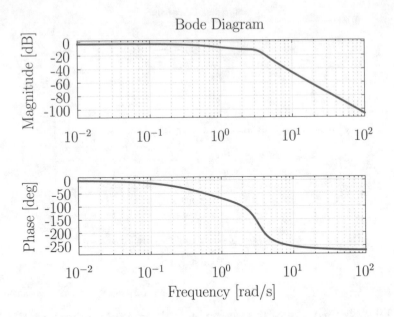

Figure 3.93: Bode plot of the system, created by *Octave*

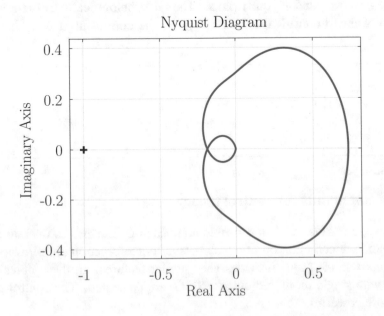

Figure 3.94: Nyquist plot of the system, created by *Octave*

47 Example : A very short demonstration

Using the transfer function given by

$$G(s) = \frac{4 + 4.8\,s}{5.5 + 17.5\,s + 14.5\,s^2 + 3.5\,s^3 + s^4}$$

a few of the commands in the control package are illustrated.

- Use `tf()` to build transfer function. Often the coefficients for the numerator and denominator polynomials are specified.

- Find description and examples of `bode()` and `nyquist()` above.

- Use the command `tfdata()` to determine the coefficients of the numerator and denominator polynomials. The answers are given a cells, since the control package can work with systems with multiple inputs and outputs. Thus use `den = den{1}` to obtain the coefficients as vector.

- Use the command `pole()` to calculate the poles of the system, i.e. the zeros of the denominator. For the examined example the real part of all poles are negative, thus the system is stable. This is confirmed by the result of `isstable(G)`, generating the answer 1.

- With `pzmap()` a figure with the location of the zeros and poles of the transfer function is generated. See Figure 3.95(a).

```
% generate the transfer function
G = tf([4.8 4],[1 3.5 14.5 17.5 5.5]);
[num, den] = tfdata(G);          % extract numerator and denominator
num = num{1}; den = den{1}       % convert to vectors
poles = pole(G)                  % poles of the transfer function
isstable(G)                      % system is stable
pzmap(G); legend off             % graph poles and zeros
-->
den =    1.0000    3.5000   14.5000   17.5000    5.5000


poles =   -1.0000 + 3.1623i
          -1.0000 - 3.1623i
          -1.0000 +      0i
          -0.5000 +      0i


ans =    1
```

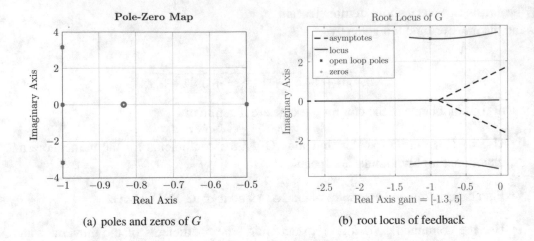

(a) poles and zeros of G (b) root locus of feedback

Figure 3.95: Graphical results for a transfer function

48 Example : A feedback and root locus example

For a given transfer function $G(s)$ examine the feedback system shown below.

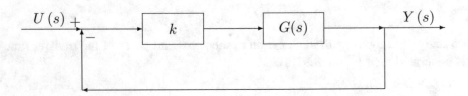

To determine the transfer function of the feedback system examine

$$
\begin{aligned}
Y(s) &= k\,G(s)\,(U(s) - Y(s)) \\
(1 + k\,G(s))\,Y(s) &= k\,G(s)\,U(s) \\
Y(s) &= \frac{k\,G(s)}{1 + k\,G(s)}\,U(s)\,.
\end{aligned}
$$

Thus the zeros of $1 + k\,G(s)$ are to be examined for the stability of the feedback system. With the command `rlocus()` the poles of this feedback system can be generated for different values of the parameter k. For $-1.2 \le k \le 5$ use

```Octave
rlocus(G,0.2,-1.3,5); legend('location','northwest')
```

to generate Figure 3.95(b). Since all poles have negative real part, the feedback system is stable for all feedback parameters $-1.3 \le k \le 5$.

The setup in MATLAB is slightly different, i.e. the feedback is differently placed.

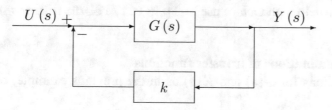

The calculation

$$
\begin{aligned}
Y(s) &= G(s)\,(U(s) - k\,Y(s)) \\
(1 + k\,G(s))\,Y(s) &= G(s)\,U(s) \\
Y(s) &= \frac{G(s)}{1 + k\,G(s)}\,U(s)
\end{aligned}
$$

shows that the poles of the feedback system are identical to the *Octave* version. With MATLAB Figure 3.95(b) is generated by `rlocus(G,linspace(-1.3,5,20))`. ◊

49 Example : A feedback example

As an example consider a feedback system where the individal transfer functions are given by

$$
K(s) = \frac{s+1}{s} \quad \text{and} \quad P(s) = \frac{1}{s+2} \,.
$$

Standard results for transfer functions imply that the transfer function for this feedback system is given by

$$
T(s) = \frac{P(s)}{1 + K(s)\,P(s)} = \frac{\frac{1}{s+2}}{1 + \frac{s+1}{s}\frac{1}{s+2}} = \ldots = \frac{s}{s^2 + 3\,s + 1} \,.
$$

The command `feedback()` combines the given systems P and K to a new system with transfer function $T(s)$.

```
K = tf([1 1],[1,0]);    P = tf(1,[1 2]);
T = feedback(P,K)
-->
Transfer function 'T' from input 'u1' to output ...

            s
  y1:  -------------
        s^2 + 3 s + 1
Continuous-time model.
```

To access the builtin help with *Octave* use `help @lti/feedback`, or with MATLAB use `help feedback`   ◇

50 Example : Combinations of transfer functions
Use the same functions for $G(s)$ and $K(s)$ as in the previous example, but in a slightly different setup.

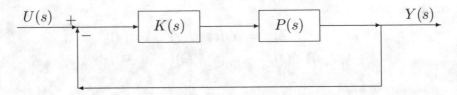

Some algebraic operations lead to the transfer function of this feed back system.

$$
\begin{aligned}
Y(s) &= K(s)P(s)\,(U(s) - Y(s)) \\
(1 + K(s)P(s))\,Y(s) &= K(s)P(s)\,U(s) \\
Y(s) &= \frac{K(s)P(s)}{1 + K(s)P(s)}\,U(s) \\
&= \frac{\frac{s+1}{s^2+2\,s}}{1 + \frac{s+1}{s^2+2\,s}}\,U(s) = \frac{s+1}{s^2+3\,s+1}\,U(s)
\end{aligned}
$$

With MATLAB/*Octave* the new transfer function is readily computed.

- Use `mtimes()` to multiply the two transfer functions.

- Followed by a call of `feedback()` to generate the new transfer function $T(s)$.

```
K = tf([1 1],[1,0]);    P = tf(1,[1 2]);
T = feedback(mtimes(P,K),1)
-->
Transfer function 'T' from input 'u1' to output ...
          s + 1
  y1:   --------------
        s^2 + 3 s + 1
Continuous-time model.
```

Similarly the addition of transfer functions can be performed by calling `plus(P,K)`.
                                                                                      ◇

Bibliography

[AbraSteg] M. Abramowitz and I. A. Stegun. *Handbook of Mathematical Functions.* Dover, 1972.

[Bevi69] P. R. Bevington. *Data Reduction and Error Analysis for the Physical Sciences.* McGraw–Hill, New York, 1969.

[BiraBrei99] A. Biran and M. Breiner. *Matlab 5 für Ingenieure. Systematische und praktische Einführung.* Addison–Wesley, 3rd edition, 1999.

[DownClar97] D. Downing and J. Clark. *Statistics, The Easy Way.* Barrons's, 1997.

[GandHreb95] W. Gander and J. Hřebíček. *Solving Problems in Scientific Computing Using Maple and MATLAB.* Springer–Verlag, Berlin, second edition, 1995.

[Grif18] D. F. Griffiths. *An Introduction to Matlab.* University of Dundee, 2018.

[HansLitt98] D. Hanselmann and B. Littlefield. *Mastering Matlab 5.* Prentice Hall, 1998.

[Hans11] J. S. Hansen. *GNU Octave.* Packy Publishing, Bimingham, 2011.

[Hind93] A. C. Hindmarsh and K. Radhakrishnan. *Description and Use of LSODE, the Livermore Solver for Ordinary Differential Equations.* NASA, 1993.

[Hock05] R. Hocking. *Methods and Applications of Linear Models: Regression and the Analysis of Variance.* Wiley Series in Probability and Statistics. Wiley, 2005.

[HuntLipsRose14] B. R. Hunt, R. L. Lipsman, and J. M. Rosenberg. *A Guide to MATLAB: For Beginners and Experienced Users.* Cambridge University Press, New York, NY, USA, third edition, 2014.

[Kove20] P. D. Kovesi. MATLAB and Octave functions for computer vision and image processing. Available from: <https://www.peterkovesi.com/matlabfns/>.

[MoleVanLoan03] C. Moler and C. van Loan. Nineteen dubious ways to compute the exponential of a matrix, twenty-five years later. *SIAM Review*, 45(1), 2003.

[MontPeckVini12] D. Montgomery, E. Peck, and G. Vining. *Introduction to Linear Regression Analysis.* Wiley Series in Probability and Statistics. Wiley, 2012.

© The Author(s), under exclusive license to Springer Fachmedien Wiesbaden GmbH, part of Springer Nature 2022
A. Stahel, *Octave and MATLAB for Engineering Applications*, https://doi.org/10.1007/978-3-658-37211-8

[DLMF15] N. I. of Standards. NIST Digital Library of Mathematical Functions, at http://dlmf.nist.gov, 2015.

[Quat10] A. Quateroni, F. Saleri, and P. Gervasio. *Scientific Computing with MATLAB and Octave*. Springer, third edition, 2010.

[Rivl69] T. J. Rivlin. *An Introduction to the Approximation of Functions*. Blaisdell Publishing Company, 1969. Republished by Dover.

[Seyd00] R. Seydel. *Einführung in die numerische Berechnung von Finanz–Derivaten*. Springer, 2000.

[Stah08] A. Stahel. Numerical Methods. lecture notes, BFH-TI, 2008.

[Stah99] W. A. Stahel. *Statistische Datenanalyse*. Vieweg, 2. auflage edition, 1999.

[Stew13] I. Stewart. *Seventeen Equations that Changed the World*. Profile Books Limited, 2013.

[Thom98] R. B. Thompson. Global Positioning System: The Mathematics of GPS Receivers. *Mathematics Magazine*, 71(4):260–269, 1998.

[Wilm98] P. Wilmott. *Derivatives, the Theory and Practice of Financial Engineering*. John Wiley&Sons, 1998.

List of Figures

© The Author(s), under exclusive license to Springer Fachmedien Wiesbaden GmbH, part of Springer Nature 2022
A. Stahel, *Octave and MATLAB for Engineering Applications*, https://doi.org/10.1007/978-3-658-37211-8

List of Tables

© The Author(s), under exclusive license to Springer Fachmedien Wiesbaden GmbH, part of Springer Nature 2022
A. Stahel, *Octave and MATLAB for Engineering Applications*, https://doi.org/10.1007/978-3-658-37211-8

Index

© The Author(s), under exclusive license to Springer Fachmedien Wiesbaden GmbH, part of Springer Nature 2022
A. Stahel, *Octave and MATLAB for Engineering Applications*, https://doi.org/10.1007/978-3-658-37211-8

Printed in the United States
by Baker & Taylor Publisher Services